Augmented Cognition:
A Practitioner's Guide

Edited by
Dylan D. Schmorrow
Kay M. Stanney

Editorial Board

Editors
Dylan D. Schmorrow, Medical Service Corps, U.S. Navy
Kay M. Stanney, Design Interactive, Inc.

Associate Editor
Leah M. Reeves, Potomac Institute for Policy Studies

Advisory Board
Ali Ahmad, Design Interactive, Inc.
Ami Bolton, Strategic Analysis, Inc.
Roberto Champney, Design Interactive, Inc.
Ben Clegg, Colorado State U.
Joseph Cohn, Medical Service Corps, U.S. Navy
Andrew Cowell, Pacific Northwest National Laboratory
Julie Drexler, U. of Central Florida, Institute for Simulation and Training
Jack Gelfand, State U. of New York at Oswego
David Graeber, The Boeing Company
Jeffrey Grubb, Naval Aviation Human Systems Integration Laboratory
Leland Kollmorgen, TLK, Inc.
Ben Lawson, Naval Aerospace Medical Research Laboratory
James Lewis, IBM Corporation
Santosh Mathan, Honeywell International
Dennis K. McBride, Potomac Institute for Policy Studies
Denise Nicholson, U. of Central Florida, Institute for Simulation and Training
James Patrey, Office of Naval Research
Michael B. Russo, Tripler Army Medical Center
Roy Stripling, Office of Naval Research
Maria Thomas, Army Aeromedical Research Laboratory
Karl Van Orden, Naval Health Research Center
Glenn Wilson, Physiometrex, Inc.

Augmented Cognition:

A Practitioner's Guide

**Edited by
Dylan D. Schmorrow
Kay M. Stanney**

Published by

**Human Factors and
Ergonomics Society**

P.O. Box 1369, Santa Monica, CA 90406-1369 USA
310/394-1811, Fax 310/394-2410
http://hfes.org, info@hfes.org

To the dedicated pioneers whose efforts advanced augmented cognition
from a vision to an established field of study.

Copyright 2008 by Human Factors and Ergonomics Society

Individual readers of this book and nonprofit libraries acting for them are permitted to make fair use of the material in it, such as to make a copy for use in teaching or research. Permission is granted to quote excerpts from the book in scientific works with the customary acknowledgment of the source, including the author's name and book title, year, and page(s). Permission to reproduce any chapter or substantial portion (more than 500 words) thereof, or any figure or table, must come from the Human Factors and Ergonomics Society (HFES). Republication or systematic or multiple reproduction of any material in this book is permitted only under license from HFES. Address inquiries and notices to Communications Director, Human Factors and Ergonomics Society, P.O. Box 1369, Santa Monica, CA 90406-1369 USA; 310/394-1811, fax 310/394-2410.

To order additional copies of this book, go to http://www.hfes.org/Publications/ProductDetail.aspx?ProductID=90, or call or write to HFES at the address above.

ISBN 978-0-945289-33-3

Library of Congress Cataloging-in-Publication Data

Augmented cognition: A practitioner's guide / edited by Dylan D. Schmorrow, Kay M. Stanney.
   p. cm.
  Includes index.
  ISBN 978-0-945289-33-3
  1. Human-computer interaction. 2. User interfaces (Computer systems) 3. Neuroergonomics. 4. Cognitive science. I. Schmorrow, Dylan, 1967- II. Stanney, Kay M.
  QA76.9.H85A7885 2008
  004.01'9--dc22

2008030667

Contents

Preface ix
    Dylan D. Schmorrow and Kay M. Stanney

Icon Key xv

Chapter 1: Brain Sensors and Measures 1
    Gabriele Gratton, Arthur F. Kramer, and
    Monica Fabiani

Introduction 1
Ideal Characteristics of Physiological Brain Sensor and Measures
    for Augmented Cognition 1
Physiological Brain Sensors and Measures 4
Measures of Neuronal Activity 5
Hemodynamic Methods 13
Measures of Incidental Behavior Related to Brain Processing .. 16
Lessons Learned 20
Parting Message 22
Acknowledgments 21
References 21

Chapter 2: Functional Near-Infrared (fNIR) Sensors ... 27
    Evan D. Rapoport, Erin M. Nishimura, Colby Raley,
    Traci H. Downs, and J. Hunter Downs III

Introduction 27
Scenario 27
Background on fNIR 30
General Approach and Associated Toolkit 31
Lessons Learned 34
Best Practice 36
Design Guidelines 38
Parting Message 39
References 40

Chapter 3: Sensor Integration to Characterize
    Operator State 41
    Thomas Schnell, Blaze M. Keller, and Todd J. Macuda

Introduction 41
Scenario 43
General Approach and Associated Toolkit 44
Sensor Integration on the Computerized Airborne Research
    Platform (CARP) 46
CATS Flight Tests 59
Lessons Learned 68

Best Practices 70
Design Guidelines 72
Parting Message 72
Appendix: Explanation of Special Terms 73
References 74

Chapter 4: Cognitive State Estimation in Mobile Environments 75
*Michael C. Dorneich, Santosh Mathan,
atricia May Ververs, and Stephen D. Whitlow*

Introduction 75
Scenario 79
General Approach and Associated Toolkit 80
Advanced Warfighting Experiment 92
Lessons Learned 98
Best Practices 103
Design Guidelines 106
Parting Message 107
References 107

Chapter 5: A Mitigation Framework for Enhancing Situation Awareness 112
*Sven Fuchs, Kelly S. Hale, Chris Berka, and
Joseph Juhnke*

Introduction 112
Scenario 113
General Approach and Associated Toolkit 115
Lessons Learned 133
Best Practices 136
Design Guidelines 137
Parting Message 140
References 141

Chapter 6: Methodology, Methods, and Metrics for Testing and Evaluating Augmented Cognition Systems 144
Frank L Greitzer

Introduction 144
Scenario 145
General Approach 147
Lessons Learned 165
Best Practices 167
Guiding Principles 170
Parting Message 171
References 171

Chapter 7: Engineering Control System Theory in the Behavioral Sciences ... 175
Peter M. Young and Patricia A. Aloise-Young

Introduction	175
Scenario	176
General Approach and Associated Toolkit	177
Lessons Learned	187
Best Practices	191
Design Guidelines	192
Parting Message	194
References	195

Chapter 8: Design Platform Methodology for Augmented Cognition ... 196
Mark Austin and Colby Raley

Introduction	196
Design Platforms	198
Handling System Complexity in Design	204
General Approach to Platform-Based Design	210
Scenario: Platform-Based Design of Automobile Dashboards Enhanced by Augmented Cognition	215
Design Guidelines for Platform-Based Design	220
Parting Message	220
References	221

Chapter 9: Practical Considerations for Developing Augmented Cognition Applications ... 225
Mark St. John and David A. Kobus

Introduction	225
General Approach	226
Best Practices	240
Parting Message	241
References	241

About the Authors ... 245

Index ... 253

Preface

The hope is that, in not too many years, human brains and computing machines will be coupled together very tightly, and that the resulting partnership will think as no human brain has ever thought and process data in a way not approached by the information-handling machines we know today. (Licklider, 1960, p. 5)

Prior to the formalization of the field of **augmented cognition** in the early 21st century, the world experienced a unique confluence of revolutions in science and technology, including the cognitive revolution, biomedical revolution, and computer revolution. These revolutions coincided at the end of the last century with the **Decade of the Brain**, which was a multibillion-dollar investment by the U.S. government to better understand the human brain. Augmented cognition was born out of this union of scientific and technical advancements.

Augmented cognition is a burgeoning field whose roots can be found in the musings of early visionaries such as W. R. Ashby (1956), J. C. R. Licklider (1960), and D. C. Engelbart (1963). What they envisioned was a time when human and computer were so closely intertwined that their symbiosis was empowering to both entities, creating a mutually "productive and thriving partnership," (Licklider, p. 4) in which "intellectual power, like physical power, can be amplified" (Ashby, 1956, p. 272). Their forethoughts are now becoming a reality through the field of augmented cognition, which is redefining human-computer interaction and enabling humans to harmonize with information and technology.

There were many stepping-stones along the way to modern-day augmented cognition, including automated adaptive instruction (Patel, Scott, & Kinshuk, 2001), adaptive automation (Rouse, 1988; Scerbo, 1996) and brain-machine interfaces (Lebedev & Nicolelis, 2006). To achieve adaptive instruction, a learner's behavior is monitored during interactive instruction, which enables the development of optimized and personalized instruction. Thus, under this paradigm, adaptation takes the form of *reconfigurable pathways* (Hartenstein, 2001) whereby the pathways essentially are set according to patterns of use and do not depend on the specific state of a given user (Linton & Schaefer, 2000).

Adaptive automation aims to enable the level of automation of a system to be tied more closely to a user's needs at any given time based on the user's specific cognitive load (Parasuraman, 1993). Adaptive automation efforts were some of the first to leverage physiological data to drive system modification in the form of task allocation (Pope, Bogart, & Bartolome, 1995; Prinzel, Scerbo, Freeman, & Milkulka, 1997; Wilson, Lambert, & Russell, 2000). Under this paradigm, adaptation takes the form of automation, and the level of task automation can be modified in real time based on the user's physiology.

Brain-machine interfaces provide a direct communication link (i.e., via direct brain implants or noninvasive neuroimaging techniques such as electroencephalography) between a user's brain activation patterns and a computational device, thereby allowing for control of this device (Lebedev & Nicolelis, 2006). Under this model, adaptation takes the form of *actuator control*, whereby neuronal signals are translated into commands that reproduce a user's intent in artificial actuators.

Each stepping-stone has brought us closer to the ideal pursuit of augmented cognition: using modern neuroscientific tools and methodologies to determine a person's cognitive state in real time in order to adapt information, technology, and the environment to meet the user's needs. Under this paradigm, adaptation takes the form of optimized human-technology couplings, whereby user interaction is precisely calibrated to the capabilities and limitations of both the user and the computing device.

The quintessential first formal foray into the contemporary field of augmented cognition was the Defense Advanced Research Projects Agency's (DARPA's) **Augmented Cognition Program,** which later became the **Improving Warfighter Information Intake Under Stress Program.** The name change was needed to be able to distinguish the DARPA project from the explosion of other augmented cognition–related projects that were emerging during this same time across the world. The DARPA program funded a set of interconnected studies from 2001 to 2006 that provided advances in augmented cognition technology, including physiological sensors, mitigation strategies, and intelligent architectures (e.g., Schmorrow & Kruse, 2005; Schmorrow & McBride, 2004; Schmorrow, McBride, Worcester, & Patrey, 2001; Schmorrow, St. John, Kobus, & Morrison, 2005).

As participants in these programs, we can attest that this practitioner's guide captures not only first-hand, hard-earned lessons of the DARPA program but also the worldwide advances within this new field of study during this same period. The chapter authors were augmented cognition pioneers; they established this field of study through their investigations of human-systems integration and their efforts to develop tools and technologies that help *remove the burdens of technology.* Their words provide valuable guidance and advice on how to study and practice in this field successfully.

About the Contents

In Chapter 1, Gratton, Kramer, and Fabiani review different types of **physiological measures of brain activity** that can be and have been used to direct adaptation in closed-loop augmented cognition systems. The authors first consider the ideal characteristics of physiological brain sensors that can drive system adaptation. Next, different types of brain sensors are discussed, including neuronal and hemodynamic methods, as well as the benefits of coupling these measures with incidental behavior related to brain processing, such as eye movements. Taken together, this information can be used to assist practitioners in determining which brain sensors are best suited to the needs of their particular application.

In Chapter 2, Rapport, Nishimura, Raley, Downs, and Downs review in detail one of the hemodynamic methods discussed by Gratton and colleagues, that of **functional near-**

Human Factors and Ergonomics Society

Fiftieth Anniversary

P.O. Box 1369

Santa Monica, CA

90406-1369 USA

TEL 310 / 394-1811

FAX 310 / 394-2410

info@hfes.org

http://hfes.org

November 5, 2008

Joseph Juhnke
Tanagram Partners
125 north halsted street
Chicago, IL 60661

Dear Joseph:

We would like to thank you for your contribution to *Augmented Cognition: A Practitioner's Guide*. Through your contributions, we have been able to compile what we feel is certain to become the quintessential reference book for this emerging field. We hope that you will take a peek at the dedication, as we have dedicated the book to all you pioneers, whose efforts advanced augmented cognition from a vision to an established field of study.

Enclosed is a complimentary copy of the book. In addition, you are entitled to a 40% discount on the purchase of additional copies. If you wish to place an order, please complete and return the attached form with payment by check (payable to Human Factors and Ergonomics Society) or credit card (MasterCard, VISA, or American Express) to the address below.

Again, please accept our gratitude for your assistance with the book.

Sincerely,

Dylan Schmorrow, Ph.D.
Email: schmorrow@yahoo.com

Kay Stanney, Ph.D., C.H.F.P.
President
Design Interactive, Inc.
Email: kay@designinteractive.net

infrared imaging (fNIR) sensors. The authors present challenges in developing fNIR sensor technology, particularly issues associated with developing a noninvasive, comfortable, and portable functional system that measures hemodynamic changes associated with the brain in real time. They provide lessons learned and design guidelines that can guide practitioners who elect to use fNIR as a sensor in closed-loop augmented cognition applications.

The works of Gratton et al., Rapoport et al., and others empower practitioners to choose from a vast suite of sensor options, which in turn need to be integrated to provide a coordinated picture of the neural mechanisms contributing to operator state. In Chapter 3, Schnell, Keller, and Macuda describe their efforts in tackling **sensor integration.** Schnell and colleagues developed the Cognitive Avionics Tool Set (CATS), which includes software modules that allow multisensory data collection, data synchronization, artifact removal, querying, classification, and visualization of mass quantities of sensor data through which to characterize cognitive state. They provide many lessons learned to guide practitioners in how best to approach sensor integration. They particularly stress the need to tag and synchronize data in real time during data collection and warn that if synchronization is left to postprocessing, it can impose a huge burden on the practitioner.

In Chapter 4, Dorneich, Mathan, Ververs, and Whitlow discuss the challenges inherent in mobile **cognitive state classification,** including the ability to collect robust and clean signals, the ability to create a mobile computing and data-processing infrastructure, the ability to classify cognitive state reliably, and the ability to experimentally assess the accuracy and specificity of algorithms in a mobile operational setting. Dorneich and colleagues describe their efforts in the context of a field evaluation in which they tested the ability to classify cognitive workload level in an unconstrained, free-play operation with dismounted soldiers executing missions in an urban environment. They also provide a framework to assist practitioners in selecting, assessing, and optimizing cognitive state classification techniques.

Although physiological sensors may be looked at as the "brains" of an augmented cognition system, mitigation strategies are the "brawn." Mitigation strategies are the tools by which we realize seamless, supportive, and highly effective human-system interaction. Mitigations aim to boost human performance through reduced workload, increased throughput, improved quality of decisions, enhanced situation awareness, and other such augmentations. As Greitzer (Chapter 6) states, "The focus of augmented cognition mitigation approaches is to define, demonstrate, and exploit neuroscience and behavioral measures that support inferences about (operator) cognitive state that prescribe the nature and timing of mitigation."

In Chapter 5, Fuchs, Hale, Berka, and Juhnke present a framework that can be used to direct such **dynamic, real-time adaptive system mitigation.** This framework provides an event-based approach to evaluating cognitive processes in real time—by measuring physiological reaction (via sensors such as those described in Chapters 1 and 2) to specific good/bad performance events—and then configuring mitigation strategies on the fly to optimize operator state. Design guidelines are provided for guiding the implementation of mitigation strategies, specifically those associated with enhancing situation awareness. The strategies provided by Fuchs et al. can assist practitioners in identifying what, how, and when to mitigate.

In Chapter 6, Greitzer tackles the challenge of developing valid **evaluation methodologies, metrics, and measures to assess the impact of mitigation strategies,** particularly external validity (applicability to the target operational context) and internal validity (reliability of performance measures). The guidance provided by Greitzer can assist practitioners in the selection and use of methodologies for modeling cognitive state, instantiating adaptive mitigation strategies, and employing effective performance measurement/analytic techniques.

Augmented cognition systems close a feedback loop around a human operator, and, given that humans are subject to a great deal of variation and uncertainty, there is a need for a robust approach to guarantee the stability of such closed-loop systems. In fact, as Young and Aloise-Young point out in Chapter 7, all such feedback loops experience a natural tension between stability and performance, and these authors contemplate the **application of mathematical systems and control systems theory to closed-loop augmented cognition systems.** They take readers through a series of control system models at various levels of abstraction and demonstrate how the nature of the model changes with the application task at hand. Further, they provide several examples of how a well-designed Proportional-Integral-Derivative (PID) controller can deliver closed-loop stability and good performance. Practitioners can adopt the methodologies provided in this chapter as part of an overarching strategy for iteratively designing and evaluating augmented cognition systems, with the aim of achieving stability, robustness, and solid performance.

The technologies required to implement augmented cognition systems are still in their infancy and will evolve over the next decade and beyond. Therefore, augmented cognition system architectures will have to support ever-changing technology components. In Chapter 8, Austin and Raley discuss how **platform-based design techniques,** which aim to achieve efficiency and customization in the creation of design iterations, can accelerate the transition of augmented cognition into commercial applications by enabling system designs that are simultaneously modular and highly integrated. They describe a variety of techniques for handling system complexity in the design and enhancement of iterative design via system modularity. They also propose a methodology for the definition, architecture-level design, and personalized customization of an augmented cognition application.

In Chapter 9, St. John and Kobus provide **guidelines for developing augmented cognition applications for operational tasks,** from developing operationally relevant augmentation models through an iterative design process of increasingly realistic proxy tasks. They provide further guidelines for choosing psychophysiological measures that suit the operational context. Practitioners can use the five-step process provided in this chapter to address a variety of constraints arising from the complexities of the operational task, capabilities and limitations of real users, and nature of the operational environment and artifacts used to accomplish real-world tasks.

Taken together, the chapters in this Practitioner's Guide represent the first comprehensive, professional publication dedicated to formalizing the study and practice of augmented cognition. It can serve equally well as a reference for the seasoned expert in the field of augmented cognition or as a primer for anyone eager to join this emergent field of study. The guide is built on the collective efforts of hundreds of dedicated scientists and engineers, who worked diligently over the past decade to establish the science, tools, and technologies that make up the field. The collective work of these dedicated pioneers is vast and can be found in a wide range of disparate professional journals and general publications. This guide pulls a wide array of this available information into a single source and would not have been possible without the contributions of the augmented cognition pioneers who established this field of study. We thus dedicate this guide to their pioneering efforts to advance augmented cognition from a vision to an established field of study.

The Practitioner's Guide is the story of a new beginning. It is a manifesto of the augmented cognition pioneers' early intentions, principles, theories, and practices. These pioneers, now augmented cognition settlers, will continue to contribute to this field of study through various professional journals and assorted scientific publications and books. Professional organizations, such as the Human Factors and Ergonomics Society, and international symposia, such as the Human-Computer Interaction International biannual conference series, will lead the way in maintaining and celebrating the works of the augmented cognition settlers.

Although there are sure to be many other publications in the future that will help shape the study and practice of augmented cognition, it is our hope that readers will save and treasure this guide and the information in these chapters. This Practitioner's Guide is truly the first of its kind for the field of augmented cognition and is meant to serve as a foundational publication for generations of scientists and practitioners to come.

Dylan D. Schmorrow, Medical Service Corps, U.S. Navy
Kay M. Stanney, Design Interactive, Inc.

References

Ashby, W. R. (1956). *An introduction to cybernetics.* London: Chapman & Hall.

Engelbart, D. C. (1963). A conceptual framework for the augmentation of man's intellect. In P. W. Howerton (Ed.), *Vistas in information handling* (pp. 1–29). Washington, DC: Spartan Books.

Hartenstein, R. (2001). A decade of reconfigurable computing: A visionary retrospective. In W. Nebel & A. Jerraya (Eds.), *Proceedings of the Conference on Design, Automation and Test in Europe* (2001): *Design, Automation, and Test in Europe* (pp. 642–649). Piscataway, NJ: IEEE Press.

Lebedev, M. A., & Nicolelis, M. A. (2006). Brain-machine interfaces: Past, present and future. *Trends in Neuroscience, 29,* 536–546.

Licklider, L. C. R. (1960). Man-computer symbiosis. *IRE Transactions on Human Factors in Electronics, v.HFE-1,* 4–11.

Linton, F., & Schaefer, H.-P. (2000). Recommender systems for learning: Building user and expert models through long-term observation of application use. *User Modeling and User-Adapted Interaction, 10,* 181–207.

Parasuraman, R. (1993). Effects of adaptive function allocation on human performance. In D. J. Garland & J. A. Wise (Eds.), *Human factors and advanced aviation technologies* (pp. 147–157). Daytona Beach, FL: Embry-Riddle Aeronautical University Press.

Patel, A., Scott, B., & Kinshuk. (2001). Intelligent tutoring: from SAKI to Byzantium. *Kybernetes, 30*(5/6), 807–819.

Pope, A. T., Bogart, E. H., & Bartolome, D. S. (1995). Biocybernetic system validates index of operator engagement in automated task. *Biological Psychology, 40,* 187–195.

Prinzel, L. J., Scerbo, M. W., Freeman, F. G., & Milkulka, P. J. (1997). Behavioral and physiological correlates of a biocybernetic, closed-loop system for adaptive automation. In M. Mouloua & J. M. Koonce (Eds.), *Human-automation interaction: Research and practice* (pp. 66–75). Mahwah, NJ: Erlbaum.

Rouse, W. B. (1988). Adaptive aiding for human/computer control. *Human Factors, 30,* 431–443.

Scerbo, M. W. (1996). Theoretical perspectives on adaptive automation. In R. Parasuraman & M. Mouloua (Eds.), *Automation and human performance: Theory and applications* (pp. 37–63). Mahwah, NJ: Erlbaum.

Schmorrow, D., & Kruse, A. (2005). Session overview: Foundations of augmented cognition. In D. D. Schmorrow (Ed.), *Foundations of augmented cognition,* (pp. 441–445). Mahwah, NJ: Erlbaum.

Schmorrow, D., & McBride, D. (Eds.) (2004). Augmented cognition (special issue). *International Journal of Human-Computer Interaction, 17*(2), 127–130.

Schmorrow, D., McBride, D., Worcester, L., & Patrey, J. (2001). Augmented cognition: New design principles for human-computer symbiosis. In *Proceedings of the 37th Annual International Applied Military Psychology Symposium.* Prague, Czech Republic: Ministry of Defence of the Czech Republic Olomouc.

Schmorrow, D., St. John, M., Kobus, D., & Morrison, M. (2005). Overview of the DARPA Augmented Cognition Technical Integration Experiment. In D. D. Schmorrow (Ed.), *Foundations of augmented cognition* (pp. 446–452). Mahwah, NJ: Erlbaum.

Wilson, G. F., Lambert, J. D., & Russell, C. A. (2000). Performance enhancement with real-time physiologically controlled adaptive aiding. In *Proceedings of the 14th Triennial Congress of the International Ergonomics Association and Human Factors and Ergonomics Society 44th Annual Meeting* (pp. 3-61–3-64). Santa Monica, CA: Human Factors and Ergonomics Society.

Icon Key

ICON KEY
📁 Valuable Information
✏️ Test Your Knowledge
💻 Applied Exercise
📖 Review

The icons used in each chapter are provided to help the reader understand the significance and utility of the information discussed and to indicate how this information might be used.

The "Valuable Information" icon calls out key points that are discussed and encourages readers to focus their attention.

The "Test Your Knowledge" icon highlights portions of the chapter that provide checkpoints to test the reader's comprehension of knowledge and facts discussed.

The "Applied Exercise" icon points out opportunities for readers to think more about the concepts and apply them to problems that they are studying in their own laboratories.

The "Review" icon indicates where high-level principles are reiterated to refresh memories and provide opportunities for more reflection.

Brain Sensors and Measures

Gabriele Gratton, Arthur F. Kramer, and Monica Fabiani
University of Illinois at Urbana-Champaign

> *Physiological sensors are a key component of the augmented cognition closed loop. They need to be informative, reliable, and practical. Here we review the potential utility of several measures of brain function, including neuronal measures, neurovascular signals, and measures of incidental behavior.*

Introduction

In this chapter we review and discuss the use of a subset of physiological measures of brain activity in augmented cognition. **Augmented cognition is conceptualized here as a closed-loop system** in which information about an operator's brain and body states and functions is used to regulate the flow of information between humans and computer-based systems in order to achieve maximum cognitive performance. Within this context, **physiological measures of brain activity can be used to infer the operator's brain states, information-processing strategies, and future performance.**

Ideal Characteristics of Physiological Brain Sensors and Measures for Augmented Cognition

Many different types of physiological measures of brain activity can be and have been used to direct adaptation in closed-loop augmented cognition systems. What characteristics would make a particular brain measure suitable for augmented cognition? First, **measures should be sensitive to different brain states and/or processes.** They should provide the rest of the system with critical information about the *readiness* of the operator and the *availability of resources* to process incoming information. This issue is complex, as it implies a theory of mind and brain—that is, a theory of how cognitive states and processes are related to readiness and resource availability—and how these states and processes are manifested by brain measures. Our current knowledge of these issues, though extensive, is far from complete, and answers to basic questions still need to be obtained. For instance, how does readiness manifest itself in functional brain measures? Are there separate or specific forms of "readiness" (and resources) for different tasks and/or states, or is it a general function?

Everyday experience and years of research in human factors engineering and related fields indicate that tasks differ in the way they interfere with one another (e.g., Wickens, 2002), and that training, level of expertise, and/or cognitive strategies may play a major role in determining the extent of interference (e.g., Ruthruff, Johnston, Van Selst, Whitsell, & Remington, 2003). Similarly, readiness can be specific for particular tasks (e.g., an operator may be ready or primed to receive visual rather than auditory feedback) but may also involve some general state that is common to a variety of tasks (e.g., a fatigued operator may be less capable of coping with a variety of tasks). Thus, brain measures to be used in augmented cognition should provide information about different states and processes. Another way of expressing this concept is that ideally, measures of brain function should be not only sensitive but also rich—in the sense that they need to provide information about as many states and processes as possible.

Second, the **measures must be reliable.** This means that inferences that are made about states and processes on the basis of a particular brain measure in one instance can be extended to other, similar instances (test-retest reliability). This is of crucial importance in applied cases. In a real operating environment one cannot typically afford a high error rate and a low level of accuracy in the estimation of brain function. An unreliable system may not only lead to fatal mistakes but also lower its acceptance. Reliability is sometimes difficult to estimate, as it includes not only properties of the measurement itself but also the concrete instantiation of the technology in each practical case.

Third, the **measures must be practically usable.** Most, if not all, current brain imaging methods (and related recording systems and sensors) are designed largely for medical and/or research applications and are therefore not readily portable to a real-world environment without substantial engineering developments (see Chapter 4 for a notable exception). An assessment of how close a measure may be to applied deployment includes a number of parameters. One of these is *invasivity*. The level of invasivity of a measure is a relative concept. It ranges from the possibility that a measurement system may pose risks or harm to the operator (which is typically unacceptable) to the possibility that a measurement system may induce various levels of discomfort in the operator (without any risk or long-term consequences). Thus, even measures that are just uncomfortable may have limited acceptability for practical uses and distract operators from their normal tasks. In a manner similar to reliability, invasivity/discomfort may not necessarily be inherent to a technique but may often be determined by the way a given technique is instantiated in practical cases.

Another aspect that may limit the practical use of a measure is the *interaction that a particular measurement system has with the environment* in which the operator works.

Some measurement systems can operate only in very special environments (e.g., environments with very small magnetic fields or minimal vibration). For example, the majority of recording systems will work correctly only within specific (and often narrow) ranges of temperature, humidity, shielding from electromagnetic interferences (such as those produced by elevators), and so on. In addition, some techniques require the operator to remain still or in a supine position, again making the measurement system hardly compatible with normal operating environments in such cases. Other recording systems are very large and cumbersome to use and cannot be practically employed in many applied settings.

Two additional practical considerations should also be mentioned. The first is *cost*. Although the exact determination of which cost is acceptable may be based on the specific application, it is clear that very expensive machinery may be impractical for extensive field applications. Similarly, once data are recorded, it is important to factor in the cost of specialized, and often extensively trained, personnel associated with deriving the appropriate measures through data analysis. It should be noted, however, that costs could be much diminished with technological improvements and mass production, especially if the analysis aspects of physiological measurements could be automatized and provide indices that could be interpreted by a closed-loop system or by non-specialized personnel. Therefore, initially high costs for the development of new methods (which necessarily include the costs associated with the research and development needed for deployment) may not be a significant issue, whereas projected long-term costs may be more important.

The second consideration is the *technical compatibility among different sensors*. In assessing the pros and cons of different methodologies, it should be understood that different techniques can be combined to provide a richer and/or more reliable set of measures of brain states and processes. This is the case particularly if different measures can be recorded concurrently without interference. Interference can occur because of different factors: It may be necessary to physically place devices at the same location; they may produce electromagnetic fields (or influence already present fields) so as to seriously disrupt the measurements, and so on.

The level of compatibility of different measurement systems can also vary depending on the specific instantiation of each of them: A particular pair of measures can be made compatible through the adoption of appropriate materials. For example, nonferrous materials could be used in a magnetic resonance environment. Similarly, a remote eye tracker could be coupled with scalp-recorded electroencephalography (EEG). Finally, other specific hardware (e.g., shielding) or software (e.g., off-line artifact removal) solutions may be available or could be developed to address various compatibility problems. In the remainder of this chapter we review different sensor techniques, with a particular emphasis on issues of sensitivity, reliability, and practicality.

Test Your Knowledge
What characteristics make brain measures particularly suitable for augmented cognition applications? Make a list and cite specific examples for each (different from those mentioned in the text and, if possible, specific to your own application).

Applied Exercise
Think of a practical situation in which two or more recording devices should be used concurrently (e.g., EEG recordings and eye tracking). Pick an environment (e.g., the cockpit of an aircraft) and then list all the possible environmental and operator-related limitations posed by such an environment. Propose solutions or alternatives for each.

Physiological Brain Sensors and Measures

Up to the 1920s, no method was available for studying brain function noninvasively in humans, apart from the observation of behavior or introspection. This changed with the introduction of the EEG by Berger in 1929. During the last 50 years there has been enormous and accelerating advancement in the development of methods for studying the human brain in vivo and for inferring brain states and processes. Many techniques have emerged, some leading to major breakthroughs in the understanding of brain function. These include measures of the electromagnetic activity produced by the brain, such as event-related brain potentials (ERPs; Fabiani, Gratton, & Federmeier, 2007; Luck, 2006) and magnetoencephalography (MEG; Hari, Levanen, & Raij, 2000); measures of the blood flow and hemodynamic changes that follow brain activity, such as O_{15}-positron emission tomography (PET; Posner, Petersen, Fox, & Raichle, 1988) and functional magnetic resonance imaging (fMRI; Frackowiak et al., 2003); and measures of changes in brain transparency, such as near-infrared spectroscopy (NIRS; Villringer & Chance, 1997; see Chapter 2) and the event-related optical signal (EROS; Gratton & Fabiani, 1998, 2001, 2007).

Notwithstanding these advancements, much remains to be learned about brain states and processes and their relation to cognitive performance. This is a complex and very active area of research within the domain of cognitive neuroscience. In this chapter we do not describe any specific model or theory of performance in cognitive neuroscience but, rather, focus on the type of information that can be obtained with different techniques for studying brain function, and how this information potentially can be useful for augmented cognition. We briefly review both the sensors (i.e., the specific hardware used to record brain data) and measures (i.e., the specific parameters derived from the recording through data analysis, which are used to drive adaptation in augmented cognition applications).

Further, for reasons of space and focus, we limit our discussion to three types of measures:

a. Measures of neuronal activity (EEG/ERPs, MEG, and EROS)
b. Measures of hemodynamic phenomena associated with neuronal activity (fMRI, PET, and NIRS)
c. Measures of incidental behavior related to brain processing (eye movements).

Other physiological and behavioral measures are covered elsewhere in this book. Each of these measures is discussed in terms of the type of information it reveals about brain states and processes, the quality of this information, and practical aspects related to its application to augmented cognition. We emphasize methods that, for theoretical and practical reasons, are expected to be more amenable to field application (EEG, ERPs, EROS, NIRS, and eye movements), and only briefly discuss other methods (MEG, fMRI, and PET).

Measures of Neuronal Activity

Measures of neuronal activity provide information about the mass activity of neurons in the brain. Given the density and number of neurons in the brain, it is impossible to obtain measures of individual neurons without placing probes very close to the brain and increasing the spatial resolution to microns. This would require invasive recording methods, which cannot be used in augmented cognition paradigms and, in fact, are permissible in humans only when justified by surgical necessities. However, measures of mass activity can be very informative about brain function, and it can be argued that they may provide a good statistical estimate of brain activity—something that would be difficult to obtain with single-unit measures.

All the measures we are considering here are related to **ionic exchanges that occur during neuronal activity** (depolarization and hyperpolarization of neuronal membranes). These ionic exchanges engender two types of phenomena: (a) changes in the electromagnetic fields and currents around cortical areas, which propagate to the surface of the scalp, and (b) volumetric and density changes in different tissue compartments (intra- and extracellular compartments as well as cell membranes), which result in changes in the overall transparency of the tissue to light (in particular in the near-infrared range). All these changes are practically simultaneous with neuronal activity. Therefore, **these measures can provide a timely and relatively faithful representation of the activation of large numbers of neurons in the cortex**. In principle, these measures could provide feedback about brain states and processes with very short delays (on the order of seconds and even shorter), mostly related to the analysis of the information obtained from the brain. For this reason, measures of neuronal activity hold great promise for augmented cognition.

Electrophysiological Measures

As mentioned, electrophysiological measures include EEG, ERPs, and MEG. EEG is the measurement of differences in electric potentials at the surface of the scalp resulting from electrical activity occurring inside the brain. The brain and head tissues are sufficiently conductive to allow for the propagation of electrical fields at the surface of the head. However, the skull has low conductivity with respect to intracranial tissue, which has a practical effect equivalent to that of measuring the intracranial potentials from a great distance (such as > 50 cm away, even though the thickness of the skull is

< 1 cm). As a result, (a) the potentials measured at the scalp are greatly attenuated with respect to those that are measurable inside the skull (i.e., they scale down from millivolts to microvolts); and (b) the potentials are spatially smeared, so that it is practically impossible to isolate the effects of areas located in very close proximity to one another (e.g., less than a few centimeters apart). From this issue ensues the **major limitation of EEG-derived measures: They have low spatial resolution** (compared with other techniques available for measuring brain activity).

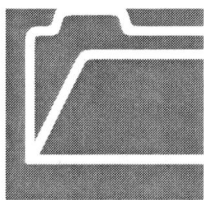

Valuable Information

SInce the initial recordings made by Berger (1929), it was clear that the EEG is very sensitive to the overall level of activation of the brain. When people are actively engaged in tasks, the EEG shows low-amplitude oscillations, which tend to be dominated by relatively high frequencies (> 20 Hz). These oscillations are called β *rhythms*. When people are calm or drowsy, larger oscillations with a frequency between 8 and 12 Hz appear (α *rhythms*). Even slower and larger oscillations are present when subjects are asleep or in a coma.

These findings can be of significant utility for augmented cognition, because in a closed-loop system, they can be informative about whether an operator is in a state that is appropriate for complex decision making or for operating complex machinery (e.g., Keckluno & Akersteot, 1993). Information of this type can be obtained very reliably (as EEG measures, if obtained with appropriate procedures, have a very high signal-to-noise ratio), with minimum computational effort and at relatively low cost, because the recording of EEG requires an inexpensive apparatus. Furthermore, EEG sensors are minimally invasive, as the EEG is produced spontaneously by the subject and no additional stimulation is required.

Finally, methods for recording EEG using telemetry (i.e., wireless links with external systems) have been available for a long time (Walter et al., 1967). For basic EEG recordings, measuring only the general level of alertness, few leads (3 to 5) are needed, further facilitating their recording.

In recent years, researchers have focused on deriving more complex measures from EEG. They have studied features of the EEG that could indicate the occurrence of specific types of brain activities and therefore may be signaling the presence of particular types of brain states or processes. An example is the occurrence of 4–6-Hz activity at frontal locations (also called *theta rhythm*), which is considered to be indexing a state of active cognitive engagement in a particular task (e.g., Canolty et al., 2006).

Another example is the use of high-frequency EEG using ratios of spectral power, which has been explored in relationship to attention (Sing et al., 2006). Finally, EEG measures (such as the *mu rhythm*) have already been used in an area related to augmented cognition: mental prosthesis. In that case the intent is to allow the brain of a "locked-in" patient to communicate directly with a machine without the mediation of muscular outputs (Wolpaw, McFarland, Neat, & Forneris, 1991).

Review

The EEG can be used to measure the general level of arousal of the brain with good signal-to-noise ratio, low cost, low invasivity, and relatively good portability, making it suitable for augmented cognition applications. Other uses of EEG for augmented cognition are being investigated (e.g., Erdogmus et al., 2005; St. John, Kobus, Morrison, & Schmorrow, 2004). However, the low spatial resolution of EEG may limit the use of this tool for more complex classifications of brain states and processes, such as those required to assess whether an operator may be primed to process one type of input over another.

Test Your Knowledge

What brain states accompany fast and slow EEG rhythms? What are likely to be the major advantages and disadvantages of EEG in simulator and field applications?

Valuable Information

The **ERP** is an application of EEG to the investigation of scalp electrical brain activity that is time-locked to external or internal events. Typically, with this term we indicate specific patterns of brain activity characterized by a particular morphology (or wave shape)—including a specific temporal relationship with external triggering events, a particular spatial distribution over the scalp, and a characteristic way of responding to specific experimental manipulations (Donchin, Ritter, & McCallum, 1978; Fabiani, Gratton, & Federmeier, 2007).

This framework is based on the assumption that ERP patterns, or components, reflect the activation of patches of cortical tissue (not necessarily contiguous) involved in specific, elementary information-processing transactions (Donchin et al., 1984). Therefore, the elicitation of a particular ERP component may manifest the occurrence of a particular brain state or process—and therefore may be very useful in the context of augmented cognition.

Research conducted during the last 40 years has led to the identification of a set of approximately 20 ERP components, each thought to be related to a particular set of brain activities and corresponding cognitive processes. For example, the **P300** (first described by Sutton, Braren, Zubin, & John, 1965) is a potential with a latency exceeding 300 ms from stimulation, positive over centroparietal scalp areas, and elicited by attended stimuli, in particular when they occur relatively rarely (such as once every 10 s or so; Polich & Kok, 1995). The P300 could be particularly useful in augmented cognition because it is very large and can be recognized, at least in some cases, on individual trials with some level of confidence (Fabiani, Gratton, Karis, & Donchin, 1987).

Various theories have been proposed about the nature of the cognitive process that is manifested by P300 (e.g., context updating; Donchin, 1981; Donchin & Coles, 1988; context closure: Verleger, 1988) and about its brain generators (e.g., hippocampus: McCarthy & Wood, 1987; angular gyrus: Yamazaki, Kamijo, Kiyuna, Takaki, &

Kuroiwa, 2001; a network comprising frontal and parietal regions: Linden, 2005). However, for the purposes of augmented cognition, the occurrence of a P300 could be used to signal that a particular eliciting stimulus was within the spotlight of attention of an operator (see the use of **P300 as a communication tool in mental prosthesis:** Farwell & Donchin, 1988) and even the extent to which the event was expected. This could be useful to inform a closed-loop system about which stimuli within a set the operator is paying attention to. This information could be used to redirect future information delivery so as to maximize the probability that information is processed appropriately.

The P300 is also sensitive to the relative distribution of attentional resources to different tasks—when they are conducted in parallel (Kramer & Parasuraman, 2007; Sirevaag, Kramer, Coles, & Donchin, 1989). Specifically, **the amplitude of the P300 elicited by secondary task probes can be used to determine the amount of central processing resources that are NOT used in the primary task** and therefore are available for other tasks (e.g., Kramer, Wickens, & Donchin, 1985). This information can be used to determine whether the operator may be capable of processing additional information related to either the primary or the secondary task. This type of information could be of significant value in augmented cognition.

Other ERP components could be valuable in augmented cognition. For instance, the error-related negativity (**ERN**) is a component that is elicited when subjects are aware of having made an error (Falkenstein, Hohnsbein, Hoormann, & Blanke, 1991; Gehring, Goss, Coles, Meyer, & Donchin, 1993). The ERN is derived from the subtraction of correct from incorrect trials, and studies of the ERN typically average across a number of trials. However, in principle, measurement of the ERN on single trials (by means of appropriate reference algorithms; see next paragraph) could be used to determine moment-by-moment error awareness. This type of information could be of significant utility for augmented cognition purposes in feedback loops, as well as in the selection and training of operators.

Similarly, several other ERP components could be potentially useful. For instance, the lateralized readiness potential (**LRP**) could be employed to detect initial activation of incorrect responses or response biases in advance of the presentation of a stimulus (Gratton, Coles, Sirevaag, Eriksen, & Donchin, 1988). This could be used to influence and correct biases and thus prevent costly mistakes.

A problem with several ERP components is their **relatively low signal-to-noise ratio;** that is, the relative magnitude of the signal or component of interest compared with the background noise within a single trial recording. This makes it difficult, and sometimes nearly impossible, to detect their occurrence on a single-trial basis. Some ERP components are so small (relative to the background noise present in the EEG) that their identification—even on data averaged over a large number of trials—is dependent on comparing or subtracting different conditions.

Both the ERN and the LRP are defined as the difference between two paired conditions. Thus, in principle, they are not identifiable on individual trials. This problem can be circumvented by developing an algorithm that compares the actual waveform observed on a particular trial with an expected waveform (or a set of different expected waveforms), which could be taken to represent different scenarios related to particular brain states and processes. The expected waveform(s) may be built so as to take into account the context within which a particular trial occurs (for instance, taking into account the trial type, or the sequence of preceding trials, etc.). The algorithm would then determine the probability that a particular observed waveform belongs to a particular class—and therefore the probability that a particular brain state or process (identified by the occurrence of the ERP component) has occurred.

An example of this logic is **the single-trial classification of P300 used in interrogative polygraphy** to determine whether a subject has previously encountered a particular stimulus or event (e.g., Farwell & Donchin, 1991). Note that even in such a case, a reliable level of discrimination between different brain states (with error rates less than 5%) can be obtained only through repeated presentations of the stimulus material (as well as of distracters). This means that **single-trial classification of ERP data is likely to provide only probabilistic statements about specific brain states or processes**. The quality of these probabilistic statements will depend on the relative size of the signal compared with that of the noise. The signal size is related to the amplitude of a given ERP component, which is typically on the order of a few microvolts, and that of the noise to the background oscillations of the EEG, which can be 5 to 10 times larger. This apparently dismal situation can be improved by using filters and pattern recognition algorithms (e.g., Fabiani et al., 1987).

For instance, if the ERP activity of interest is most prominent in the 3–6-Hz range, eliminating other frequencies from the EEG may greatly improve its identification. Further, given that ERP components possess specific scalp distributions that the background noise may not share, utilization of information from multiple electrodes can also help in separating signal from noise. For instance, we (Fabiani et al., 1987; Gratton, Coles, & Donchin, 1989) showed that using frequency filtering and spatial distribution information may greatly increase the reliability of single-trial estimation of P300 amplitude. However, by and large, single-trial classification of brain states and processes based on ERPs may be constrained by **the amount of information transmitted through ERPs and EEG, which may be limited by the spatial resolution of the methods.**

Another issue to consider about ERPs is whether it is useful or necessary to **use discrete probes to derive information about brain states and processes** using this approach. For instance, we have mentioned the use of the P300 elicited by secondary-task probes to determine the amount of resources left free from a primary task. This can be extremely useful information for augmented cognition, as the amount of additional information delivered to an operator may be adapted on the basis of these data.

The use of probes, however, generates two problems. First, it reduces the temporal resolution of the technique, because information about the brain state can be obtained only when a probe is presented. If elicitation of a P300 requires relevant probes to be presented as an infrequent event in a secondary input stream (i.e., once every 10 seconds or so), information about available resources would be obtained only at this rate (assuming that a single-probe presentation is sufficient to provide this information). Thus the temporal resolution of this approach is at best 10 seconds.

Further, this approach introduces a secondary stream of stimuli and another task—which themselves may influence the amount of available resources and interfere with the normal performance of the operator. To avoid these types of problems, it is preferable that the brain activity that is measured be either related to events embedded in the operator's task or, even better, distinguishable from the normal ongoing brain activity without necessary reference to a particular stimulus. The latter may require a good signal-to-noise ratio and is likely to be possible only for a small subset of ERP activities.

Review
ERPs have several features that make them a useful tool for augmented cognition. They can provide important summary information about brain states and processes and are relatively inexpensive and unobtrusive to record. Potentially, ERPs can provide rapid feedback (within a few hundred milliseconds), although this may vary depending on the specific implementation. Their major limitation is their signal-to-noise ratio, especially for smaller or subtracted components, which may hinder their practical effectiveness so that only a few types of states and processes can be detected with confidence on a trial-by-trial basis.

Applied Exercise
To acquire expertise in ERP recording and analysis, for both naive and more advanced users, several summer schools are available, including ERP boot camp (University of California, Davis; *http://www.erpinfoorg/bootcamp*) and Neuroscan School (*http://63.134.192.29/school.cfm*).

Test Your Knowledge
What are the main characteristics of ERP measures? What are some of the ERP components that may be of interest for augmented cognition? What are the advantages and disadvantages of averaging across trials to obtain a summary ERP? What are likely to be the major advantages and disadvantages of ERPs in field applications?

Valuable Information
In addition to electric fields, it is possible to record magnetic fields associated with brain activity by using **MEG** (see Hari, Levanen, & Raji, 2000, for a review). The propagation of magnetic fields is little influenced by the skull's resistivity, which improves the spatial resolution of the technology. However, the magnetic fields produced by the brain are very small and require a bulky and expensive apparatus for recording.

Specifically, MEG recording involves the use of superconductive quantum interference devices (SQUIDs), which need to be kept at extremely low temperature (just a few degrees Kelvin). This involves a large and expensive cooling and insulating apparatus. In addition, magnetic fields are very small compared with environmental sources. Thus, MEG requires special recording conditions, such as a magnetically shielded room and the absence of large sources of magnetic fields in the neighboring environment.

Review
MEG is very useful for cognitive neuroscience studies and some medical applications. However, because of the requirements and constraints associated with MEG recording, at present MEG appears to be of little practical utility for direct field applications in augmented cognition and therefore is not discussed further.

Fast Optical Signal (EROS)

As mentioned earlier, neuronal activity is associated with changes in the transparency of brain tissue (Frostig, Lieke Ts'o, & Grinvald, 1990; Rector, Carter, Volegov, & George, 2005). These changes can be measured noninvasively with a technology called *EROS* (Gratton, Corballis, Cho, Fabiani, & Hood, 1995; Gratton & Fabiani, 2007; see Chapter 2). This technique is based on **passing near-infrared (NIR) light through areas of the cortex and measuring changes in either the amount of light moving though the brain (intensity) or the time taken by photons to move through the brain (time-of-flight)**. Although mostly opaque to visible light, head and brain tissues are much more transparent to NIR light of wavelengths between 700 and 900 nm (Wilson, Patterson, Flock, &Wyman, 1989). At these wavelengths, the major obstacle to the movement of light is the high scattering coefficient of most head tissues. As a consequence, if a point of the head is illuminated with this type of light (*source*), light will diffuse in a random fashion inside the head. If a *detector* is located at some distance (between 2 and 6–7 cm from the source) on the surface of the head, it may pick up a portion of the photons coming from the source.

Interestingly, it can be shown both mathematically and empirically that the photons traveling between the source and the detectors are likely to travel within a curved-spindle volume, reaching a maximum depth that is roughly proportional to the distance between source and detector (Gratton, Sarno, Maclin, Corballis, & Fabiani, 2000). Therefore, at appropriate source-detector distances, this volume may encompass regions of the cortex. If a change in transparency occurs in these areas, parameters of the light reaching the detector (intensity and/or time-of-flight) may be affected. Therefore, changes in transparency associated with brain activity can be measured with optical methods, as demonstrated by extensive work conducted in our lab (for a review, see Gratton & Fabiani, 2007) and replicated in other labs (e.g., Franceschini & Boas, 2004; Tse, Tien, & Penney, 2006; Wolf et al., 2000).

Given that **changes in transparency are practically simultaneous with electrical activity**, optical measures have a latency that is consistent with that of ERPs (e.g., Gratton et al., 1997); they also show oscillations that are consistent with EEG signals (Zhao, Valle-Inclan, & Hackley, 2006). In addition, as we discuss later in this chapter, optical measures can be sensitive to other changes related to hemodynamic phenomena (Villringer & Chance, 1997). However, the neuronal changes are much more rapid than the vascular changes.

Compared with electrical (and even magnetic) measures, EROS is more localized and does not require the use of source-modeling algorithms. This is partly because photon propagation through the head is limited, so that activity in a particular brain area is visible only through sources and detectors located relatively close to it (at a distance of not more than 2–3 cm). However, the spatial resolution can be greatly increased by using time-of-flight measures (which are especially sensitive to local variations in transparency) and by using a high density of source-detector pairs (also called *recording channels,* or *optodes*), combined with appropriate mathematical procedures for combining data (*pi detectors;* Wolf et al., 2000). Empirical evidence shows that through use of this approach, the spatial resolution of EROS is clearly less than 1 cm and perhaps as good as about 5 mm (Gratton & Fabiani, 2003).

The combination of spatial and temporal resolution of EROS means that **this technique has the potential to provide a great deal of information about brain function**. However, it should be kept in mind that the relatively limited spread of photons through the brain also causes an important limitation of EROS: **It is impossible, with the current methodology, to use this technique to monitor the activity of deep brain structures** (such as the hippocampus or basal ganglia).

Two major types of optical recording instruments are available. The simplest method is based on light sources of fixed (or slowly oscillating) intensity: the *continuous-wave,* or CW, method. This method can provide only measures of the amount of light moving between the source and the detector (intensity). A more complex methodology is based on the use of rapidly oscillating sources: the *frequency-domain,* or FD, method. This approach provides measures of both intensity and photons' time-of-flight.

Most current studies of EROS are based on FD technology, in contrast with most of the studies of slow (hemodynamic) optical signals (NIRS, reviewed later in this chapter; also see Chapter 2), which are based on CW technology. However, it appears that both the FD and the CW technologies are in fact capable of revealing both fast (Gratton et al., 1995) and slow (Franceschini & Boas, 2004) optical signals, although the signal-to-noise ratio and spatial and temporal resolution of different devices may vary widely. It should be noted, however, that few optical recording machines currently on the market can record data with sufficient temporal resolution to reveal fast optical signals.

Review

Optical measures are potentially useful for augmented cognition because they can be informative about a variety of brain states and processes, as demonstrated by research conducted in our lab and elsewhere. Other advantages of the optical technology include its cost (which, though higher than that for EEG recording, is relatively low compared with those for other brain-imaging methods), small size of the equipment (again, compared with fMRI and MEG), and compatibility with other technologies (optical recordings can be performed simultaneously with most other imaging methods with little or no interference).

For EROS to be useful in augmented cognition, however, it should be demonstrated that it is possible to classify brain states and processes on the basis of single-trial information. As we have seen with ERPs, this classification depends in large part on the possibility of identifying a signature signal in the midst of a varying background of noise. Up to now, all EROS studies have been based on the computation of averages, often of a large number of trials. This is because the EROS signal is small, with changes in the photon time-of-flight on the order of picoseconds, or 10^{-12} s. However, filtering methods (Maclin, Gratton, & Fabiani, 2003) and appropriate recording techniques, such as the use of optimal wavelengths (Gratton et al., 2006) and optimal oscillation frequencies (for FD methods, see Maclin, Low, Fabiani, & Gratton, 2007), may greatly improve the signal-to-noise ratio of EROS. The possibility of using EROS to classify brain activity on the basis of single-trial data is under investigation in our lab.

Test Your Knowledge
What types of light wavelengths are used for EROS recordings? What are the main advantages and disadvantages of EROS for field applications? What are the main neuronal measures available for noninvasive recording of brain activity? How do they compare?

Applied Exercise
Pick a familiar applied scenario and imagine recording brain activity from an operator with each of the techniques listed in this section. Make a list of all potential problems and possible solutions. Identify whether each solution requires the development of new hardware or software or modifications of current hardware and software.

Hemodynamic Methods

Brain activity is associated with localized changes in oxygenation level and blood flow. Because the location (and sometimes the timing) of these changes differs depending on the types of brain states and processes, measurement of these neurovascular changes can be important for augmented cognition.

During the last 30 years, several techniques have been developed to image in vivo localized changes in hemodynamic parameters. Here we consider three methods that have been most commonly applied to cognitive neuroscience and/or have significant potential for augmented cognition: O_{15}PET, fMRI, and NIRS.

These techniques are based on substantially different physical principles, but they share the target phenomenon they aim at visualizing: the vasodilation and consequent increase in concentration of oxy-hemoglobin that occurs in cortical areas where neurons are active (sometimes also labeled the *blood-oxygenation-level dependent*, or **BOLD,** signal). This phenomenon appears to be general and is observable not only in humans but also in a variety of animals (Ogawa, Lee, Kay, & Tank, 1990). In fact, early anaerobic neurovascular responses are very localized – within a few hundred microns from the active area—so that they can pinpoint the areas where neurons are active with great resolution when invasive methods (Frostig et al., 1990) or high-field MR methods (Uğurbil et al., 1999) are used. However, the BOLD signal typically obtained in vivo is less localized (on the order of mm) and also relatively slow in its development over time, so that it takes several seconds for it to unfold (with a peak 4–7 s after the neuronal activity has occurred). This limits the temporal resolution of hemodynamic methods. For augmented cognition purposes, this means that one can get information about a particular brain state or process only a few seconds after it has occurred (but see Sigman, Jorbet, Lebihan, & Dehaene, 2007, for the development of much faster fMRI temporal resolution).

Positron emission tomography **(PET)** is a method to image the location of certain types of radioactive isotopes within the body. The radioisotopes can be introduced into the bloodstream intravenously (making PET impractical as a field application) or through inhalation. These isotopes then diffuse through the body and reach the target organ of interest, where a portion of the radioisotopes will decay (β decay) and emit positrons, which interact with electrons in the environment, annihilating them and producing pairs of opposite γ rays. The location where the β decays occur can be imaged based on tomographic principles.

By appropriately selecting isotopes as well as the chemical substance in which they are embedded and the procedures used for imaging, it is possible to use PET to visualize a number of target organs and physiological or anatomical parameters. Of interest here is the use of PET for imaging brain activity. The most common method is to use water containing the O_{15} isotope, which is carried, at least initially, to those areas in the body with the largest amount of blood flow. O_{15} is a rapidly decaying isotope, so that, even though water ends up diffusing through the tissue, most of the radioactivity is emitted while O_{15} is still inside the blood vessels. This means that the O_{15}PET signal can be used to visualize blood flow. Furthermore, because of the rapid decay of the radioisotope, measurements can be taken relatively rapidly (within a few minutes) and can be repeated relatively quickly. **This makes it possible to compare changes in blood flow occurring in different conditions.** This is a critical step for generating sensitive measures with good spatial resolution.

Review

O_{15}PET has played a significant role in cognitive neuroscience (e.g., Posner et al., 1988). It was one of the first widely used neuroimaging methods capable of resolving areas of activation at a resolution of less than 2 cm. It has proven to be capable of localizing a number of brain areas that are involved in specific tasks. This information could, in principle, be used to identify a variety of brain states and processes, a useful datum for augmented cognition.

However, the temporal resolution of O_{15}PET is limited, as measurements can be taken only every few minutes and cannot be repeated frequently because of the use of ionizing radiation. Finally, PET is a very expensive technique and is impractical to use outside research and medical environments. As a consequence, O_{15}PET cannot be considered a good candidate for augmented cognition applications. However, it remains important because of historical and theoretical considerations.

Several of the problems that exist with O_{15}PET are alleviated or improved using **fMRI**. fMRI is currently the most commonly used tool in functional neuroimaging. Magnetic resonance technology is based on the magnetic properties—specifically, the spin—of the nuclei of certain atoms, most notably hydrogen. Through the use of a combination of a strong, stable magnetic field and weaker magnetic pulses, some of these nuclei are made to spin at a particular frequency and in a particular orientation. This spinning creates magnetic fields of their own, which can be measured with appropriate detectors.

These induced magnetic fields tend to disappear (or relax) at a speed that is partly influenced by the environment in which the atoms are located. Thus, by measuring parameters of the relaxation time, it is possible to infer some properties of the internal milieu of living individuals in a fundamentally noninvasive manner.

Through modulations of additional magnetic fields (called *gradient fields,* which modify the frequency of the magnetic signal at different locations), it is possible to produce two- and three-dimensional images of these properties, with a spatial resolution of a few millimeters (and even less, in some cases). Interestingly, the presence of deoxy-hemoglobin (but much less of oxy-hemoglobin) is capable of influencing the strength of the MR signal. Variations in the concentration of deoxy-hemoglobin are therefore reflected by variations in the strength of the MR signal. (This signal, however, is influenced by a number of other factors.)

Review
Compared with O_{15}PET, BOLD-fMRI measures do not require the introduction of contrast agents into the body. They can be repeated frequently, so that most fMRI studies produce images of the brain every 2–3 s. As with PET—and in fact even more so—a variety of brain processes and states can be differentiated using fMRI. Although expensive, an MR scanner is substantially cheaper than a PET imager. The fMRI signal, though small, appears to be reliable (Noll et al., 1997).

However, the scanners used for producing fMRI images are very bulky; they weigh several tons and require specialized environments. They produce huge magnetic fields that are incompatible with a large number of working devices. Finally, a substantial number of people present counterindications for MR, such as pacemakers, hearing aids, and claustrophobia. For these reasons, fMRI is likely to remain an experimental or medical technique that can be used to derive principles and ideas for augmenting cognition, rather than as a tool for field applications.

Valuable Information
These latter problems (bulkiness and compatibility issues) can be addressed by using slow optical measures to image hemodynamic changes. This can be done because oxy- and deoxy-hemoglobin have distinctive absorption spectra in the NIR range. It is therefore possible, using spectroscopic and tomographic principles, to measure in vivo the concentration of these substances in the body with NIR light with NIRS technology (Villringer & Chance, 1997; see Chapter 2). As for the fast optical signal, the major limitation for the measurement of hemodynamic changes with optical imaging is the limited penetration of NIR photons into the head tissues. This makes it impossible to derive measures from deep structures.

With NIRS it is feasible (and in fact already done in practice) to obtain almost continuous estimates of changes in the concentration of oxy- and deoxy-hemoglobin in the cortex with a resolution of a few centimeters (Villringer & Chance, 1997). Recent data show that this spatial resolution could be improved using relatively dense optode arrays (Gratton, Goodman-Wood, & Fabiani, 2001). Although this resolution is inferior to that of fMRI, it is still sufficient to differentiate cortical regions on the scale of Brodmann areas. NIRS has therefore been proposed as a tool for augmenting cognition (Fabiani, Schmorrow, & Gratton, 2007). Among its most attractive aspects are its relatively low cost and invasivity, as well as the possibility of coupling it with a variety of other techniques.

Interestingly, the same devices can be used to record simultaneously the EROS and NIRS signals (Gratton et al., 2001). It is also possible to record both these signals as well as EEG and ERPs simultaneously (Gratton et al., 2001). As for EEG, telemetric techniques (Hoshi, 2003) and even remote recording methods are currently being developed.

Measures of Incidental Behavior Related to Brain Processing

Valuable Information: Eye Tracking
Over the past half century, the examination of eye movements and other oculomotor features has served an important role in the understanding of human attention, visual search, and memory in both basic laboratory studies of these processes and simulated and real-world transportation, medical, and industrial systems. Compared with the brain indices reviewed earlier, eye tracking is more portable and lower in cost and yet can provide an index of brain states and processes, albeit not a direct one.

Early studies focused on the use of video-based eye-tracking systems to examine the effect of aircraft cockpit designs on pilot performance, the manner in which individuals inspect photographs and paintings, and the study of automobile drivers (Fitts, Jones, & Milton, 1950; Mourant & Rockwell, 1970; Yarbus, 1967). Although these studies yielded important information concerning the manner in which individuals extract and process relevant features of a visual environment, eye movement recording was cumbersome, and the analysis of data was extremely time-consuming (given the requirement to visually inspect many frames of videotape). More recent studies have employed both head-mounted and remote eye-tracking equipment that can enable the recording of eye position and eye movements with high degrees of temporal and spatial precision, as well as semiautomated scoring and analysis of eye movement data (see Jacob & Karn, 2003, for a description of eye-tracking technology).

Several types of eye movements have been examined in the aforementioned studies: **saccadic movements, pursuit movements, and vergence movements.** Vergence shifts are movements in which the left and right eyes rotate in opposite directions, moving the point of regard in depth relative to the observer. Pursuit movements are those in which the eyes travel smoothly and in conjunction so as to maintain fixation on a moving object. Of most interest to researchers studying augmented cognition are saccades, ballistic movements that rapidly shift the observer's gaze from one area of interest to another. Saccadic eye movements are typically 30 to 50 ms in duration and can reach velocities of 500°/s. They tend to occur 3–4 times/s in normal scene viewing and are separated by fixations that are generally 200–300 ms in duration.

Saccades can be broadly classified as *reflexive, voluntary,* or *memory-guided.* Reflexive saccades are visually guided movements programmed automatically in response to a transient signal at the saccade target location. Voluntary saccades are programmed endogenously to a location not marked by a visual transient. Memory-guided saccades are made to a cued or previously attended location, but only after delay (Pierrot-Deseilligny, Milea, & Müri, 2004).

The use of eye movements, primarily saccades, to study human performance and cognition has often been predicated on the **eye-mind hypothesis.** This hypothesis states than **an individual is attending to the location of a fixation** (Just & Carpenter, 1980). Although this is often the case, it is known from several decades of research that attention can be focused on one area of the visual field while the eyes are focused elsewhere (Posner, 1980). Knowledge of such dissociation has been acquired in experiments in which individuals are required to fixate in the center of a display and attend to locations in the visual periphery. Areas of the visual field that are attended to in such studies show performance benefits when stimuli are presented in attended locations.

A second caution with respect to the eye-mind hypothesis is that fixation on a particular area or object in the visual field does not always imply adequate processing to detect or remember the fixated event. For example, Strayer, Drews, and Johnston (2003) showed that conversation on a hands-free cell phone resulted in poor recognition of road signs in a simulated driving task, even when drivers had fixated on the road signs. Indeed, automobile accidents have been attributed to "looked but failed to see" instances; that is, the failure to appreciate the urgency of a potential collision situation despite fixating on another vehicle or on a pedestrian (Langham, Hole, Edwards, & O'Neil, 2002).

Despite the demonstration that attention can be allocated independently of eye position, it generally appears to be the case that there is a functional relationship between attention and eye movements, such that the processing of visual stimuli is very efficient when locations to be attended to are the same as the position to which saccades are directed (Deubel & Schneider, 1996; Hoffman & Subramaniam, 1995). Indeed, attention appears to be covertly directed to locations in advance of overt eye movements or saccades (Henderson, Pollatsek, & Rayner, 1987; Peterson, Kramer, & Irwin, 2004).

Several different **measures of oculomotor behavior** have been employed in the study of human performance and cognition. *Fixation durations,* the durations between saccades, are generally acknowledged to provide a measure of the difficulty of extracting information from the fixated object or region, the informativeness of the fixated information, and/or preparation or planning for subsequent saccades (Bellenkes et al., 1997; Loftus & Mackworth, 1978). Multiple consecutive fixations on a single area, object, or instrument are usually grouped together and referred to as a *gaze duration.*

Measures of gaze duration are particularly useful when individuals need to scan large objects (e.g., a ship or aircraft) or complex multifunctional displays such as those found in a modern-day aircraft, to extract task-relevant information. The number or frequency of fixations in different areas of interest is also a measure of eye movement behavior that has proven useful in laboratory, simulator, and real-world studies of human perception and cognition. Fixation frequency is usually employed as a measure of the importance of a particular area of interest or instrument, such as the directional gyro in an aircraft, the road ahead of an automobile, or a suspect region in luggage during airport baggage screening (Bellenkes et al., 1997; McCarley et al., 2004a, 2004b; Mourant & Rockwell, 1970).

The eye movement measures that we have discussed thus far are recorded as individuals freely scan a picture, scene, or instrument panel in a vehicle or industrial system. However, in recent years **other methods of utilizing eye movement measures also have been developed.** These procedures, called *moving-window* or *gaze-contingent displays* (Reingold, Loschky, McConkie, & Stampe, 2003), entail changes in displays contingent on an individual's eye position or eye movements. For example, those who study reading have employed such displays to examine the type of information that is used from the visual periphery to guide eye movements. In this case the area of high-resolution information surrounding a fixation is systematically varied. This research has determined that the perceptual span (i.e., the region of space, in letters, that can be discriminated) extends 3–4 characters to the left of a reader's fixation and approximately 14–15 characters to the right of fixation (McConkie & Rayner, 1975). (Interestingly, this asymmetry is reversed when reading Hebrew; see Pollatsek, Bolozky, Well, & Rayner, 1981.) Thus, perceptual span appears to develop as a function of reading experience to help to efficiently guide in the extraction of relevant information.

More recently, McCarley and colleagues used a variant of the gaze-contingent paradigm to examine the role and nature of **memory processes** in guiding visual search behavior. In the version of the contingent search paradigm developed by McCarley and colleagues (2003; see also Boot et al., 2004), a subject is initially presented with a fixation cross and one search item on a computer display. At this point, there is only one possible search item to examine. During the initial saccade to the first search item, a second search item is presented. When new search items are presented during saccades, saccadic suppression prevents the new items from acting as abrupt onsets and capturing attention.

Gaze-contingent presentation of potential saccade targets also eliminates the ability to plan multiple saccades (and therefore prospective memory), thereby providing a relatively pure measure of the capacity of retrospective memory in dynamic search tasks. If the fixated item is the target, the subject is to respond with the appropriate key press and the trial is terminated. If the fixated item is a distractor, search continues until the target is found or all the items have been presented. After the first saccade, the subject still has only one potential saccade target. As he or she makes the next saccade, a new item is added to the display and the first item remains on the screen. From this point on, three items are always present in the display: the currently fixated item and two potential saccade targets. One of the potential saccade targets is always a new item that has not yet been identified, and the other is an old item that has already been examined (the *decoy*). Decoys reappear in the same location and orientation as originally presented, and except for the third event, when there is only one previously examined item, decoys are randomly chosen from the list of examined items.

Critical to this paradigm is that the subject is forced to choose between making a saccade to a new item or to a decoy. If visual search is guided by a memory that is perfect, then a new item will always be chosen. If there is no memory, or memory has failed, the subject will choose to make a saccade to a decoy half the time. McCarley and colleagues (2003) examined how long memory lasts during visual search by looking at the probability that a decoy was reexamined based on the number of intervening fixations since the decoy was first examined (*lag*). Memory for the last two previously fixated items is quite good; most of the saccades go to the new item. However, with additional lags, performance eventually reaches chance levels (generally by the fourth or fifth lag), which provides an estimate of the size of the memory buffer that supports memory for previously visited locations in search.

Follow-up studies with this gaze-contingent paradigm have revealed the following:

a. The memory representations that underlie visual search are largely implicit (Boot et al., 2004).
b. Background layout and landmarks boost memory for previously inspected locations (Becic, Kramer, & Boot, 2007a; Peterson et al., 2004).
c. Substantial individual differences exist in the size of the memory buffer that supports efficient visual search (Becic, Kramer, & Boot, 2007b; Kramer et al., 2006).
d. Memory representations are largely spatial and appear to contain little information about the identity of previously inspected objects (Beck, Peterson, Boot, Vomela, & Kramer, 2006; Beck, Peterson, & Vomela, 2006).
e. Both prospective and retrospective memory support visual search (Peterson et al., 2007).

Review

Eye movement research has a long history of supporting the study of human cognition and performance in laboratory, simulator, and real-world tasks and systems. Future development of eye movement techniques will depend on both further development of models of oculomotor control that take into account attention, memory, and other cognitive processes, and on the coordination and integration of eye-movement-tracking technology with other techniques such as ERPs and optical and fMRI.

Test Your Knowledge

What types of eye movements can be measured, and what are their characteristics?

What is the mind-eye hypothesis? Is the eye movement hypothesis supported by recent studies?

Lessons Learned

In this chapter we have reviewed both neuronal and hemodynamic methods that could be used as sensors in closed-loop augmented cognition applications. Table 1.1 summarizes the lessons learned from our review of these methods.

Table 1.1. Lessons Learned in Determining the Suitability of Physiological Sensors for Closed-Loop Augmented Cognition Systems

Lesson Learned	Example of Lesson in Action
Neuronal methods provide timely information with good temporal resolution, but their spatial resolution is variable.	EEG can be used to measure the brain's general level of arousal with good signal-to-noise ratio, low cost, low invasivity, and relatively good portability. However, low spatial resolution may limit the use of EEG for more complex classifications of brain states and processes, such as those required to assess whether an operator may be primed to process one type of input over another. EROS may provide an alternative measure of neuronal activity with good spatial and temporal resolution.
Hemodynamic methods have excellent spatial resolution and a relatively high signal-to-noise ratio; however, the need for contrast agents and/or bulky equipment makes these methods ill-suited for augmented cognition applications, with the exception of NIRS measured, which can be obtained with a low-profile system.	NIRS can provide almost continuous estimates of changes in the concentration of oxy- and deoxy-hemoglobin in the cortex with a resolution of a few centimeters, which is sufficient to differentiate cortical regions on the scale of Brodmann areas, providing a relatively low cost and low-invasivity solution for augmented cognition.
Coupling methods may provide the best solution.	The same devices can be used to record simultaneously the EROS and NIRS signals, as well as EEG and ERPs. Coupling methods can provide complementary information. For example, EEG measures electrical activity and NIRS measures blood oxygenation levels in the brain, thereby providing complementary sources of information about brain function.
Measures of incidental behavior can provide indirect information regarding human attention, visual search, and memory function, thereby complementing information provided by neuronal and hemodynamic methods.	Eye tracking can provide critical information about a subject's attentional states in a portable and low-cost solution.

Parting Message

Physiological brain measures are critical elements in the closed loop used in augmented cognition. They can provide crucial information about brain states and processes. To be usable in applied settings, they should be informative about a number of different states and processes, provide reliable data, and respond to a number of practical requirements. It is important to note, however, that knowledge about brain physiology and function derived from all these methods, as well as from more invasive methods used in nonhuman studies, can provide invaluable data to inform theory of mind as well as the design and implementation of augmented cognition loops.

We have considered three types of neuronal methods: EEG/ERPs, MEG, and EROS. They typically provide timely information with good temporal resolution, but their spatial resolution is variable (with EROS and MEG providing the highest spatial resolution and EEG/ERPs the lowest). Also, in many cases, their signal-to-noise ratio is low. However, they are relatively easy to adapt to practical settings (with the exception of MEG). Some of these methods are sufficiently well developed for use in applied settings, such as the use of EEG to measure operator alertness, whereas others are being investigated for similar applications. Both EEG/ERPs and EROS appear to be promising techniques for future development.

Hemodynamic methods (O_{15}PET, BOLD-fMRI, and NIRS) are among the most commonly used techniques in cognitive neuroscience. Although informative in terms of spatial resolution (providing the potential ability to measure a variety of brain states and processes) and possessing relatively high signal-to-noise ratios, PET and fMRI are impractical for most applied settings. However, NIRS may preserve some of the advantages of these technologies in a low-profile system that could have extensive practical applications.

Measures of incidental behavior, such as eye tracking, can also provide critical information about a subject's attentional states. Combining multiple technologies may further enrich the armamentarium of augmented cognition, providing the theoretical possibility of monitoring in real time a variety of states and processes of the brain.

Test Your Knowledge
List neuronal and hemodynamic measures of brain function. How do neuronal and hemodynamic measures compare with each other for applied use?

Acknowledgment

The authors wish to acknowledge the support of ONR MURI to Arthur Kramer (Gabriele Gratton and Monica Fabiani, co–principal investigators) in the preparation of this chapter.

References

Becic, E., Kramer, A. F., & Boot, W. R. (2007a). Age-related differences in the use of background layout in visual search. *Aging, Neuropsychology, and Cognition, 14*, 1–17.

Becic, E., Kramer, A. F., & Boot, W. R. (2007b). Age-related differences in visual search in dynamic displays. *Psychology & Aging, 22*(1), 67–74.

Beck, M. R., Peterson, M. S., Boot, W. R., Vomela, M., & Kramer, A. F. (2006). Explicit memory for rejected distractors during visual search. *Visual Cognition, 14*(2), 150–174.

Beck, M. R., Peterson, M. S., & Vomela, M. (2006). Memory for where, but not what, is used during visual search. *Journal of Experimental Psychology: Human Perception and Performance, 32*(2), 235–250.

Bellenkes, A. H., Wickens, C. D., & Kramer, A. F. (1997). Visual scanning and pilot expertise: The role of attentional flexibility and mental model development. *Aviation, Space, and Environmental Medicine, 68*, 569–579.

Berger, H. (1929). Über das Elektroenkephalogramm des Menschen [On the human electroencephalogram]. *Archive Psychiatrie Nerven, 87*, 527–570.

Boot, W. R., McCarley, J. S., Kramer, A. F., & Peterson, M. S. (2004). Automatic and intentional memory processes in visual search. *Psychonomic Bulletin and Review, 11*, 854–861.

Canolty, R. T., Edwards, E., Dalal, S. S., Soltani, M., Nagarajan, S. S., Kirsch, H. E., Berger, M. S., Barbaro, N. M., & Knight, R. T. (2006). High gamma power is phase-locked to theta oscillations in human neocortex. *Science, 313*, 1626–1628.

Deubel, H., & Schneider, W. X. (1996). Saccade target selection and object recognition: Evidence for a common attentional mechanism. *Vision Research, 36*, 1827–1837.

Donchin, E. (1981). Surprise! ... surprise? *Psychophysiology, 18*, 493–513.

Donchin, E., & Coles, M. G. (1988). Is the P300 component a manifestation of context updating? *Behavioral & Brain Sciences, 11*, 357–427.

Donchin, E., Heffley, E., Hillyard, S. A., Loveless, N., Maltzman, I., Ohman, A., Rosler, F., Ruchkin, D., & Siddle, D. (1984). Cognition and event-related potentials II. The orienting reflex and P300. *Annals of the New York Academy of Sciences, 425*, 39–57.

Donchin, E., Ritter, W., & McCallum, C. (1978). Cognitive psychophysiology: The endogenous components of the ERP. In E. Callaway, P. Tueting, & S. H. Koslow (Eds.), *Event-related brain potentials in man* (pp. 349–411). New York: Academic Press.

Erdogmus, D., Adami, A., Pavel, M., Tian, L., Mathan, S., Whitlow, S., & Dorneich, M. (2005). Cognitive state estimation based on EEG for augmented cognition. *2nd International IEEE EMBS Conference on Neural Engineering* (pp. 566–569).

Fabiani, M., Schmorrow, D., & Gratton, G. (Eds.). (2007). Special issue of *IEEE EMB Magazine on Optical Imaging, 26*(4).

Fabiani, M., Gratton, G., & Federmeier, K. (2007). Event related brain potentials. In J. Cacioppo, L. Tassinary, & G. Berntson (Eds.), *Handbook of psychophysiology* (3rd ed., pp. 85–119). New York: Cambridge University Press.

Fabiani, M., Gratton, G., Karis, D., & Donchin, E. (1987). Definition, identification, and reliability of measurement of the P300 component of the event related brain potential. In P. K. Ackles, J. R. Jennings, & M. G. H. Coles (Eds.), *Advances in psychophysiology* (Vol. 2, pp. 1–78). Greenwich, CT: JAI Press.

Falkenstein, M., Hohnsbein, J., Hoormann, J., & Blanke, L. (1991). Effects of crossmodal divided attention on late ERP components. II. Error processing in choice reaction tasks. *Electroencephalography and Clinical Neurophysiology, 78*, 447–455.

Farwell, L. A., & Donchin, E. (1988). Talking off the top of your head: A mental prosthesis utilizing event-related brain potentials. *Electroencephalography & Clinical Neurophysiology, 70*, 510–523.

Farwell, L. A., & Donchin, E. (1991). The truth will out: Interrogative polygraphy ("lie detection") with event-related brain potentials. *Psychophysiology, 28*, 531–547.

Fitts, P. M., Jones, R. E., & Milton, J. L. (1950). Eye movements of aircraft pilots during instrument-landing approaches. *Aeronautical Engineering Review, 9*, 24–29.

Frackowiak, R. S., Friston, K. J., Frith, C., Dolan, R., Price, P., Zeki, S., Ashburner, J., & Penny, W. D. (2003). *Human brain function*. New York: Academic Press.

Franceschini, M. A., & Boas, D. A. (2004). Noninvasive measurement of neuronal activity with near-infrared optical imaging. *NeuroImage, 21*, 372–386.

Frostig, R. D., Lieke, E. E., Ts'o, D. Y., & Grinvald, A. (1990). Cortical functional architecture and local coupling between neuronal activity and the microcirculation revealed by in vivo high-resolution optical imaging of intrinsic signals. *Proceedings of the National Academy of Sciences, 87,* 6082–6086.

Gehring, W. J., Goss, B., Coles, M. G. H., Meyer, D. E., & Donchin, E. (1993). A neural system for error detection and compensation. *Psychological Science, 4,* 385–390.

Gratton, G., & Fabiani, M. (2003). The event related optical signal (EROS) in visual cortex: Replicability, consistency, localization and resolution. *Psychophysiology, 40,* 561–571.

Gratton, G., & Fabiani, M. (2007). Optical imaging. In R. Parasuraman & M. Rizzo (Eds.), *Neuroergonomics: The brain at work* (pp. 65–81). Cambridge, MA: Oxford University Press.

Gratton, G., Brumback, C. R., Gordon, B. A., Pearson, M. A., Low, K. A., & Fabiani, M. (2006). Effects of measurement method, wavelength, and source-detector distance on the fast optical signal. *NeuroImage, 32,* 1576–1590.

Gratton, G., Coles, M. G. H., & Donchin, E. (1989). A procedure for using multielectrode information in the analysis of components of the event-related potentials: Vector filtering. *Psychophysiology, 26,* 222–232.

Gratton, G., Coles, M. G. H., Sirevaag, E., Eriksen, C. W., & Donchin E. (1988). Pre and post-stimulus activation of response channels: A psychophysiological analysis. *Journal of Experimental Psychology: Human Perception and Performance, 11,* 331–344.

Gratton, G., Corballis, P. M., Cho, E., Fabiani, M., & Hood, D. (1995). Shades of gray matter: Noninvasive optical images of human brain responses during visual stimulation. *Psychophysiology, 32,* 505–509.

Gratton, G., & Fabiani, M. (1998). Dynamic brain imaging: Event-related optical signal (EROS) measures of the time course and localization of cognitive-related activity. *Psychonomic Bulletin & Review, 5,* 535–563.

Gratton, G., Fabiani, M., Corballis, P. M., Hood, D. C., Goodman-Wood, M. R., Hirsch, J., Kim, K., Friedman, D., & Gratton, E. (1997). Fast and localized event-related optical signals (EROS) in the human occipital cortex: Comparisons with the visual evoked potential and fMRI. *NeuroImage, 6,* 168–180.

Gratton, G., Goodman-Wood, M. R., & Fabiani, M. (2001). Comparison of neuronal and hemodynamic measure of the brain response to visual stimulation: An optical imaging study. *Human Brain Mapping, 13,* 13–25.

Gratton, G., Sarno, A. J., Maclin, E., Corballis, P. M., & Fabiani, M. (2000). Toward noninvasive 3-D imaging of the time course of cortical activity: Investigation of the depth of the event-related optical signal (EROS). *NeuroImage, 11,* 491–504.

Hari, R., Levanen, S., & Raij, T. (2000). Timing of human cortical functions during cognition: Role of MEG. *Trends in Cognitive Science, 4,* 455–462.

Henderson, J. M., Pollatsek, A., & Rayner, K. (1987). Effects of foveal priming and extrafoveal preview on object identification. *Journal of Experimental Psychology: Human Perception & Performance, 13,* 449–463.

Hoffman, J. E., & Subramaniam, B. (1995). The role of visual attention in saccadic eye movements. *Perception & Psychophysics, 57,* 787–795.

Hoshi, Y. (2003) Functional near-infrared optical imaging: Utility and limitations in human brain mapping. *Psychophysiology, 40*(4), 511–520.

Jacob, J. K., & Karn, K. S. (2003). Eye-tracking in human-computer interaction and usability research: Ready to deliver the promises. In J. Hyönä, R. Radach, & H. Deubel (Eds.), *The mind's eye: Cognitive and applied aspects of eye movement research* (pp. 574–605). Amsterdam: North-Holland.

Just, M., & Carpenter, P. A. (1980). A theory of reading: From eye fixations to comprehension. *Psychological Review, 87,* 329–354.

Keckluno, G., & Akersteot, T. (1993). Sleepiness in long distance truck driving: An ambulatory EEG study of night driving. *Ergonomics, 36,* 1007–1017.

Kramer, A. F., & Parasuraman, R. (2007). Neuroergonomics: Applications of neuroscience to human factors. In J. Cacioppo, L. Tassinary, & G. Berntson (Eds.), *Handbook of psychophysiology* (3rd ed., pp. 704–722). New York: Cambridge University Press.

Kramer, A. F., Boot, W. R., McCarley, J. S., Peterson, M. S., Colcombe, A., & Scialfa, C. T. (2006). Aging, memory and visual search. *Acta Psychologica, 122,* 288–304.

Kramer, A. F., Wickens, C. D., & Donchin, E. (1985). Processing of stimulus properties: Evidence for dual-task integrality. *Journal of Experimental Psychology: Human Perception and Performance, 11,* 393–408.

Langham, M., Hole, G., Edwards, J., & O'Neil, C. (2002). An analysis of "looked but failed to see" accidents involving parked police cars. *Ergonomics, 45,* 167–185.

Linden, D. E. J. (2005). The P300: Where in the brain is it produced and what does it tell us? *The Neuroscientist, 11,* 563–576

Loftus, G. R., & Mackworth, N. H. (1978). Cognitive determinants of fixation location during picture viewing. *Journal of Experimental Psychology: Human Perception & Performance, 4,* 565–572.

Luck, S. J. (2006). *An introduction to the event-related potential technique.* Cambridge: MIT Press.

Maclin, E., Gratton, G., & Fabiani, M. (2003). Optimum filtering for EROS measurements. *Psychophysiology, 40,* 542–547.

Maclin, E. L., Low, K. A., Fabiani, M., & Gratton, G. (2007). Improving the signal-to-noise ratio of event related optical signals (EROS) by manipulating wavelength and modulation frequency. *Special issue of IEEE EMBM, 26*(4), 47–51.

McCarley, J. S., Kramer, A. F., Wickens, C. D., Vidoni, E. D., & Boot, W. R. (2004a). Visual skills in airport security inspection. *Psychological Science, 15,* 302–306.

McCarley, J. S., Vais, M. J., Pringle, H. L., Kramer, A. F., Irwin, D. E., & Strayer, D. L. (2004b). Conversation disrupts visual scanning and change detection in complex traffic scenes. *Human Factors, 46,* 424–436.

McCarley, J. S., Wang, R. F., Kramer, A. F., Irwin, D. E., & Peterson, M. S. (2003). How much memory does oculomotor search have? *Psychological Science, 14*(5), 422–426.

McCarthy, G., & Wood, C. C. (1987). Intracranial recordings of endogenous ERPs in humans. *Electroencephalography and Clinical Neurophysiology Supplement, 39,* 331–337.

McConkie, G., & Rayner, K. (1975). The span of the effective stimulus during a fixation in reading. *Perception & Psychophysics, 17,* 578–586.

Mourant, R. R., & Rockwell, T. H. (1970). Mapping eye movement pattern to the visual scene in driving: An exploratory study. *Human Factors, 12,* 81–87.

Noll, D. C., Genovese, C. R., Nystrom, L. E., Vazquez, A. L., Forman, S. D., Eddy, W. F., & Cohen, J. D. (1997). Estimating test-retest reliability in functional MR imaging II: Application to motor and cognitive activation studies. *Magnetic Resonance in Medicine, 38,* 508–517.

Ogawa, S., Lee, T. M., Kay, A. R., & Tank, D. W. (1990). Brain magnetic resonance imaging with contrast dependent on blood oxygenation. *Proceedings of the National Academy of Sciences, 87,* 9868–9872.

Peterson, M. S., Beck, M. R., & Volema, M. (2007). Visual search is guided by prospective and retrospective memory. *Perception & Psychophysics, 69*(1), 123–135.

Peterson, M. S., Kramer, A. F., & Irwin, D. E. (2004). Covert shifts of attention precede involuntary eye movements. *Perception & Psychophysics, 66*(3), 398–405.

Pierrot-Deseilligny, C., Milea, D., & Müri, R. (2004). Eye movement control by the cerebral cortex. *Current Opinions in Neurology, 17,* 17–25.

Polich, J., & Kok, A. (1995). Cognitive and biological determinants of P300: An integrative review. *Biological Psychology, 41,* 103–46.

Pollatsek, A., Bolozky, S., Well, A. D., & Rayner, K. (1981). Asymmetries in the perceptual span for Israeli readers. *Brain and Language, 14,* 174–180.

Posner, M. I. (1980). Orienting of attention. *Quarterly Journal of Experimental Psychology, 32*, 3–25.

Posner, M. I., Petersen, S. E., Fox, P. T., & Raichle, M. E. (1988). Localization of cognitive operations in the human brain. *Science, 240*, 1627–1631.

Rector, D. M., Carter, K. M., Volegov, P. L., & George, J. S. (2005). Spatio-temporal mapping of rat whisker barrels with fast scattered light signals. *Neuroimage, 26*, 619–627.

Reingold, E. M., Loschky, L. C., McConkie, G. W., & Stampe, D. M. (2003). Gaze-contingent multiresolutional displays: An integrative review. *Human Factors, 45*, 307–328.

Ruthruff, E., Johnston, C. J., Van Selst, M., Whitsell, S., & Remington, R. (2003). Vanishing dual-task interference after practice: Has the bottleneck been eliminated or is it merely latent? *Journal of Experimental Psychology: Human Perception and Performance, 29*, 280–289.

Sigman, M., Jorbet, A., Lebihan, M. & Dehaene, S. (2007). Parsing a sequence of brain activations at psychological times using fMRI. *Neuroimage, 35*(2), 655–668.

Sing, H., Kautz, M., Thorn, D., Hall, S., Redmond, D., & Russo, M. (2006) High frequency EEG as a potential indicator of alertness, attention, and cognition. *Aviation, Space, and Environmental Medicine, 77*(3), 186.

Sirevaag, E. J., Kramer, A. F., Coles, M. G., & Donchin, E. (1989). Resource reciprocity: An event-related brain potentials analysis. *Acta Psychologica, 70*, 77–97.

St. John, M., Kobus, D. A., Morrison, J., & Schmorrow, D. (2004). Overview of the DARPA Augmented Cognition Technical Integration Experiment. *International Journal of Human-Computer Interaction, 17*, 131–149.

Strayer, D. L., Drews, F. A., & Johnston, W. A. (2003). Cell phone–induced failures of visual attention during simulated driving. *Journal of Experimental Psychology: Applied, 9*, 23–32.

Sutton, S., Braren, M., Zubin, J., & John, E. R. (1965). Evoked potential correlates of stimulus uncertainty. *Science, 150*, 1187–1188.

Tse C.-Y., Tien, K.-R., & Penney, T. B. (2006). Event-related optical imaging reveals the temporal dynamics of right temporal and frontal cortex activation in pre-attentive change detection. *NeuroImage. 29*, 314–320.

Uğurbil, K., Hu, X., Chen, W., Zhu, X. H., Kim, S. G., & Georgopoulos, A. (1999). Functional mapping in the human brain using high magnetic fields. *Philosophical Transactions of the Royal Society London B. Biological Science, 354*, 1195–1213.

Verleger, R. (1988). Event-related potentials and cognition: A critique of the context updating hypothesis and an alternative interpretation of P3. *Brain & Behavioral Sciences, 11*, 343–427.

Villringer, A., & Chance, B. (1997). Non-invasive optical spectroscopy and imaging of human brain function. *Trends in Neuroscience, 20*, 435–442

Walter, W. G., Cooper, R., Crow, H. J., McCallum, W. C., Warren, W. J., Aldridge, V. J., Storm Van Leewen, W., & Kamp, A. (1967). Contingent negative variation and evoked responses recorded by radiotelemetry in free-ranging subjects. *Electroencephalography and Clinical Neurophysiology, 23*, 197–206.

Wickens, C. D. (2002). Multiple resources and performance prediction. *Theoretical Issues in Ergonomics, 3*, 159-177.

Wilson, B. C., Patterson, M. S., Flock, S. T., & Wyman, D. R. (1989). Tissue optical properties in relation to light propagation models and in vivo dosimetry. In B. Chance (Ed.), *Photon migration in tissues* (pp. 25–42). New York: Plenum.

Wolf, U., Wolf, M., Toronov, V., Michalos, A., Paunescu, L. A., & Gratton, E. (2000). Detecting cerebral functional slow and fast signals by frequency-domain near-infrared spectroscopy using two different sensors. Paper presented at OSA Meeting in Optical Spectroscopy and Imaging and Photon Migration, April 2–5, Miami, FL.

Wolpaw, J. R., McFarland, D. J., Neat, G. W., & Forneris, C. A. (1991). An EEG-based brain-computer interface for cursor control. *Electroencephalography and Clinical Neurophysiology, 78*, 252–259.

Yamazaki, T., Kamijo, K., Kiyuna, T., Takaki, Y., & Kuroiwa, Y. (2001). Multiple dipole analysis of visual event-related potentials during oddball paradigm with silent counting. *Brain Topography, 13*, 161–168.

Yarbus, A. L. (1967). *Eye movements and vision.* New York: Plenum Press.

Zhao, J., Valle-Inclan, F., & Hackley, S. A. (2006). Fast optical signal detected in the prefrontal lobe with near-infrared spectroscopy during sleep. *Psychophysiology, 43*, 110.

Chapter 2

Functional Near-Infrared (fNIR) Sensors

Evan D. Rapoport[1], Erin M. Nishimura[1], Colby Raley[2], Traci H. Downs[1], and J. Hunter Downs III[1]
[1]Archinoetics, LLC
[2]Strategic Analysis, Inc.

Here we describe proper practices for developing hardware and software for physiological sensors that integrate seamlessly with augmented cognition systems.

Introduction

In this chapter, we cover the development and integration of **functional near-infrared imaging (fNIR)** sensors, which are **physiological sensors** that can be used to drive adaptation (e.g., mitigations, see Chapters 5 and 6) in augmented cognition systems. The integration of physiological sensors for an adaptive or information system are also discussed. For illustrative purposes, detailed examples and specific references are included based on the authors' experience developing an **fNIR** system (called **OTIS**). The chapter includes background on this relatively new **functional brain-imaging** technology as well as the OTIS system, including its hardware and software components.

The design process for developing the OTIS system within a closed-loop framework is described, covering the iterative process used for prototyping and testing, lessons learned along the way, and guidelines developed. Several considerations and requirements arose throughout the iterative development process. Some of these will be discussed in this chapter, including **rapid placement and calibration processes**, software and **signal-processing modules**, and **easy integration methods** for collaborators and other third parties to utilize.

Scenario

Many appropriate scenarios for implementing augmented cognition systems utilize physiological sensing technologies. A primary application is that of an overloaded operator who must constantly multitask. Surrounded by a number of computer monitors, speakers, telephones, and other people in the workplace, the operator is confronted with a large volume of data, which requires a high cognitive workload to process everything. Such an individual may face a great deal of stress and fatigue as a result. These factors may reduce the person's ability to process all incoming information adequately.

In a dynamic operational environment like this, a computer is unlikely to have enough information about the environment to create accurate models of the person's cognitive and physiological states. As a result, some tasks may back up and/or be performed inadequately because of operator overload. A closed-loop augmented cognition system that uses physiological sensors could directly measure the impact of the tasks on the person's capabilities, thus enabling the system to adapt, in real-time, to the changing performance of the human.

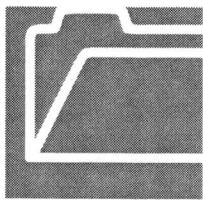

Valuable Information
A closed-loop augmented cognition system uses information about the physiological state of the operator to modify, in real-time, the appearance and/or functionality of a user interface such that it adapts to the operator's present cognitive capabilities. A system that is not closed-loop in this manner could fail to adapt to the operator's changing needs and abilities determined by changes in physiological state.

One example of an operator in this position would be an air traffic controller. The operator must monitor the status of numerous inbound and outbound flights while communicating with pilots, ground crew, and others in the control tower. Environmental factors such as poor weather conditions and darkness can cause additional concerns. The safety of thousands of people and expensive aircraft relies on the proper execution of the operator's responsibilities, which could create a stressful environment in which failure may result in serious consequences.

Given such intense job requirements, physiological sensors would be useful in providing critical information about the operator that can directly monitor cognitive states and may even indicate signs of fatigue, stress, or other potential detriments to performance. For example, if the air traffic controller were looking at flight maps, an fNIR sensor could detect high load on areas of the brain responsible for spatial working memory. Recognizing that new visual and spatial tasks could cause overload and errors, the system would modify any new incoming messages, with one option being to provide oral messages to the operator, addressing his verbal working memory system, which is less engaged and thus more receptive to new information.

Test Your Knowledge
Answer these questions to test your knowledge of scenarios for implementing closed-loop augmented cognition systems: Identify a candidate scenario in a domain with which you are familiar. (1) What are the challenges and demands for the operator? (2) What physiological information about the user (cognitive workload, heart rate, stress, fatigue, etc.) would reflect his or her current workload and/or ability to perform his/her tasks satisfactorily? (3) Based on that information, how might you modify the user interface to "close the loop" and make this system an adaptive augmented cognition system?

Another recent application for human-systems technologies is improving the design process that produces user interfaces for heavily-tasked operators. In a user-centered design (UCD) process, an interface is evaluated and modified based on the workload (among other factors) it requires to complete a given task. Traditional user workload measures (e.g., Situation Awareness Global Assessment Technique, or SAGAT, and NASA's Task Load Index, or NASA TLX) are based on observation of external signs of user performance and subjective reporting of a user's internal state after the performance of tasks (Endsley, 1988). It is difficult, if not impossible, to modify user interaction based on such post hoc analysis.

Using real-time physiological sensors in designing and evaluating human-systems interaction (HSI) allows objective and consistent measures that can supplement existing techniques of workload assessment. This also enables researchers to reduce variability in their experimental processes because the sensors can record data continuously and at precise intervals across many user interactions or multiple experiments without interrupting users.

A specific example of such experimentation is an advanced human factors effort aimed at improving the usability of future releases of the Tactical Tomahawk Weapons Control System (TTWCS). In these experiments, each user's cognitive state was measured using the TTWCS Tool for Interface Design Evaluation with Sensors (T-TIDES) toolkit, which incorporated electrocardiography (EKG), galvanic skin response (GSR), and electroencephalography (EEG), as well as traditional psychological analysis measures (see Figure 2.1). Initial results from these experiments provided researchers with significant insight into required system improvements for the useful incorporation of physiological measures into the usability testing process. Although traditional HSI experimentation is structured and rigorous, it utilizes an entirely different process and time scale than is required for physiological measures. These studies also gave a preview of the considerations for designing systems to support the collection and analysis of objective measures of cognitive workload.

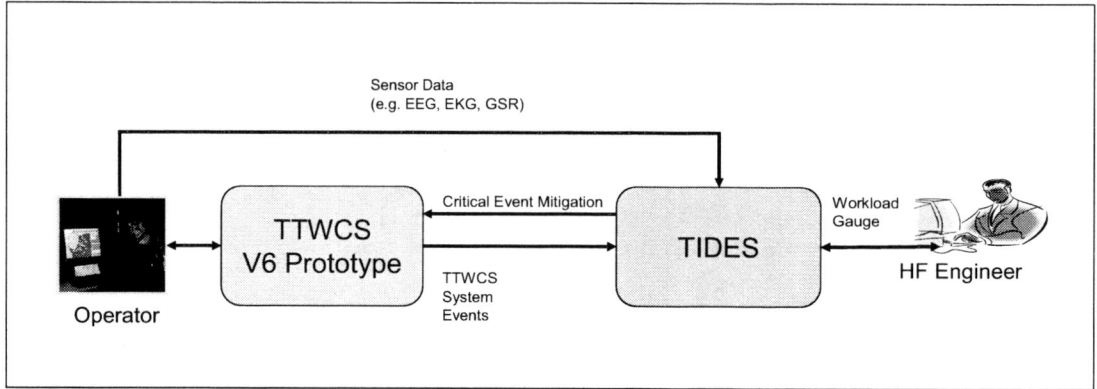

Figure 2.1. T-TIDES tool architecture from Lockheed Martin Advanced Technology Laboratories.

Background on fNIR

Near-infrared spectroscopy (NIRS) was first used by Jobsis (1977) as a noninvasive way to monitor oxygen sufficiency. NIRS has since been explored for use in various clinical applications, and most recently its potential in functional brain imaging has gained attention. It is in this latter capacity, functional near-infrared imaging (fNIR), that this current chapter is focused.

Near-infrared sensing techniques exploit the known properties of the interaction between tissue and near-infrared (NIR) light. Oxy- and deoxy-hemoglobin, present in all living human tissue, have distinct absorption spectra of NIR light. The absorption properties of oxy- and deoxy-hemoglobin at light wavelengths between 60 0nm and 900 nm allow for the determination of oxygenation levels based on the scattering response of this range of light through tissue (Cope, 1991; Schmidt, 1999).

Currently, three fundamental types of NIR instruments are used for probing the neurovascular response: (a) continuous intensity, (b) time resolved, and (c) intensity modulated (frequency). A qualifier to each of these systems is its measurement geometry, which can be either transmission geometry or a reflective geometry.

Continuous wave or continuous intensity imagers measure attenuation changes in the tissue by recording changes in the output intensity (I) of continuous input NIR light (I_0; see Cope, 1991; Siegel, Marota, & Boas, 1999). However, these are only relative changes. To determine quantitative changes, repeated measurements must be taken with a change in only one of the following variables: intensity, additional wavelengths, or source-detector spacing between each measurement.

Time-resolved systems measure the temporal distribution of output light ($I(t)$) resulting from an ultrafast pulse (< 10 picoseconds) of NIR light (Schmidt 1999). Measurements at multiple wavelengths enable estimations of the concentrations of different absorbers. In terms of capabilities, time-resolved systems are equivalent to intensity-modulated systems.

Intensity-modulated (frequency) systems measure the detected intensity ($I(t)$), the phase shift ($\Phi(t)$), and the modulation depth ($M(t)$) output from the tissue from an intensity-modulated (at radio frequencies[RF]) light source. Heterodyning is used to facilitate the measurement by reducing the RF to lower frequencies (typically audio), simplifying analog-digital conversion as well as decreasing phase errors (Chance, Cope, Gratton, Ramanujam, & Tromberg, 1998).

Lagging several seconds after the onset of neuronal activity, a hemodynamic and metabolic response lead to increases in blood flow and an increase in the concentration of oxy-hemoglobin in the area of the brain required for that activity. These regional rises in oxygenation and the amount of signal change can be measured by fNIR imaging in the specified area as an assessment of task performance and task demand.

Augmented cognition researchers have experimented with and refined numerous sensing technologies, including direct brain measures (such as fNIR or EEG) and indirect physiological measures (for example heart rate, EKG, pulse oximetry, posture, GSR, temperature, EOG, pupilometry, and gaze tracking; see Chapter 1). Measuring the same functional brain imaging signal as does functional magnetic resonance imaging (fMRI), fNIR systems have the distinct advantage of small size and low cost. These benefits allow for the deployment of fNIR systems in more natural environments; users do not need to be confined to a room, and there are no constraints on the tasks or additional test equipment used. Another benefit is that as an optical imaging technology, fNIR (unlike EEG) is not affected by electrically noisy environments (e.g., cockpits or control rooms), and thus its signal is more robust to such interference.

Test Your Knowledge
Functional near-infrared is a brain-imaging technology that measures what kind of changes in the cortex? What are its advantages and disadvantages in assessing task demands? There are many options for sensing cognitive workload and activity, ranging from invasive, direct neurological measures, such as implanted electrodes, all the way to indirect physiological measures such as galvanic skin response. What are the advantages and disadvantages of the different technologies? What types of sensors are better suited to specific environments and users?

General Approach and Associated Toolkit

Our challenge in developing fNIR sensor technology was to create a noninvasive, comfortable, portable, and functional brain-imaging system that measures hemodynamic changes in the brain in real time for use in closed-loop systems. Functional near-infrared imaging technology was relatively new and had not yet been implemented in this manner. Additionally, the system needed to monitor any location on the cortex, which meant that the sensor would need to be able to collect data from areas of the head covered by hair.

After testing several sensor and hardware versions to meet specified requirements, we developed the resulting system, OTIS (see Figure 2.2) as a continuous-wave fNIR system to monitor relative neural oxy- and deoxy-hemoglobin concentrations in the tissue underlying the sensors (Cope, 1991). As discussed in the previous section, fNIR imaging detects the increase in the concentration of oxy-hemoglobin in the area of the brain corresponding to the cognitive activity being performed. Thus, fNIR technology offers the promise of accurate, real-time cognitive monitoring in a portable and potentially wearable package that is robust to electronic noise and can be used in natural environments (Nishimura, Stautzenberger, Robinson, Downs, & Downs, 2007).

Figure 2.2. The OTIS fNIR System.
A two-channel sensor is placed with optical probes, which extend from the sensor, against the skin. This sensor is attached to an electronics enclosure, which handles analog processing and digitization of the optical signal (Nishimura, 2007).

The OTIS system was developed with a focus on its role in closed-loop systems. Design constraints and considerations were created with this and other end uses in mind. As a result, the current system (hardware and software) functions in many real-world environments as a means of detecting changes in cognitive activity. OTIS integrates well with third-party software and hardware through custom application programming interfaces (API); it also provides turnkey solutions for use in experiments and usability analyses that require data on cognitive workload.

The OTIS system is composed of an electronics enclosure, up to two sensors, and a software bundle. Raw optical data undergo initial processing through the hardware components before being sent via Ethernet for additional processing through the custom software applications. In addition to calculating relative oxy- and deoxy-hemoglobin concentrations through a modified Beer-Lambert equation (Franceschini et al., 2002), plotting results, and logging data, the software applications have several features for rapid calibration of the system to ensure signal quality and localization of placement over targeted brain areas.

Rapid calibration tools should ensure that sensors are properly placed to prevent inaccurate physiological data from entering the closed-loop system. OTIS's calibration tools include several measures that test the quality of the signal through simple, self-contained steps. Following these steps helps reduce noise that can enter into the fNIR signal from such factors as ambient light and poor sensor-to-skin contact. These tests are crucial because this type of noise can obliterate the optical signal and render the oxy- and deoxy-hemoglobin measures useless.

One of the primary components of the OTIS software package is the Signal Quality tool (Figure 2.3), which is part of the rapid calibration process. Physiological signals (heartbeat, respiration, etc.) unavoidably affect oxy- and deoxy-hemoglobin concentrations and are dominant oscillations that become overlaid onto the unfiltered fNIR signal. In the calibration process, the signal-to-noise ratio of these physiological signals—namely, the heartbeat—is used to verify the detection of a true oxy-hemoglobin concentration signal that is not affected by interruptions in the path of the optics. The Signal Quality tool (or a similar calibration process) should be run every time a sensor is placed on an individual. The Signal Quality tool displays statistics about features of the signal to confirm data integrity, without requiring the experimenter to have prior knowledge about the OTIS system or the fNIR imaging technology. Additionally, statistics are shown that may reveal hardware or connection problems, such as slow Ethernet connections or a system malfunction.

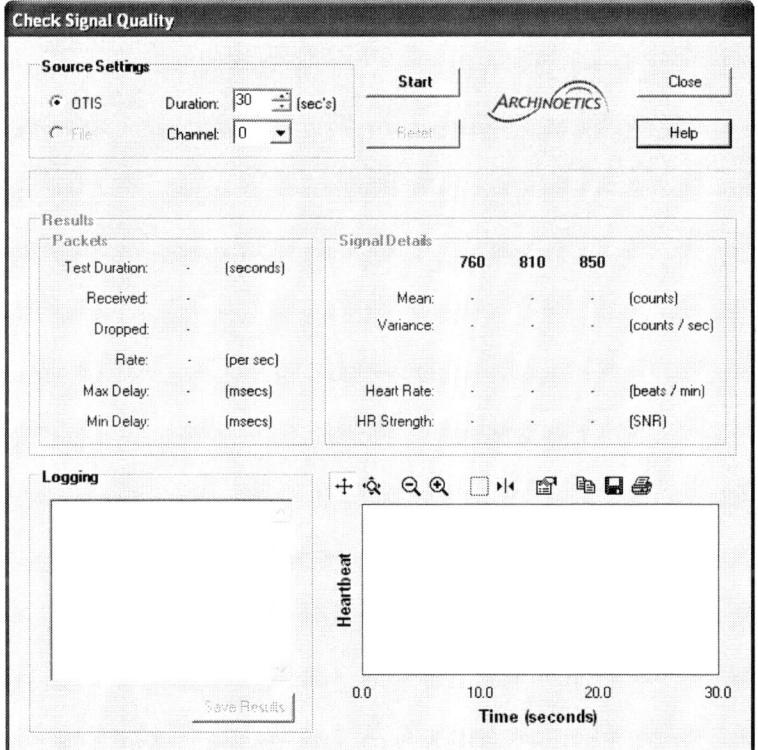

Figure 2.3. The Signal Quality Tool.

A robust calibration process should also include the use of a tool for cortical localization of the sensor over the desired area. To accomplish this in OTIS, a specialized software application executes a block design script that requests users to perform two tasks. One of the tasks (the "active" task) robustly activates the area of interest, and the other task does not (the "inactive" task). These tasks were chosen through careful literature reviews of functional brain imaging publications. Custom algorithms were then used to determine if the signal patterns for the two tasks indicated the expected differences in activity levels; a rise in the oxygenation signal was expected for the period when the active task was being performed, whereas a relatively flat signal was expected during the inactive task period. Once this pattern was achieved, it signified effective calibration. With calibration completed, in the general effort of using physiological sensors to aid in the design of interfaces, users completed task protocols, were observed and asked questions to obtain traditional psychological ratings on the interfaces, and were outfitted with the OTIS brain-sensing technology to obtain quantitative ratings during interaction.

Test Your Knowledge
What are the important components of the fNIR system discussed earlier in terms of both the hardware and the software features? What information can fNIR provide about the present state of an operator?

Lessons Learned

Through the process of integrating new physiological sensors into augmented cognition systems, we learned three primary lessons (see Table 2.1).

- First is the need to provide system developers with a rapid means of verifying the quality of the sensor's signal. Improper placement of sensors on the person can lead to inaccurate or noisy data, so establishing and maintaining data integrity are essential.
- Second is the importance of providing an easy means for third parties to receive data from physiological sensors in real time for use in their systems. Complicated integration procedures for software engineers will delay a project and inevitably create additional work for your own software team.
- Third is the benefit of a sensor simulator module that provides scripted real-time data, thus allowing the augmented cognition system to be tested thoroughly. The simulator allows for testing the entire system with a variety of physiological states of the operator, which may be difficult and timely to induce on human participants.

Additional insights regarding these lessons learned include the following.

In terms of sensor placement, working with physiological sensors—particularly new technologies—in closed-loop systems requires care in verifying that sensors are well placed and data are properly calibrated to the wearer. Depending on the application, the amount of time an individual would be willing to spend on the calibration process may vary. Streamlining the setup process so that it is fast and easy to follow will help to ensure data integrity. This may be accomplished through a wizard-type interface or through several clear steps.

As previously described with the OTIS fNIR system, a first step in using an fNIR system should be to verify that the light levels observed by the sensors' detectors fall within an acceptable range. It may prove helpful to provide a display that allows for easy visual monitoring of the values while the sensors are placed on the user.

Next, the skin-to-sensor contact should be checked through a short data collection period; we use the Signal Quality tool for this check. The signal strength of the heartbeat has a high direct correlation with the quality of this contact, so it is important to display the accuracy and signal-to-noise ratio of the heartbeat signal. If the values observed are not in acceptable ranges, sensors should be repositioned and the process repeated.

The location of the sensor on the head should be verified to make sure that readings are from the cortical area of interest. Failure to show significant differences during performance of active versus inactive tasks would indicate that a sensor is positioned over the wrong area of the brain. Executing this test during calibration ensures that the signal being measured reflects changes in the brain that the closed-loop system is using in modulating the interface.

The second lesson learned communicates the importance of providing a means for other parties to access sensor data in real time using their own software. The implementation of this lesson is beyond the scope of this chapter; however, it is crucial that the platform used and the methods for receiving sensor data are appropriate for the targeted augmented cognition applications. For example, if the sensor will need to connect to small, portable devices with low power requirements, it would be inappropriate to develop software that is processor- and memory-intensive. Moreover, developing easy, flexible ways for third-party software to receive sensor data will save time and money in the long run by reducing the number of customizations.

The third lesson learned is that there are immense benefits of a sensor simulator module that provides scripted real-time data. This should function within the same software architecture or framework as that of the actual sensor device. This will allow several things to happen:

a. The augmented cognition system may be developed and tested even when the actual sensor is not present or functional.
b. The system may be tested on physiological states of the operator that are not easily or conveniently replicable.
c. Third-party developers may be utilized effectively even if they do not have access to the physiological sensor.

Table 2.1. Lessons Learned in Designing Physiological Sensors for Closed-Loop Augmented Cognition Systems

Lesson	Purpose
Signal quality verification	Providing an easy means of ensuring a sensor's data integrity when putting it on a participant facilitates obtaining accurate readings for a closed-loop system.
Software integration	Providing an easy means of connecting to sensor software and/or hardware allows third parties to quickly begin using an fNIR system for their application.
Sensor simulation	A means of simulating different physiological states of the operator enables system designers to quickly test the impact it will have on a closed-loop system.

Two additional lessons were learned that relate to the primary lessons; these become particularly important when including multiple physiological sensors in a system. The first lesson is the importance of time-stamping for all quantitative and qualitative events. As researchers discovered, manual synchronization is inadequate for physiological sensors, as they take multiple measurements each second. Next, preplanning and "thought experiments" are important before integration events. Though the need for planning is the driving force behind pilot studies, additional work and delays could be prevented by a little mental engineering before the process starts. This would enable initial experiments to be used to refine and tweak the system setup to allow for ideal data collection, rather than to provide a forum for initial troubleshooting.

Best Practice

Developing closed-loop augmented cognition systems can be a challenging task, but following these practices may facilitate a successful implementation. The following "best practices" provide specific ways to improve your system.

Physiological monitoring, particularly when using relatively new technologies, can introduce a tremendous amount of variability into a system. Some of this variability is the result of the ever-changing physiological state of the user; however, poor practices can cause variability in the form of noisy data. Careful calibration techniques should be used to ensure the quality of the signal. This includes properly placing and localizing the sensors, verifying the presence of physiological data (such as the heart beat signal), and (if possible) executing practice trials to confirm that the desired physiological response to events is observed in the data. By eliminating unwanted noise and variability through careful calibration, you can avoid many problems of data integrity.

Furthermore, the user's comfort should be confirmed and then reconfirmed throughout data collection. Unreasonable discomfort must be avoided for reasons other than the benefit of the user; it can also cause problems for the data. First, if the user fidgets as a result of discomfort, that may introduce noise and artifacts as the sensors move. Should this occur, filters and algorithms must be able to recognize noisy data and update (or not update) the system accordingly.

Second, if the user is uncomfortable, he or she may choose to discontinue using the system. In a research study, this is typically the participant's right as granted by the informed consent form of the Institutional Review Board (IRB). Participant dropout delays the completion of a study, must be explained in reports, and can have an overall negative impact on people's perception of the study. Similarly, in an industrial, commercial, or military setting, operator discomfort may lead to the ultimate failure of the system to be accepted by the target population.

Designing an augmented cognition system to handle changes in the physiological states of an operator requires careful planning, prototyping, and, likely, several iterations. Testing the system early in the process will help to ensure that all hardware and software components connect properly; however, it is also critical to test the ability of the sensors to detect aspects of the physiology that are of interest. For example, if the system aims to monitor an operator's verbal working memory using fNIR in order to prevent auditory interruptions during verbal overload, significant testing should be done to verify that the operator's tasks do in fact activate language-processing areas (such as Broca's or Werniche's area). A literature search may support this belief, but there is no substitute for testing with the actual tasks and sensors to be used in the system.

Valuable Information
It is important to remember that most research papers report effects that are derived from cumulative results across multiple participants and multiple trials. Although such results can guide system design or help build signal-processing methods and statistical models, a closed-loop augmented cognition system does not have the same luxuries—it must function in real time while monitoring one user's data. Additionally, such systems likely will function in less controlled environments compared with environments used in many research studies. For these reasons, one should not assume that a system using physiological sensors will yield the same results as those published in research journals. It is worth repeating that it is critical to test the system with the actual sensors, tasks, and operators. This should be done as soon as possible so that changes can be made before too much time is invested in a potentially flawed approach.

Researchers attempting to integrate multiple sensors for the first time will quickly discover the importance of clear guidelines for using each type of sensor, for connecting each sensor, for involving participants, and for aiding the observer who will be completing the traditional workload measurements. All members of the integration and testing team should have a clear and consistent understanding of these processes to ensure that physiological and nonphysiological data will be comparable.

Additionally, data collected by physiological sensors are raw and unformatted and may not directly relate to the traditional, psychological measures taken during usability testing. There must be methodologies for (a) cleaning and filtering the raw sensor data into usable information regarding cognitive state and (b) determining meaningful comparisons between physiological and psychological cognitive workload measures.

Another, more general, best practice is publishing helpful information on physiological sensor systems. This includes instructions for training people to connect and calibrate sensors, as well as general allowances for space, power, and connectivity of advanced computational tools in a traditionally "low-tech" evaluation environment.

The points listed as lessons learned in the earlier section were derived from an iterative design process that incorporated input from both participants and system operators. Some of these points are specific to augmented cognition systems, though they remain consistent with standards of practice for designing systems for use by humans. More information about the iterative design process and other methods of designing for usability can be readily found in books such as *Usability Engineering* by Jakob Nielsen (1993).

Design Guidelines

In developing physiological sensors for augmented cognition systems, it is important to follow proper practices of usability engineering to ensure that sensors are appropriate for the users, environments, and applications in which they will be used. If the physiological sensor is truly intended for use in such systems, it must be designed to have minimal obtrusiveness on its wearer so as to be readily accepted when incorporated into an operational environment. Failure to meet these requirements may lead to significant resistance from users of the systems.

In developing and using fNIR technology, we have garnered several design guidelines. As you develop the technology, you must select one of the three available types of fNIR imaging techniques (continuous wave, time resolved, and intensity modulated) based on the targeted application. The continuous-wave systems have the capability for small, portable device implementations but are best for relative rather than absolute measurements of oxy- and deoxy-hemoglobin. Time-resolved and intensity-modulated systems require more complicated hardware and are slower in measurement speed, but they are more accurately calibrated for absolute measurements (Hoshi, 2003). Preprocessing techniques are also crucial in extracting useful features from the fNIR data, depending on the application. Physiologic signals can be extracted or slower cognitive changes can be isolated in real time through careful filter design.

In designing for use of the system, you must select a simple calibration task that requires the activation of the same cortical area to be used in the application (or study). This calibration task ensures that the fNIR sensor is properly placed for the detection of the monitored activity through a simple calibration procedure. Additionally, the system (or study) designer and data analyst must be aware that, as with all physiologic signals, the hemodynamic response varies among individuals—and even with changes in time for the same individual—in terms of profile and degree of signal change.

These design guidelines are summarized in Table 2.2.

Table 2.2. Design Guidelines for Guiding the Development and Use of Functional Near-infrared Imaging Sensors for Closed-Loop Augmented Cognition Systems

Purpose	Guideline
Development	Use continuous-wave techniques when relative oxygenation changes are needed and device size or portability is critical.
	Use time-resolved or intensity-modulated techniques when absolute oxy-hemoglobin measurements are required.
	Preprocessing allows for the extraction of features that are useful for the specific application.
Use	Use a simple calibration task that activates the same area of the brain as the monitored task to ensure proper placement of the sensor for the detection of targeted cognitive activity.
	Beware of individual differences and changes over time in hemodynamic responses.

Test Your Knowledge
Based on what you have learned in this chapter, outline the steps you would take to incorporate a physiological sensor into a closed-loop system. Consider a system in a domain in which you are familiar. Is fNIR, or another functional brain-imaging technology, a potentially useful technology for this application? If it is, explain what cognitive changes you would monitor and how you would ensure that you place the sensor(s) in the correct location(s).

Applied Exercise
A computational model to simulate an auditory classification task in ACT-R has been built by Son, Guhe, and Gray (2005). How could this be used to validate an fNIR system? Read the Son et al. article to learn more about such a validation exercise. What other strategies could be used for fNIR validation?

Parting Message

The development of an fNIR system (OTIS) within a closed-loop augmented cognition framework required many prototype iterations and small steps leading to a usable and useful solution appropriate for its intended use. Designing any technology within such specifications may require many trials, but clearly outlined end use operations and applications can guide and drive the design process. Incorporating such sensor technology into a closed-loop system can yield tremendous benefits for the user if proper steps are followed.

References

Chance, B., Cope, M., Gratton, E., Ramanujam, N., Tromberg, B. (1998). Phase measurement of light absorption and scatter in human tissue. *Review of Scientific Instruments, 69,* 3457–3481.

Cope, M. (1991). *The development of a near-infrared spectroscopy system and its application for non-invasive monitoring of cerebral blood and tissue oxygenation in the newborn infant.* Ph.D. thesis, Department of Medical Physics and Bioengineering, University College London.

Endsley, M. (1988). Situation Awareness Global Assessment Technique (SAGAT). In *Proceedings of the National Aerospace and Electronics Conference (NAECON)* (pp.789–795). New York: IEEE.

Franceschini, M. A., Boas, D. A., Zourabian, A., Diamond, S. G., Nadgir, S., Lin, D. W., Moore, J. B., & Fantini, S. (2002). Near-infrared spiroximetry: Noninvasive measurements of venous saturation in piglets and human subjects. *Journal of Applied Physiology, 92,* 372–384.

Hoshi, Y. (2003). Functional near-infrared optical imaging: Utility and limitations in human brain mapping. *Psychophysiology, 40,* 511–520.

Jobsis, F. F. (1977). Noninvasive, infrared monitoring of cerebral and myocardial oxygen sufficiency and circulatory parameters. *Science, 198,* 1264–1267.

Nielsen, J. (1993). *Usability engineering.* San Diego: Academic Press.

Nishimura, E. M., Stautzenberger, J., Robinson, W., Downs, T. H., & Downs, J. H. III. (2007). A new approach to fNIR: OTIS. *IEEE Engineering in Medicine and Biology Magazine, 26, 4,* 25–29.

Schmidt, F. E. W. (1999). *Development of a time-resolved optical tomography system for neonatal brain imaging.* Ph.D. thesis, Department of Medical Physics and Bioengineering, University College London.

Siegel, A., Marota, J., Boas, D. M. (1999) Design and evaluation of a continuous-wave diffuse optical tomographic system. *Optics Express, 4, 8,* 287–288.

Son I.-Y., Guhe, M., & Gray, W. D. (2005). Human performance assessment using fNIR. In *Proceedings of SPIE 5797: Biomonitoring for Physiological and Cognitive Performance During Military Operations* (pp. 158–169). Bellingham, WA: The International Society for Optical Engineering (SPIE).

Sensor Integration to Characterize Operator State

Thomas Schnell, Blaze M. Keller, & Todd J. Macuda
University of Iowa

This chapter addresses how neural and physiological sensors can be integrated into an operational environment to predict operator state.

Introduction

As operational environments such as flight crew stations, command and control centers become increasingly rich in information (e.g. visual, auditory, and tactile displays; mitigation systems; and automation systems), operator workload and performance limits may be encountered (Samel et al., 1996). Another concern is the proliferation of highly automated systems (e.g., flightdeck) that have the potential to make easy tasks easier but hard tasks harder. Adverse effects of boredom and fatigue thus become as much of a concern as do adverse effects of extremely high levels of workload.

In the field of augmented cognition, several international research groups are examining how best to measure operator state (see Chapters 1 and 2) and develop mitigating systems (see Chapters 5 and 6) to enhance operator performance and prevent hazardous conditions. The main thrust of these efforts has been to gain an understanding of the neural mechanisms contributing to operator behavior; for example, a pilot's behavior in flight (e.g., the impact of stress, fatigue, and pharmaceuticals), with a view toward developing technologies that will enhance human effectiveness in operational environments. Much of the work conducted to date has resulted in the simulation of various flight and military platforms, environments, and conditions and the measurement of neural activity using electroencephalogram (EEG) and related technologies during these simulated tasks. Participant performance on these tasks is typically correlated with neural signals, which are used to develop mitigations in near-real time during task performance.

In addition, a few studies have measured an operator's neural activity using EEG and near-infrared spectroscopy (NIRS; see Kobayashi, Kikukawa, & Onozawa, 2002: Kobayashi, Tong, & Kikukawa, 2002; Wilson, 1999, 2003, 2005). It is important to note that both simulated environments (e.g., flight test) and real environments (e.g., airborne flight test) will be necessary to continue to advance work in this field.

Although the body of work on monitoring operator state and neural activity using simulated environments is expanding, few studies have used real operational platforms. There may be several factors contributing to this scarcity of data, including limitations in recording technologies in operational environments (e.g., data synchronization, noise, electromagnetic interference), limited access to field facilities (e.g., flight test facilities), and field facilities that lack the capability to accommodate neural recording studies. (See Chapter 4 for an example of work in assessing operator state in mobile field environments.)

It is important to emphasize that conducting field testing requires considerable expense and expertise to integrate systems into real operational environments and to develop and/or maintain these platforms. There are substantial overhead costs (e.g., flight safety reviews), institutional review board (IRB) processes, training (e.g., test pilot recurrent training), and so on. The design and airworthiness certification of the test equipment is another costly item associated with flight testing. Installation of instrumentation, placement of sensors, installation of racks, cutting of access ports, placement of antennas, installation of optics, and the design of specialty systems (e.g., avionics) requires a specialized workforce that is able to incorporate certification standards into the design, so that the final apparatus can be approved (e.g., by the Designated Airworthiness Representative [DAR] for flight applications).

To expand the ability to examine the neural basis of operational performance and develop mitigations, it is necessary to create dedicated facilities for use by the broad community examining these issues. In this chapter, we summarize efforts by the National Research Council (NRC) and the University of Iowa Operator Performance Laboratory (OPL) to develop **integrated airborne neural imaging capabilities** that are now available to advance the state of research in this field. Both the OPL and Canada's NRC have established the required facilities and maintain a skilled staff to perform all testing tasks for flight applications under one roof. We discuss general approaches to physiological experiments, stressing the flight platforms developed to conduct the research. These facilities are enablers to the field of airborne augmented cognition, permitting the advancement of sensor systems, methodological approaches, and measurement techniques in the user community at large.

SENSOR INTEGRATION

Valuable Information

The development of cognitive avionics requires an airborne physiological assessment capability to elicit realistic pilot behavior to properly account for external influences such as sustained accelerations, electromagnetic noise, illumination conditions, temperature fluctuations, and an overall environment that can produce a wide range of real-world stressors. We consider our flight test platforms and the associated sensor and software architecture to be a Cognitive Avionics Tool Set (CATS) that may be useful to developers of augmented cognition systems. CATS includes software modules that allow for cognitive avionics–specific visualization, synchronization, artifact removal, querying, and classification of data that were collected in the context of airborne testing. The goal of CATS is to give researchers a hand in organizing mass quantities of data and provide an opportunity to focus on scientific aspects of the data.

In this chapter, we discuss the need for multisensory data collection, data synchronization, artifact removal, ruggedized flight-capable hardware, and flight cards that are building up from simple to complex tasks in a systems design fashion. An explanation of special terms used in the chapter may be found in the Appendix.

Scenario

The gamut of environments that are of interest to researchers range from clinical laboratory settings and flight simulators to airborne test platforms such as the ones discussed in this section. In the laboratory, researchers have the ability to control a multitude of factors that may not be controllable in an airborne setting. In terms of experimental control, flight simulator environments are more closely related to clinical laboratory settings than they are to their airborne counterparts. Clinical laboratory settings are excellent for basic research of fundamental phenomena. Flight simulators are excellent tools to evaluate the merit of cognitive avionics concepts at their early design stages, but as the designs mature, they must be tested in the ecologically valid context of real flight using an airborne physiological assessment platform.

The synchronization of data becomes extremely important in airborne testing. In clinical settings it is usually easy to execute a very tightly scripted scenario. In airborne flight tests, the scenarios may be scripted tightly, but often the scenarios need to be changed dynamically because of unforeseen influences such as air traffic control (ATC) requests, weather, noncooperating traffic, or variances in safety pilot (SP) execution of the test card. Also, space in the aircraft may be very limited, and taking notes may be difficult for a flight test engineer (FTE). CATS allows for synchronized data collection and data tagging, all but eliminating the need to take paper notes.

Valuable Information

The goal is to collect good data. If tool sets such as CATS are not used, it is easy to end up with a multitude of undocumented, nonsynchronized data files from a multitude of sensors. At best, this will bog researchers down, forcing them to sift through massive amounts of data. Worse, it is possible that incorrect scientific conclusions would be drawn based on bad or confusing data. An **integrated, multisensor data collection**

> **system** that allows for very detailed marking of data in real time, automatic synchronization, data storage, retrieval, and extraction is needed.

General Approach and Associated Toolkit

Figure 3.1 shows the NRC-FRL Bell 412 Advanced Systems Research Aircraft (ASRA), and Figure 3.2 shows the University of Iowa's Computerized Airborne Research Platform (CARP). Taken together, these flight test platforms enable researchers to study pilot state in both rotary and fixed-wing aircraft, respectively.

Figure 3.1. Bell 412 Advanced Systems Research Aircraft (ASRA) at Canada's NRC.

Bell 412 Advanced Systems Research Aircraft (ASRA)

The NRC Bell 412 ASRA was acquired in 1993 in order to expand the NRC's ability to perform research on advanced, high-bandwidth flight controllers. The Bell 412HP is a twin-engine, four-bladed, medium-utility helicopter with a "soft-in-plane" rotor system, and a gross weight of 11,900 lbs. During flight tests, the EP controls are located on the left-hand side and the SP located in the standard pilot-in-command position is on the right-hand side. The Bell 412 is fully instrumented to record a variety of flight performance data (aircraft rates, attitudes, airspeed, etc.) that can be correlated with pilot state parameters such as neural activity and other physiological markers. Additionally, ASRA's layout allows for an FTE to occupy a workstation in the cabin, who has the capability to modify system parameters and monitor performance in real time.

FRL Fly-by-Wire System

The ASRA facility is a "single-string," or simplex, full-authority, fly-by-wire (FBW) system. In general, this means the helicopter can be flown from a single flight control computer. This single-string design implies that there is little or no redundancy in the FBW system. There is one set of FBW actuators, one flight control computer, one operating system, and one set of flight control software. The simplicity of this design allows for rapid changes to be made without the added overhead of multiple coding sources and the extensive verification required for commercially certified FBW systems. However, the implication of this architecture is that flight safety is the sole responsibility of the SP. For further details regarding the ASRA facility, refer to Gubbels et al. (1997, 2000)

SENSOR INTEGRATION

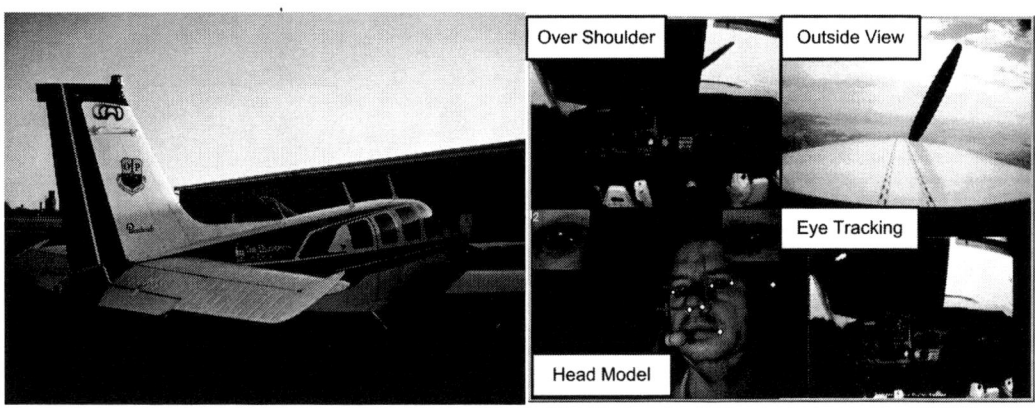

Figure 3.2. Beechcraft A-36 Bonanza Computerized Airborne Research Platform (CARP) at OPL.

Valuable Information
How is the ASRA useful for augmented cognition research? The ASRA facility has a unique capability to change the control laws of this airborne platform to allow the simulation of other helicopter types and to vary the workload experienced by the EP. Notably, the FBW systems at NRC have been used for

- testing various cockpit technologies and developing visually coupled systems,
- conducting flight controls research and development,
- handling qualities research, and
- conducting variable stability training for test pilots and FTEs.

The capacity to alter the handling qualities of the ASRA at the flip of a switch via various software coding changes offers a unique test platform for examining pilot neural activity and relating it to flight performance and pilot workload. This capability enables the EP to experience a wide range of handling qualities, from very simple to extremely difficult flight models; this provides the experimenter with the ability to cognitively load up the EP as needed for the pilot state characterization research paradigms (Taylor, 2000). Therefore, the ASRA at FRL is a real implementable "cognitive cockpit" that can be used to support the overall field of augmented cognition by providing a simulation facility that can be used to (a) validate and assess state-of-the-art recording technologies, (b) enhance data analysis techniques, (c) develop new measurement techniques, and (d) lead to an understanding of the neural mechanisms contributing to flight.

The OPL-designed CATS architecture was adapted for use in the ASRA as shown in Figure 3.3. As our pilot state characterization system is maturing, we will start to focus on augmented cognition system mitigation on the basis of operator state characterization (see Chapters 5 and 6). The adaptable FBW system could be a powerful way to demonstrate the benefits of adaptive flight control automation. This mitigation could be used to increase the level of aircraft control stability (see Chapter 7) or even autonomy when the pilot state characterization system detects that the pilot is sharing attention with a mission-related task.

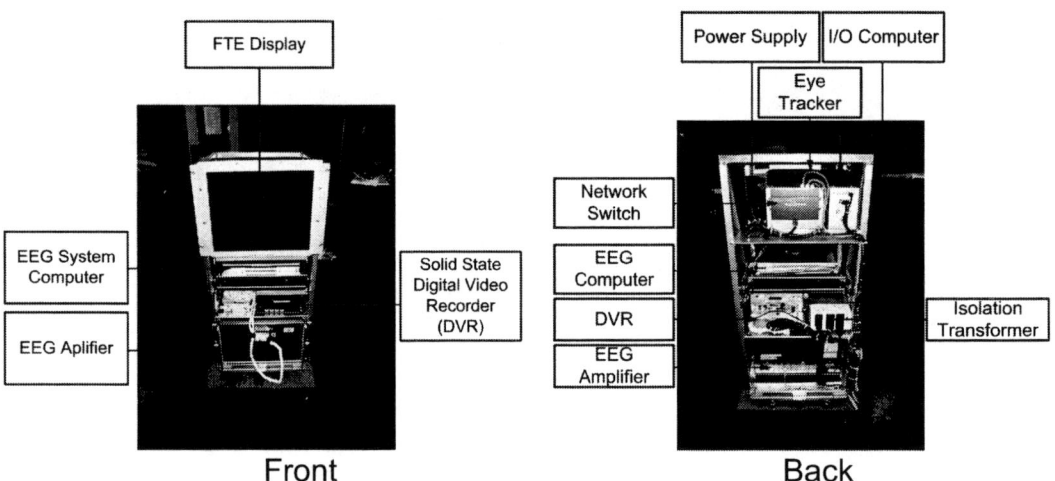

Figure 3.3. ASRA (NRC Bell 412) equipment rack with OPL CATS hardware.

Sensor Integration on the Computerized Airborne Research Platform (CARP)

The CARP is a Beechcraft Bonanza that the OPL has equipped as a flying neural imaging and physiological performance assessment capability. Work on the CARP generated massive amounts of data that needed to be organized for analysis and classification. This mass of data gave rise to the need for developing a Cognitive Avionics Tool Set. The CATS approach is to provide a rich suite of different physiological sensors and let researchers decide which sensors should be incorporated into any given system or experiment.

The extensive instrumentation of the CARP aircraft can be assessed from Figures 3.4, 3.5, and 3.6. The CARP is a flying laboratory equipped with a synchronized data collection system that can collect flight state data, eye tracking,

SENSOR INTEGRATION

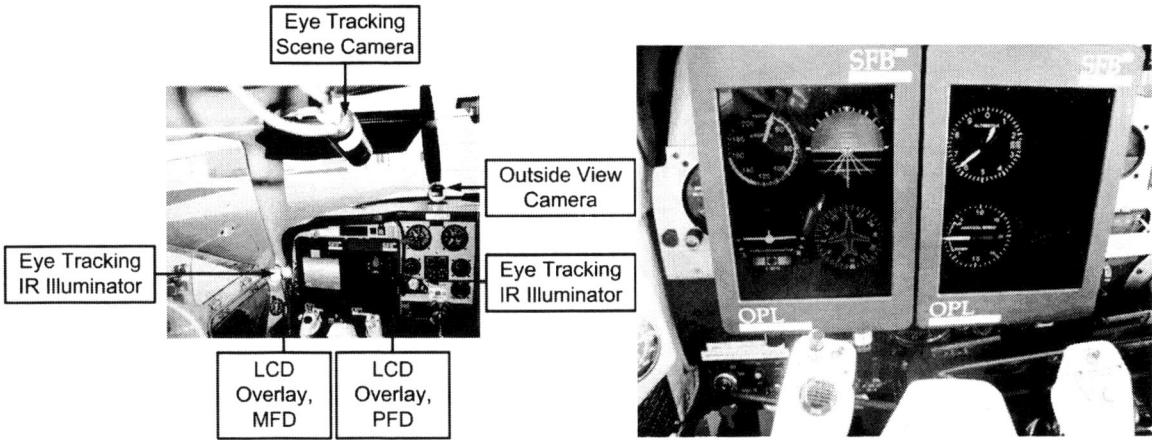

Figure 3.4. Pilot station with eye-tracking cameras (left) and dual-overlay LCD instrument displays (right).

PFD = Primary Flight Display; MFD = Multifunction Display.

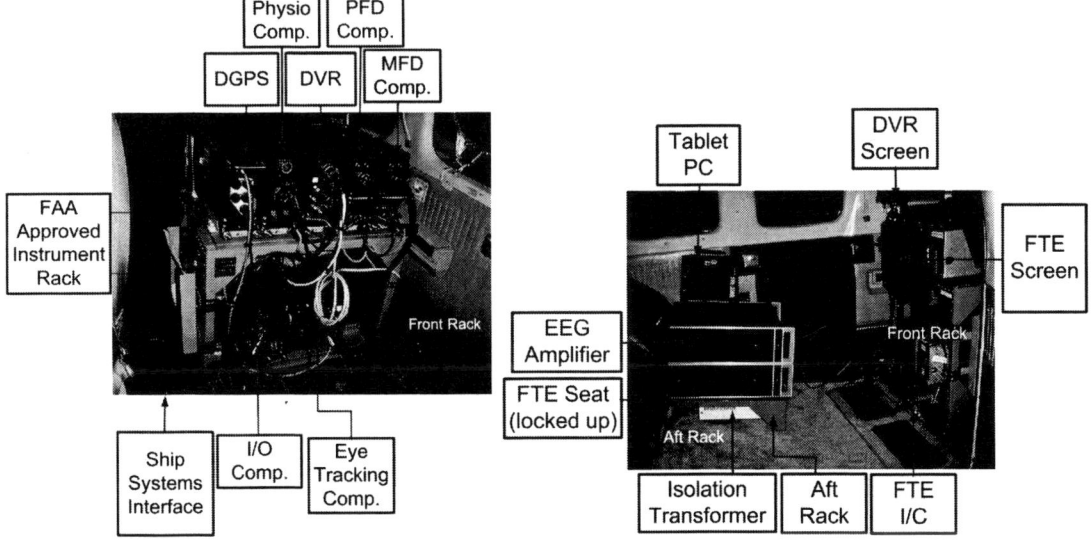

Figure 3.5. Front (right) and aft (left) instrument rack in CARP.

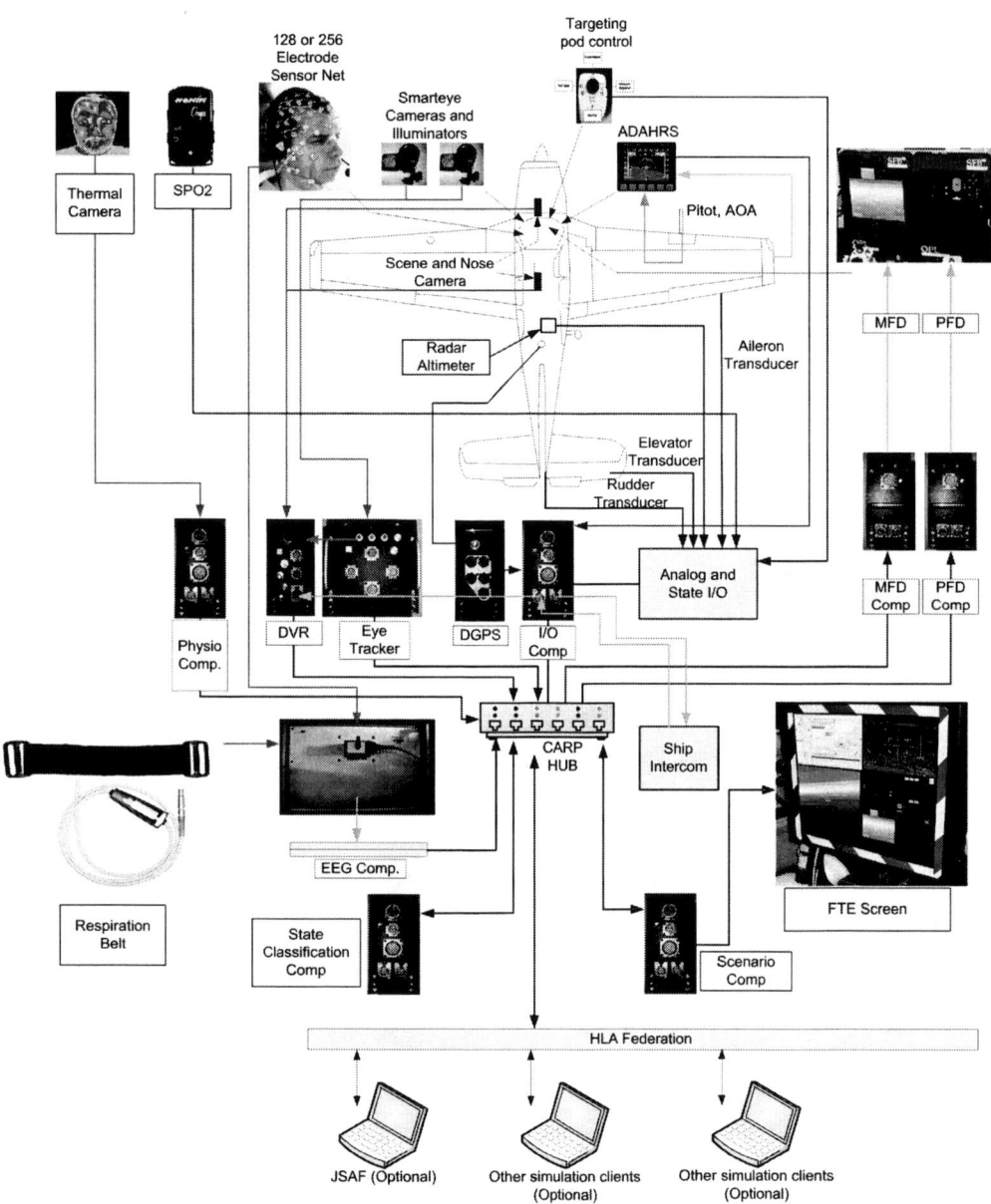

Figure 3.6. Connectivity diagram of OPL's CARP.

Dense-array EEG, facial temperature, respiration, pulse oximetry, electromyogram (EMG), and electrocardiogram (EKG). CATS architecture is completely independent of sensor manufacturers. For example, the EEG system could be exchanged for a system from another vendor with ease. In addition, the CARP has a tactile seat cueing system; an auditory alert system (programmable verbal warnings, steering commands, and sounds); an embedded event-marking system; a high-resolution, daylight-readable dual primary flight display; a multifunction display LCD overlay display at the primary flight display location (see Figure 3.4); and a third display between the control yokes.

One of the reasons LCD panels are used is to elicit pilot behavior in response to failing flight instruments. With the software-generated flight instruments, we can generate carefully calibrated indications that can be used to elicit spatial disorientation, upset, or perceived conflicts between the instrument indications and vestibular "seat-of-the-pants" perception. This would be very difficult, if not impossible, to achieve with standard electromechanical gauges, because the calibrated indications on multiple instruments of the LCD panel may follow certain decay curves that will cause the pilot to fly the aircraft into the desired entry condition for the test point. Another reason to use LCD overlay displays is their ability to show symbologies and formats used in other aircraft such as military or transport category jets.

CARP flights are flown with a crew of three. The left front seat is the pilot station for the evaluation pilot (EP), who acts as the test participant. The right front seat is the station for the safety pilot (SP), who is commander of the ship and responsible for the safety and overall success of the flight. The right aftmost seat is the station of the flight test engineer. The flight test engineer (FTE) is in charge of the experimental payload and usually acts as the test director. As such, the FTE is responsible for ensuring that the data collection system is ready prior to each sortie and that data are extracted from the aircraft after returning to base. The FTE ensures compliance with IRB protocols and is responsible for storing the data on a server with a backup stored in a secure location.

In our research sorties, the SP manages the aircraft from taxi and take-off to an initial point (IP) where the EP starts the first task. There are typically several tasks in a sortie, and in some cases, these tasks are replicated for increased statistical power. We usually execute within-subjects, repeated-measures designs with only a small number of factors under investigation. Also, we found that we get better results by performing full factorial run matrices. The FTE in the back of the aircraft will mark all events in the sortie such that removal of nontask data (taxi, takeoff, repositioning of aircraft, etc.) is automatically flagged for removal.

The CARP crew station can be configured to approximate practically any flight crew station. For example, the CARP was used to demonstrate early Synthetic Vision System concepts to the FAA. In addition, the CARP has been set up as a flying simulator that can simulate tactical fast-jet symbology set using Alion's Advanced Tactical Aircraft Simulator (ATAS). This allows the CARP platform to serve as a member in a federated Distributed Mission Operations (DMO) simulation, thus joining simulation and flight. In addition, the OPL recently procured an Aerovodochody L-29 Delfin fighter jet trainer that has been equipped similar to the CARP.

The CARP equipment is installed on two configurable racks. The front rack in Figure 3.5 (left) is installed in lieu of the standard seats in the middle row. The aft rack in Figure 3.5 (right) is installed in lieu of the left aftmost seat. The front rack contains the equipment that collects the flight state data, control surface position data, eye-tracking data (three facial cameras), audio (intercom and ATC radio), and video data (four witness cameras) and performs data concentration and synchronization; it also includes general-purpose computers. In some cases, the general-purpose computers are used to produce cueing signals for a tactile display mounted in the pilot seat or for a voice-based avionics interface system.

As noted earlier, the FTE sits in the right aftmost seat. The FTE has a display and keyboard that can be used to control all research computers from a single interface location. The FTE has access to the radio system and can make and receive radio calls to free up the flying crew, should that be necessary. A cueing and marking system is available to the FTE so that the data stream can be marked with a multitude of state descriptor markers that greatly facilitate the detection of the location of relevant segments of data during analysis.

Our philosophy is to collect data that are as close to the final analysis data as possible with as little overhead data as possible. To achieve this, we place a multitude of digital markers in the data, such that we can eliminate any nonessential data that do not belong to a specific flight card but that may have been streamed during repositioning of the aircraft.

Figure 3.6 shows the connectivity diagram of the CARP. Flight state data are acquired through a differential GPS system (Novatel), an Attitude Heading Reference System (AHRS), an Air Data Computer (ADC), and position transducers on the flight control surface linkages. A radar altimeter provides accurate digital data regarding the height of the aircraft above the surface. A three-camera Smarteye 4.0 eye tracker is used to obtain head orientation in 6 deg of freedom; facial feature data such as location of the eyebrows, corners of the eyes, nostrils, and corners of the mouth; as well as an accurate gaze vector in 3-D. A Thermovision A10 thermal camera is used to obtain facial temperature change data. A pulse oxymeter can be integrated with the data collection system. All computers are networked with an Ethernet-based local area network (LAN). The I/O computer receives data packets from each computer and synchronizes them into a global data record.

A dense-array EEG system (128 or 256 electrodes) is used in the CARP. EEG data are streamed through an IEE 1394 interface to the I/O computer for integration with all other state data, as well as recorded on a Macintosh laptop running the EEG system software as backup. Data from a respiration belt, galvanic skin response (GSR) sensor, electrocardiogram (ECG) sensor, pulse oximetry sensor (sPO_2), EMG sensor, and temperature sensor are incorporated into the I/O system through a wireless Bluetooth® data link. These sensors are worn below the flight suit, and thanks to the Bluetooth interface, no wired connections to the I/O computer are needed.

We found that refining an experimental protocol, testing the equipment, and rehearsing a sortie can be greatly facilitated by a technique that we call **aircraft-in-the-loop simulation**. In this mode, the actual aircraft is parked in front of a screen in the research hangar, thus providing a simulated view of the outside visuals. Because all flight controls are monitored with position transducers and all relevant aircraft control state data are available on the I/O computer (see Figure 3.6), it is possible to connect the aircraft to a flight model simulation. In this fashion, simulated sorties can be flown in the aircraft while they are parked safely in the research hangar.

Both the NRC and the OPL aircraft are capable of aircraft-in-the-loop simulation. This methodology allows us to test all experimental systems and run them through their paces, and it provides for a measure of the difference between simulation and flight. Both flight platforms—the ASRA and CARP—can be used in this mode (see Figure 3.7) as a low-cost tool to develop and refine flight cards and test equipment prior to a sortie. In the aircraft-in-the-loop simulation mode, all aircraft systems except the engine are active.

Valuable Information
The merged, synchronized data are continually streamed to, and stored on, the I/O computer in the aircraft. After a sortie, the data are streamed into a secure file server using an Ethernet hookup in the aircraft hangar. This means that the experimenter can generate a query suitable for his/her experimental hypothesis and receive a data set that immediately can be used for analysis.

ASRA Helicopter-in-the-Loop Simulation   CARP Airplane-in-the Loop Simulation

Figure 3.7. Aircraft-in-the-loop simulators as low-cost flight test development tools.

Test Your Knowledge
What are the essential elements of a synchronized data collection system that can collect operator state data such as eye tracking, dense-array EEG, facial temperature, respiration, pulse oximetry, EMG, and EKG?

Cognitive Avionics Tool Set (CATS)

As our team embarked on airborne cognitive avionics research, we realized very quickly that we would end up with a large amount of data that needed to be managed. Our data collection architecture has been set up to generate a single synchronized database of the many sensors that produce data at various sampling rates on the aircraft. That single data file quickly grew to dimensions that defied analysis with commercially available software packages, including those provided by the manufacturers of the sensors. Out of that need, at OPL we decided to start our own line of analysis tools, which we bundled into CATS. These tools are continually improved as our research progresses. For example, the tool has recently been updated to allow for real-time analysis of data without user intervention.

A typical data analysis of physiological data starts with an in-depth analysis of the effects within each participant. In CATS, the user (e.g., researcher, instructor) can select the data file to be included in the analysis from a file history list on the CATS desktop, as shown in Figure 3.8. By selecting the file of a single sortie or multiple sorties, the user makes an important decision about the scope of the data to be included in the analysis.

The next step typically involves an analysis of the eye movements. CATS has an eye movement analysis module, as shown in Figure 3.9. The eye movement module also serves as the visualization of the sortie. In that module, the aircraft interior is shown as a movable 3-D object consisting of a computer-aided design (CAD) model or a stitched photograph on a 3-D geometry. Real-time video can be overlaid on the outside scene, and if a head-worn scene camera is used, it is also possible to overlay a semi-transparent overlay of the pilot's view of the virtual world. The audio channel from the sortie can be monitored for quality control. For example, sometimes ATC may issue radar vectors to avoid traffic. It is important that such extraneous factors be eliminated prior to data analysis.

The eye-tracking module in CATS associates with a time line on which the experiment is being edited for analysis. Video files and data files can be cropped and selected using a method similar to that used with nonlinear video-editing tools. Narratives can be typed into the experiment time line to highlight or explain interesting events. Alternatively, the user can use the audio recorder to dictate a narrative. These narratives are tremendously useful. Our tradition is that the FTE who flew the sortie will generate this first cut of the data as well as the associated narrative. Handwritten notes from the sortie are transcribed into the timeline at that stage of the analysis.

The next step usually involves the definition of areas of interest (AOIs), as shown in Figure 3.10. The user can import a definition file for the location of the AOIs, or he/she can define new areas using the cursor to draw a polygon line into the area of interest. During the analysis, CATS will then determine the eye fixation summary statistics for each AOI.

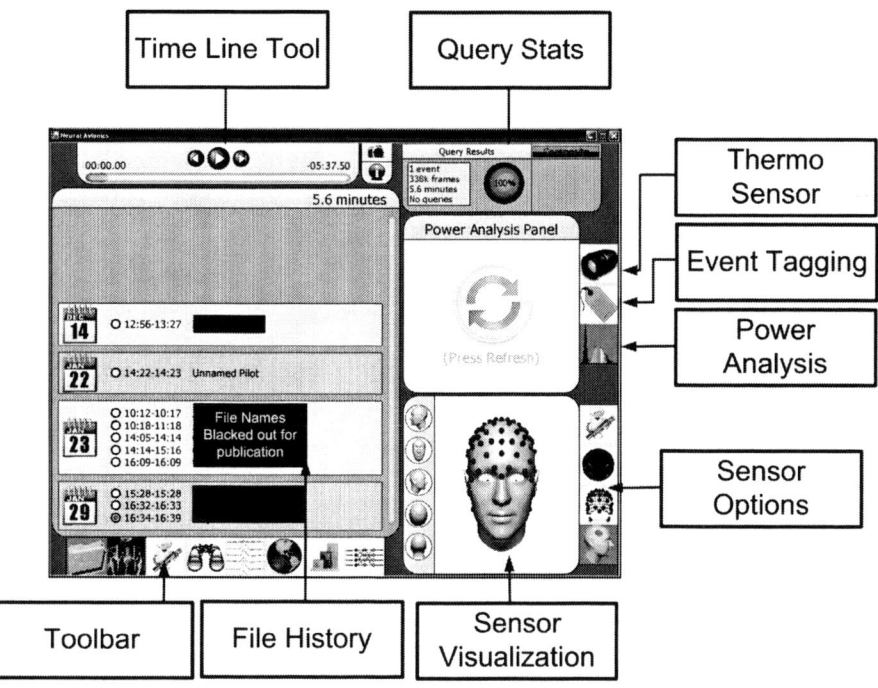

Figure 3.8. File selection page of CATS to determine which data files (usually corresponding to sortie) should be included in the analysis.

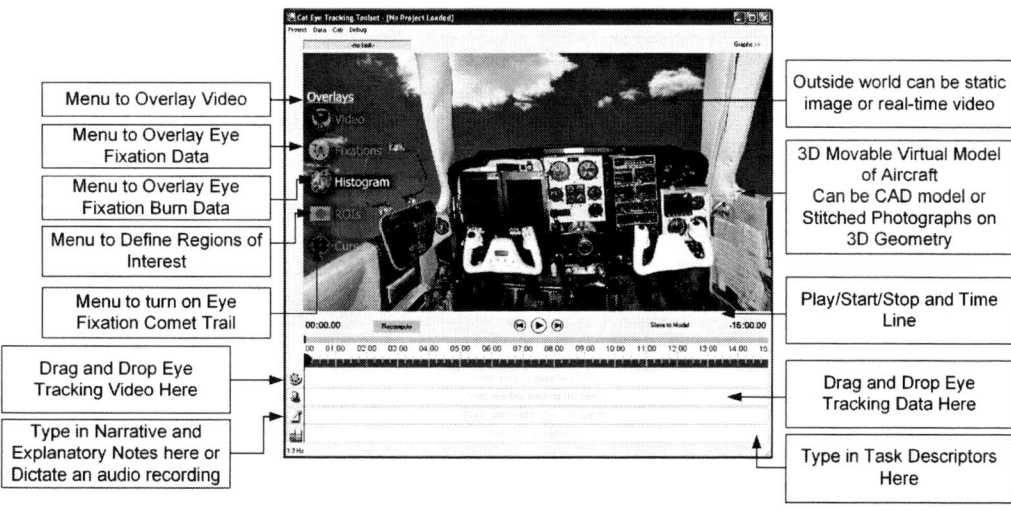

Figure 3.9. Eye-tracking module of CATS used to manage eye-tracking data.

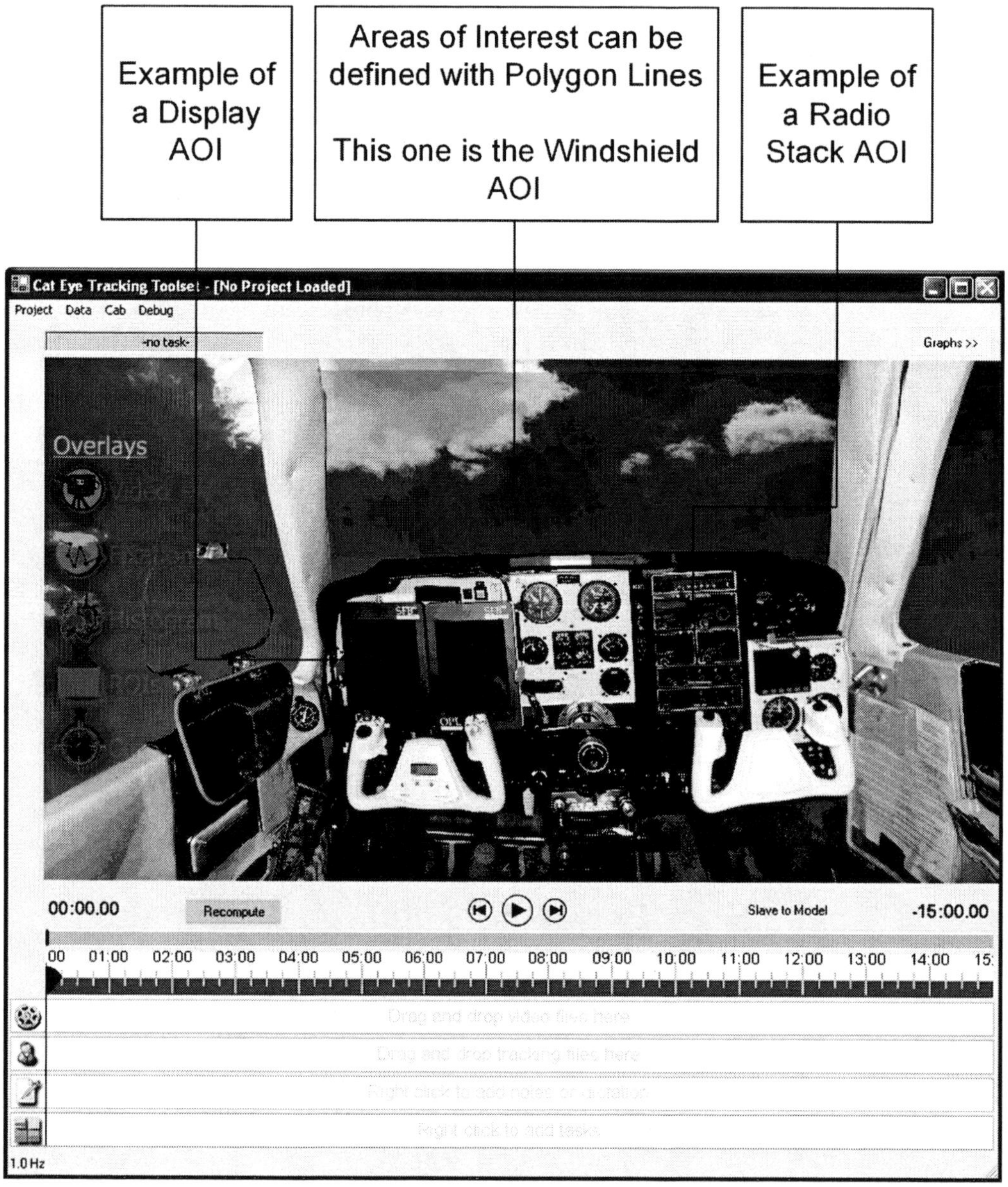

Figure 3.10. Areas of interest definition in eye-tracking module of CATS.

Many of our sorties deal with pilot-avionics interaction. Thus, we often want to see how much time a pilot is spending looking at certain AOIs. CATS offers a burn map overlay that shows a color plot of the number of eye fixation durations in a selected area. Using the Play/Stop toolbar, the user can witness in real time or step by step how the eye fixations form on an AOI for a certain flying task. Summary graphs of the eye fixation data can be generated for use in reports and presentations.

The benefit of performing the eye-tracking analysis up front is that subsequent analysis of other physiological data can be performed separately for each area of interest. For example, the user can instruct CATS to generate the EEG analysis for those time segments when the EP was looking out the front windshield, the head-up display (HUD), or any other area of interest.

In the next step in analyzing the EEG data, the user specifies how to handle artifact removal. Eye blinks, eye movements, and head movements generate artifacts on the EEG data that need to be accounted for. The Smarteye eye tracker provides knowledge about eye movements, eye blinks, and head movements. Speech generates artifacts as well, and we use filtering algorithms to remove, as best as we can, such artifacts from the EEG. It should be noted that the artifacts need to be removed for analysis of EEG data. However, the artifacts themselves convey information about eye blink frequency, head movement frequency, occurrence of speech, and so on and can thus also be treated as a source of state information.

Next, the user can select tagged events to be included in the data analysis. In CATS, data can be down-selected for analysis using a query method. For example, a query could be generated to analyze data for Pilot 1 flying Maneuver 1. Or a query could be generated to analyze data in pilots and all left-banking turns exceeding 80 deg angle of bank and an acceleration of more than 2 g.

CATS also offers a graphical query tool, as shown in Figure 3.11. Using this tool, the user can select a portion of interest of the flight using a graphical tool with photorealistic texture imagery. This is an excellent tool to extract extraneous data that is often collected during repositioning of the aircraft for the subsequent maneuver. We like to have an ability to include or exclude different flying tasks, because we often correlate (using neural network classifiers) the physiological data with self-reported workload ratings that are made by the EP at the conclusion of each task.

Next, a power spectra analysis can be performed. On the right side of the EEG analysis desktop in CATS, the user has access to 2-D and 3-D representations of the sensor net. This representation of the net (see Figure 3.12; p. 57) can be rotated, and EEG activity can be shown in a red-white-blue color-coding scheme in real time or step by step.

The EEG power can be analyzed for all channels and all frequencies that the system is capable of recording. The power analysis interface is shown in Figure 3.13 (p. 57). Power spectra are updated constantly as the time line is playing back the collected data. Using a snapshot feature, the researcher can visually compare power spectra in different tasks. The user can navigate to the "signals" tab and look at traces of EEG activity, as shown in Figure 3.14 (p. 58). To add traces to the window, the user needs to click on a sensor in the 3-D representation of the net.

Finally, CATS generates a power analysis in the clinical bands, and the resulting data can be sent to an artificial neural network (ANN) classifier, as shown in Figure 3.15 (p.

59), where the data set can be used to train the classifier for prediction of workload or other scalar data that were collected during the sortie (see Chapter 4 for other classification techniques). This scalar data could be self-reported workload scores (NASA TLX, Bedford, etc.) or derived from an a priori assumption of task difficulty. Other subjective rating scales against which one may be able to classify include situation awareness (SA) scores and drowsiness/fatigue ratings.

Test Your Knowledge
List the steps involved in a typical data analysis of an integrated set of physiological data.

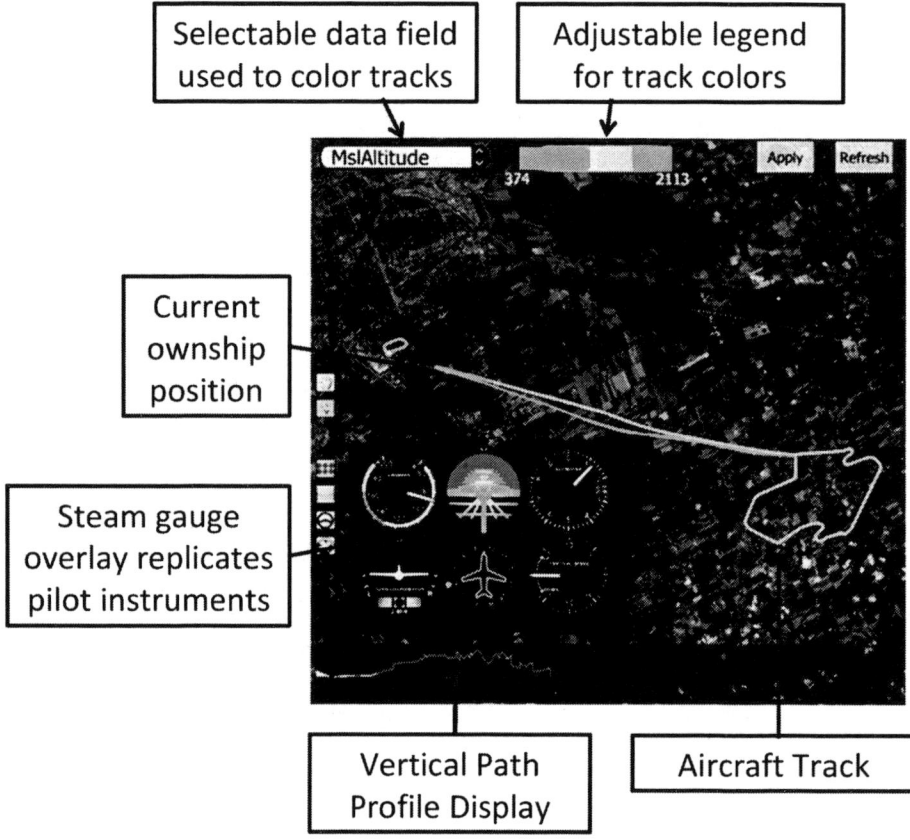

Figure 3.11. Query page of CATS used to include or exclude data records based on a user-defined query.

SENSOR INTEGRATION

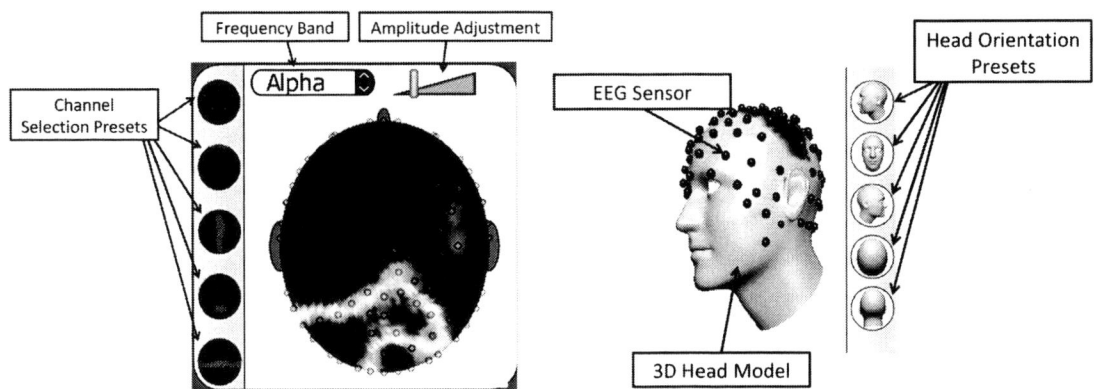

Figure 3.12. Movable 3-D sensor net view of CATS shows activity in real time.

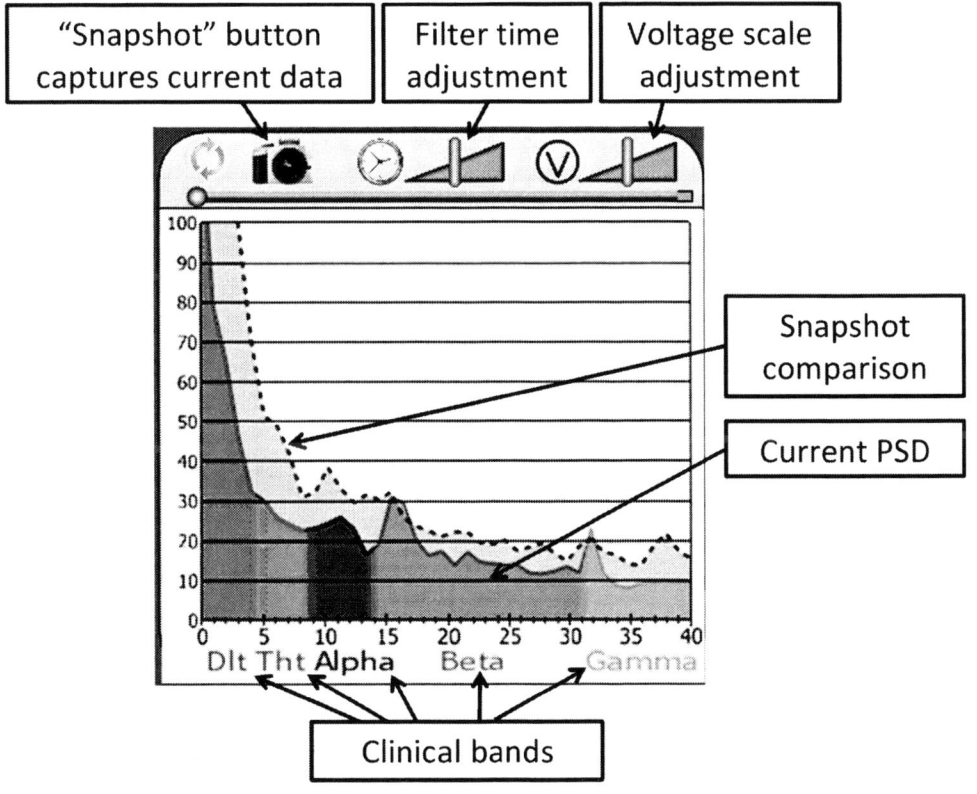

Figure 3.13. CATS EEG power analysis.

AUGMENTED COGNITION: A PRACTITIONER'S GUIDE

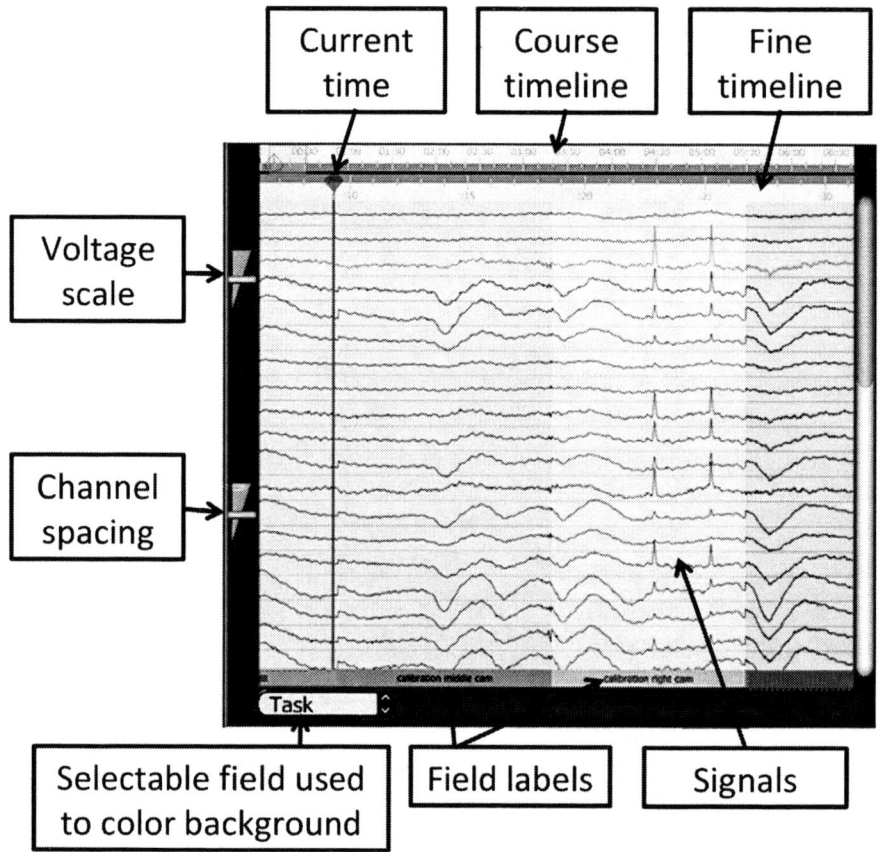

Figure 3.14. Signals page of CATS used to examine EEG signals for selected sensors.

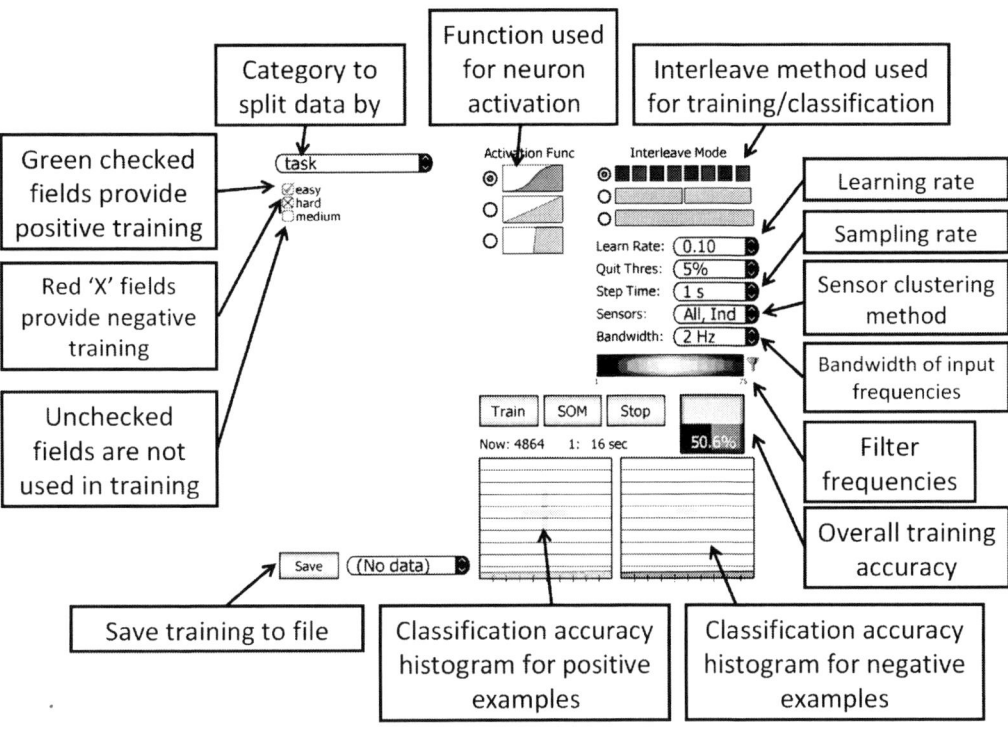

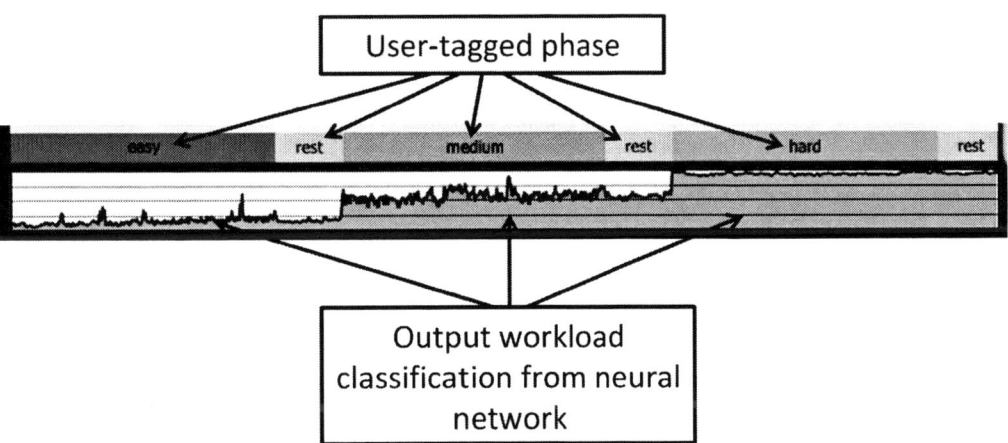

Figure 3.15. Neural network operator state classifier.

CATS Flight Tests

In the following section, we will discuss two flight tests that were performed to collect data for our CATS development. We also wanted to establish a database of experiences dealing with acceptance of neurophysiological monitoring in an aviation setting. At this time, many of our sensors are intrusive and require considerable preparation. We are using our flight test experience to continually refine our sensor suite in hopes that, eventually, we will have all sensors integrated in such a way that no preparation is needed. One concept that we pursued along those lines is called the Cognitive Pilot Helmet (CPH), a system that incorporates all relevant neurophysiological and physiological sensors in an aviator helmet.

The reader may find this section useful to see how we built up maneuvers to isolate certain pilot behaviors. The goal was first to develop simple maneuvers to collect data that could be used to develop the pilot state classification system. In a subsequent flight test phase, we intend to develop a new series of maneuvers that resemble normal phases of flight, such as take-off and climb, cruise, approach, and landing.

We conducted a flight test to separate pilot physiological data on the basis of different levels of cognitive, motor, or perceptual components of a flying task, as indicated in Figure 3.16.

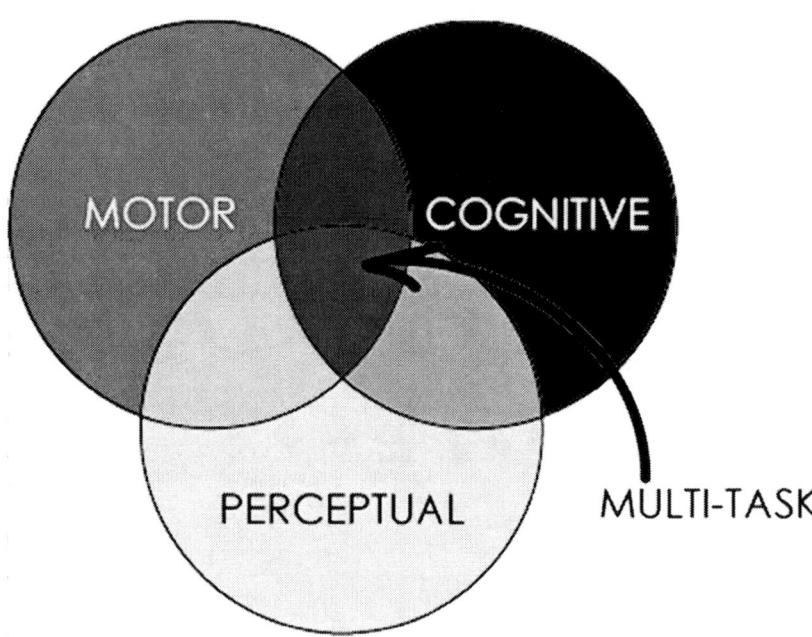

Figure 3.16. Isolating motor, cognitive, and perceptual components of operator performance.

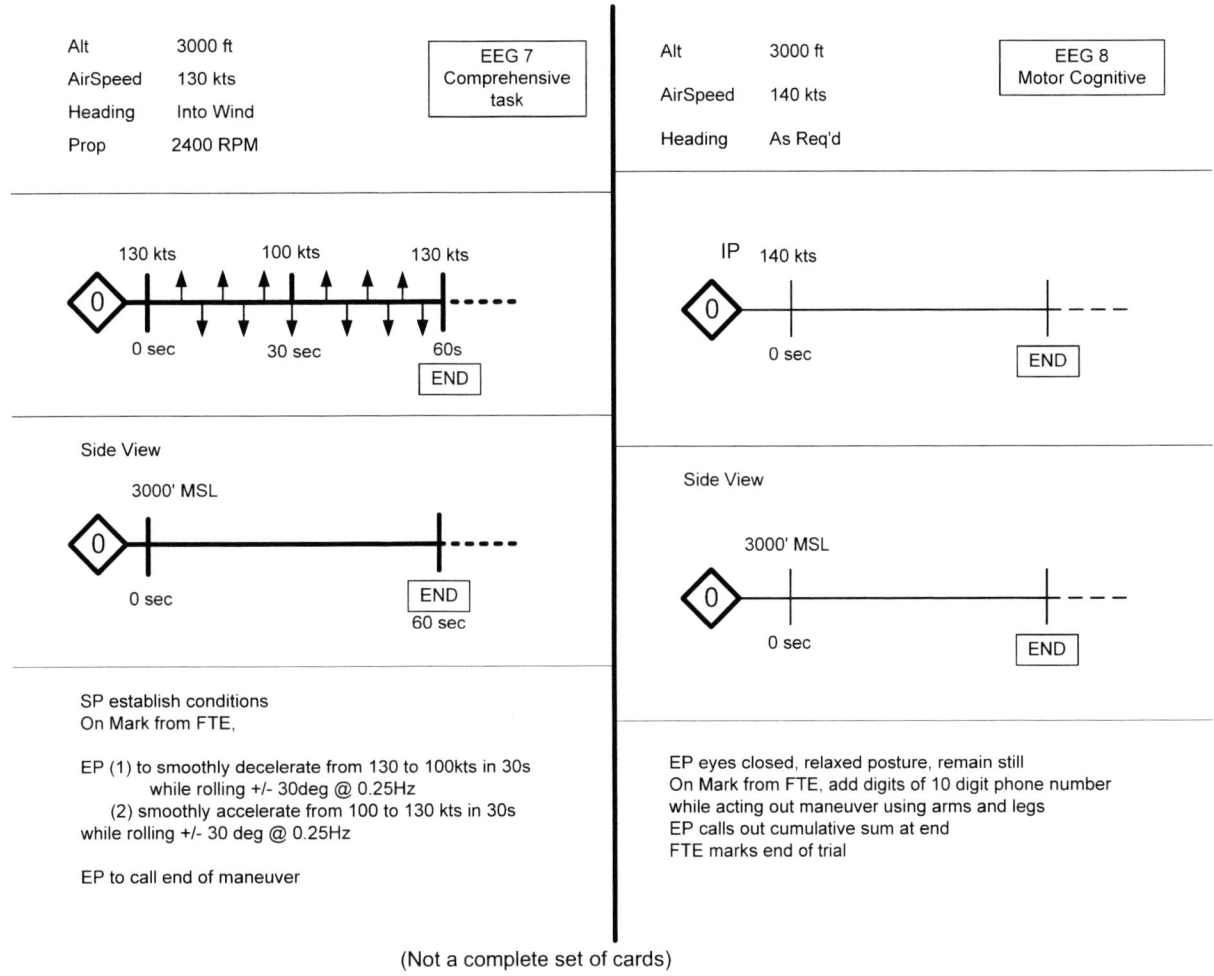

Figure 3.17. Sample flight cards used in a fixed-wing flight test.

Eight special flight maneuvers were designed to isolate motor cognitive and perceptual components of operator performance. We refer to these maneuvers as EEG-1 to EEG-8. A sample test card for EEG-7 and EEG-8 can be seen in Figure 3.17. In Maneuver EEG-1, the EP performs a manual motor task with the hands and feet to mimic the motions he/she would perform to fly a precision roll capture maneuver. For this motor task, there is virtually no cognition involved; perception is minimized because the EP has his/her eyes closed, and the intercom is isolated. There is also extremely limited vestibular and kinesthetic feedback, because the EP does not have hands and feet on the controls but operates a set of imaginary controls. The SP controls the CARP during this maneuver.

Immediately following this task, the EP takes control of the aircraft and flies the EEG-2 roll capture task. This task involves banking left and right at 30 deg angle of bank at 2 Hz while keeping the aircraft on a straight heading. This maneuver is somewhat motor intensive, and inverse coordination between aileron and rudder is required. There is little cognitive loading, because the task is simple and repetitive. The perceptual pathway is primarily visual, and there will be force feedback from the controls and vestibular feedback from the aircraft motion itself.

EEG-3 is a task that was designed to test cognitive loading. The EP is asked to add the digits in a familiar phone number and report the sum as soon as it is available. Cognitive workload in this task is high, whereas motor and perceptual loading are low. The SP controls the CARP during this maneuver.

EEG-4 is a cognitive task in which the EP listens to, remembers, and recalls a holding clearance. This task is cognitively intensive, and perception is permitted to the extent that it helps the EP remember the clearance. Typically, pilots use the altimeter and heading indicators as aids in remembering the direction and altitude of the holding clearance. The SP controls the CARP during this maneuver.

EEG-5 is a task in which we attempt to elicit EP perception of a flight maneuver flown by the SP. Specifically, the SP will fly a wingover (both left and right) and the EP will observe the maneuver by following through on the outside scene and the instruments. No motor activity is performed and cognition is minimal.

In EEG-6, the EP flies the wingover him/herself. This task has the same visual perception components as EEG-5 but with added feedback from the flight control forces. Cognition in this maneuver is higher than it was in EEG-2. Motor activity is present but relatively mild.

EEG-7 is referred to as the *comprehensive task*. This is a very difficult task because the EP needs to combine the roll capture maneuver (EEG-2) with a speed control task in which the speed is reduced from 130 to 100 kts during the first 30 s of the maneuver and then increased from 100 to 130 kts during the second 30 s of the maneuver.

EEG-8 is a maneuver in which the EP will combine maneuvers EEG-1 and EEG-3. For each of these maneuvers, the following dependent measures were obtained:

1. Dense-array EEG (128 channels
2. EKG, EMG
3. Pulse oxymetry
4. Respiration (frequency and amplitude
5. Eye movements
6. Facial feature point locations:
    a. at a minimum, eye location, eyebrow location, and corner of mouth;
    b. if possible, also mouth shape, nasolabial furrows, and wrinkles in corner of the eye
7. Control control inputs (for controller-based analysis)
8. Flight state data
9. Facial temperature (exploratory, not yet part of the OPL CARP package; integration is not particularly difficult but is relatively expensive).

For each participant, three randomized replications of the maneuvers were performed. The data were marked in flight using the CATS synchronized event marker system so that we always knew which portion of the maneuver was flown by the EP and which was flown by the SP.

Figure 3.18 shows EEG activity in two clinical bands for a participant in the flight test. Power in the clinical bands can be indicative of certain functional aspects of cognition. The delta band is for frequencies below 4 Hz. The theta band is for frequencies between 4 and 7 Hz. Physiological markers of workload, decision making, and conflict monitoring can be found in this band. The alpha band is for frequencies of 8 to 12 Hz, and the beta band is for frequencies of 13 to 20 Hz. The gamma band is for frequencies above 30 Hz; markers of perception and attention can be found in this band.

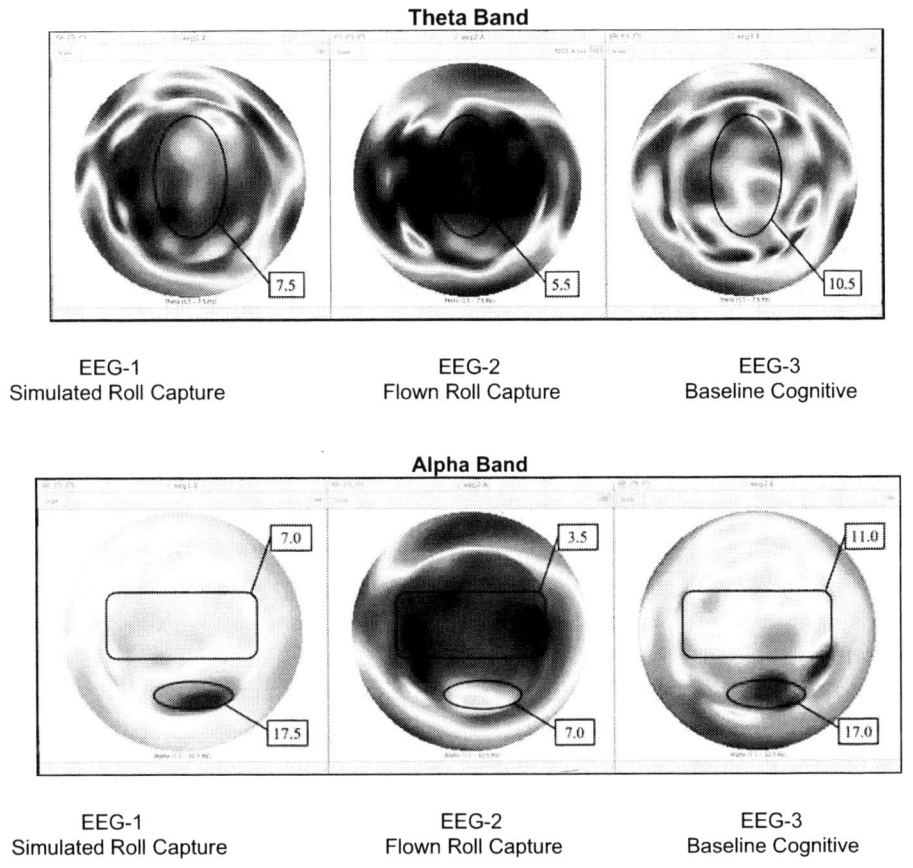

Figure 3.18. Sample EEG activity for three flight maneuvers.

Alpha activity represents a sort of "idle" state. Normally, power in this band is fairly large over the back third of the brain (mainly occipital areas) when the eyes are closed but the participant is awake. Power in the alpha band is usually reduced in those areas when a person becomes mentally busy. Also, for voluntary movements (e.g., finger tapping), EEG activity is desynchronized in alpha ranges over sensorimotor areas.

Applied Exercise
Design an operational task to separate physiological data on the basis of different levels of cognitive, motor, and perceptual components of the task. Specify which types of activities you plan to capture with alpha, beta, delta, gamma, and theta bands.

The three flight maneuvers for which we show EEG activity in Figure 3.18 are EEG-1, the simulated roll capture; EEG-2, actual flown roll capture; and EEG-3, baseline cognitive. EEG-1 involved voluntary movements of the arms and legs as if manipulating the ailerons and rudder with the eyes closed. EEG-2 involved a left/right roll capture task flown by the EP, obviously with the eyes open, and EEG-3 involved the cognitive task of adding the digits in a phone number with the eyes closed. EEG-1 and EEG-2 did not involve much cognitive workload, because the simulated and actual roll capture tasks were very simple, repetitive maneuvers. The EEG-3, however, it a is relatively difficult cognitive task to perform.

Figure 3.18 shows that the midline power in the theta band is highest for the cognitively intensive EEG-3 task. In the alpha band, we notice that power in the occipital area is highest when the eyes were closed (EEG-1 and EEG-3) and lowest when the eyes were open in the perceptually intensive task of flying the roll capture task (EEG-2). Also, we see a desynchronization in the sensorimotor areas for EEG-2 and a strong synchronization for the cognitively intensive numbers-adding task in EEG-3.

In December 2006, we repeated this experiment on NRC's ASRA, a Bell 412 helicopter with an experimental, full-authority FBW system. Integration of the CATS architecture of the ASRA was easily accomplished in a matter of a few working days (see Figure 3.3). For this experiment, seven flight tasks were designed for the ASRA sorties. The first task, FF-1, involved manual left/right movement of the cyclic FBW control stick as if banking the helicopter from straight and level to an angle of bank of about 30 deg to each side, with a full roll cycle taking about 2 s. For this maneuver, the EP had his eyes closed and the FBW controls were not actually affecting the aircraft. The SP maintained the aircraft at a straight-and-level attitude during this task, which was performed to isolate simple cyclic motor movement artifacts.

For the next maneuver, FF-2, the FBW controls were activated and the EP actually performed the roll capture maneuver with his eyes open and with a maximum angle of bank to either side of 30 deg and a roll cycle time of 2 s. The third task, FF-3, involved mental addition of 10-digit phone numbers (with eyes closed), and the fourth task, FF-4, was a memory task requiring the EP to remember an instrument holding clearance for 30 s (eyes open). The fifth task, FF-5, required the EP to observe a wingover maneuver that was flown by the SP. The goal was to give the EP a chance to visually monitor performance values (speed heading, angle of bank, g-loading, etc.) and mentally follow through this relatively complex task without any movement artifacts. The next task, FF-6, consisted of the same wingover, but this time it was flown by the EP.

This allows for isolating the effect of motor movement between the observed wingover and the flown wingover.

The last task, FF-7, was a comprehensive task that included a repeated cyclic roll capture to 30 deg of bank (left/right) at a cycle time of 4 s with a simultaneous reduction in speed from 100 kts to 70 kts during the first 30 s of the maneuver and a subsequent increase from 70 kts to 100 kts during the second 30 s of the task. This task was cognitively very demanding because it involved a cyclic banking motion with control of a speed trend. Between each task, we recorded data for a 30-s rest period (eyes closed).

In addition, we added several high-gain tasks from the ADS-33 catalog of maneuvers. Because the ASRA is a platform that allows augmentation of the flight control laws, we elected to study the effect of flight control augmentation on physiological patterns as a within-subject variable. We used a rate-damped (RD) flight control law as a representative of a flight control law that is typical for modern military helicopters and temporal rate command (TRC) as a representative of a highly augmented simple (low-workload) flight control law. In RD, the pilot was exposed to essentially normal flight behavior and workload. In TRC, hovering is accomplished by moving the cyclic to a neutral position. To move the aircraft in any direction from the TRC hover, the pilot needs only to push the cyclic in the direction that he/she wants to go, with the amplitude of the deflection commensurate to the desired speed. Clearly, TRC will lead to lower workload levels than will RD for the same maneuver.

Data were collected for eight pilots in the ASRA simulator and in flight. Data are presented for two contrasting participants: an experienced high-time military pilot and an NRC nonpilot researcher with moderate simulator experience on the ASRA (see Figures 3.19 and 3.20; pp. 66–68). Both participants were middle-aged males. The CATS tool was used to explore the clinical EEG bands for each participant and each task across the spatial extent of the 128 electrodes. We were particularly interested in locating specific regions on the scalp that conveyed significant markers of cognitive loading. If localized regions can be designated, it would be possible to design pilot state monitoring systems with EEG arrays that are sparse and perhaps integrated into the helmet or aviation headset. In describing the following results, it is important to note that EEG power should be compared only within each participant relative to task difficulty.

Figure 3.19 (pp. 67, 68) shows the beta/alpha ratio in the occipital area for both pilots and for all tasks. The beta/alpha ratio is a good indicator of cognitive workload, with higher ratios corresponding to higher levels of cognitive workload. Both participants showed the lowest beta/alpha ratio during the rest task. Consistent with the difficulty of the task, the experienced pilot exhibited the highest beta/alpha ratio during the very difficult comprehensive maneuver (bank aircraft and change speed). The novice participant showed high levels of beta/alpha for the clearance callback, the wingover, and the comprehensive task.

Our qualitative observation during task performance clearly indicated that this participant tried very hard to do well in these tasks, which were unfamiliar to him. Our Bell-412 data clearly indicated that levels of cognitive performance in experienced and novice pilots can be differentiated on the basis of EEG and flight technical data in a technically challenging environment.

Figure 3.20 shows the EEG theta power in the midline regions for the experienced and novice participants for the flight maneuvers. Midline theta power is a good indicator of cognitive expenditures for performance monitoring and working memory. However, during wakeful rest with the eyes closed, midline theta power is high as well. Thus, if one were to use this measure for real-time pilot state characterization, one may want to use some other measure, such as eye tracking, to determine if the eyes are open or closed.

The first task involved simple left/right movements with the arm, and virtually no cognitive expenditures were necessary. Task monitoring in the flying tasks involved keeping track of headings, bank angles, roll cycle times, speeds, altitudes, torque, and so on. The roll capture, wingover, and comprehensive tasks required close control of most of these parameters. The mental arithmetic (numbers-adding) task required maintenance of running sums in working memory, and the clearance callback required maintenance of holding pattern parameters in working memory for 30 s. For both pilots, midline theta power was lowest for the first task, in which the participants moved the cyclic stick left and right with their eyes closed. This task required virtually no cognitive (performance-monitoring or memory) expenditures.

For the experienced pilot, mental arithmetic required a higher expenditure (higher midline theta power) of cognitive effort than did the highly overlearned flying tasks. Based on the low midline theta power, the clearance callback appeared to be easy for this experienced instrument pilot, and he was in fact able to recall all parameters of the clearance without much effort or error. For the novice participant, we noticed that the wingover and comprehensive tasks required a higher expenditure of cognitive resources compared with the mental arithmetic task. Qualitative observation of the novice participant during these tasks clearly indicated that he was working hard to perform these unfamiliar maneuvers. We noted that the clearance callback of the novice (non-instrument-rated, nonpilot) participant was not correct; it may be that he gave up trying to remember the clearance parameters that he was supposed to remember for 30 s. His midline theta power remained relatively low in this task.

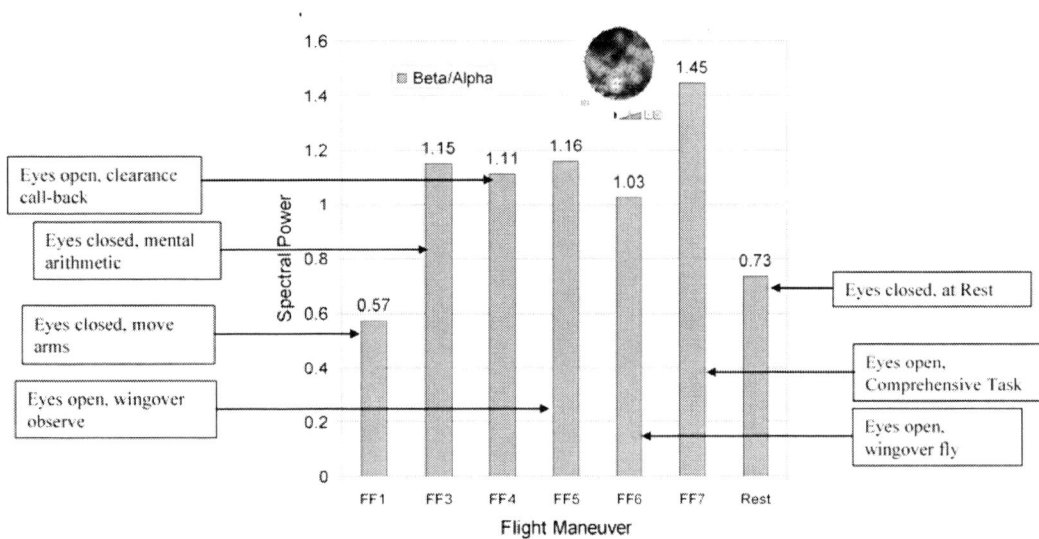

a. Experienced Pilot

Figure 3.19a. EEG beta/alpha power ratio for the experienced pilot and the novice participant, Bell 412 flight maneuvers.
Note: Topography of sensor locations is shown on the right of the bar graph. Dots indicate sensors that were included in the analysis. Shading indicates average activity over the duration of the maneuver. Anterior is oriented up. FF2 data are missing due to a data collection problem.

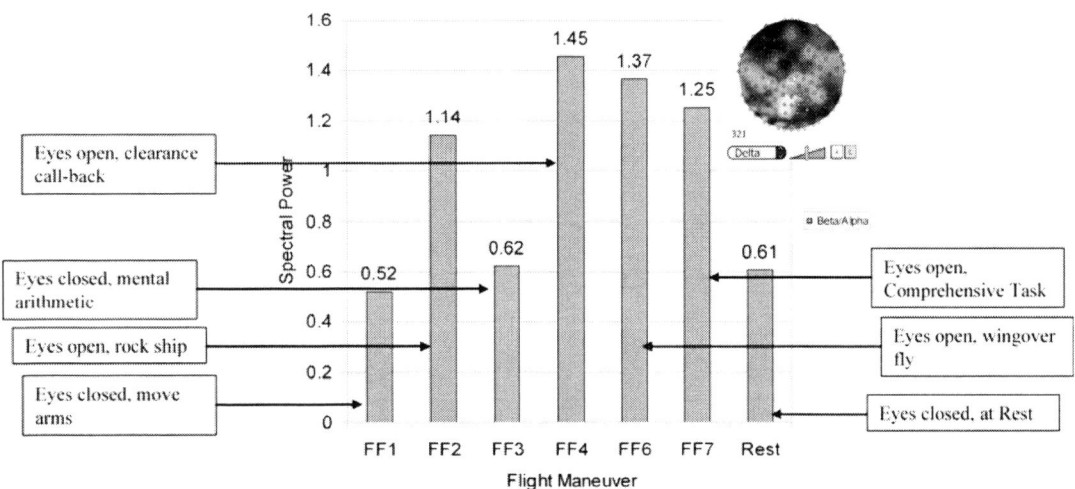

b. Novice Participant

Figure 3.19b. EEG beta/alpha power ratio for the experienced pilot and the novice participant, Bell 412 flight maneuvers.
Note: Topography of sensor locations is shown on the right of the bar graph. Dots indicate sensors that were included in the analysis. Shading indicates average activity over the duration of the maneuver. Anterior is oriented up. FF5 data are missing due to a data collection problem.

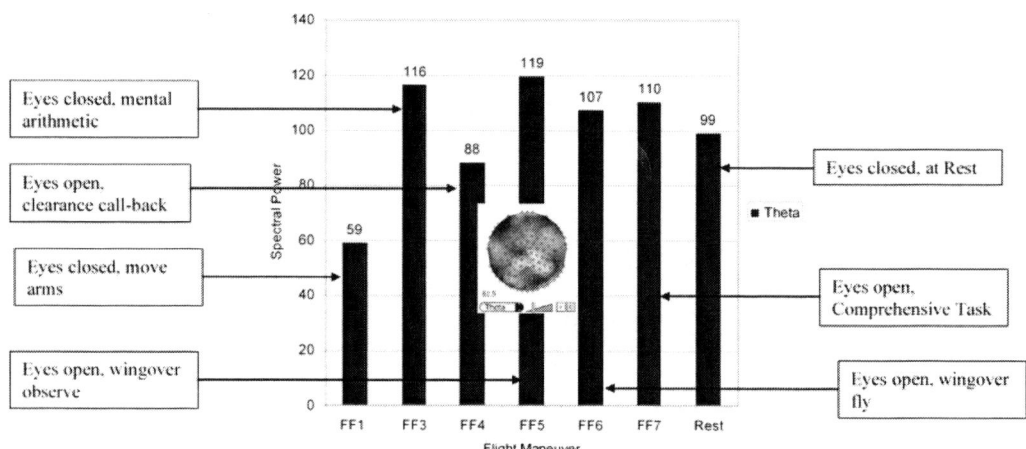

a. Experienced Pilot

Figure 3.20a. EEG midline theta power for the experienced pilot and the novice participant, Bell 412 flight maneuvers.
Note: Topography of sensor locations is shown on the right of the bar graph. Dots indicate sensors that were included in the analysis. Shading indicates average activity over the duration of the maneuver. Anterior is oriented up. FF2 data are missing due to a data collection problem.

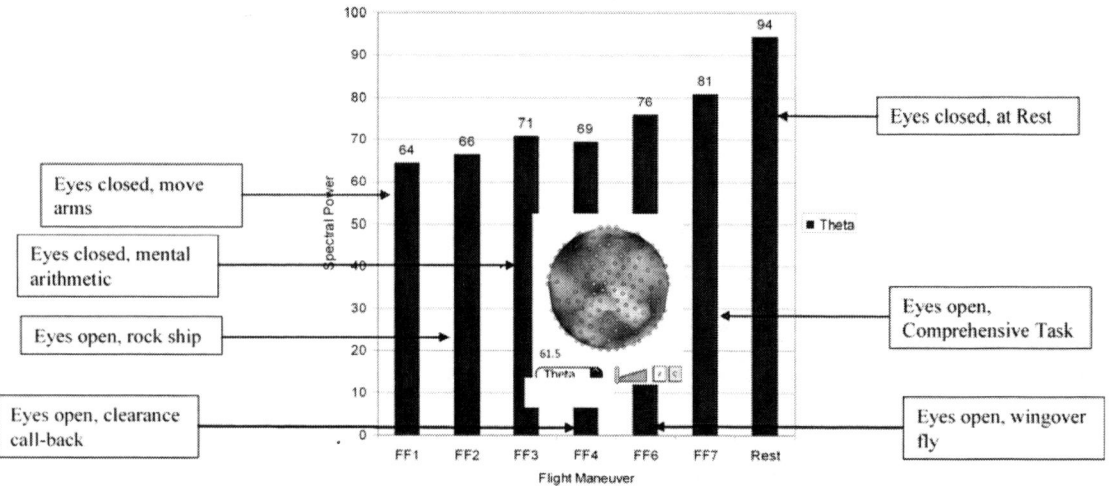

b. Novice Participant

Figure 3.20b. EEG midline theta power for the experienced pilot and the novice participant, Bell 412 flight maneuvers.
Note: Topography of sensor locations is shown on the right of the bar graph. Dots indicate sensors that were included in the analysis. Shading indicates average activity over the duration of the maneuver. Anterior is oriented up. FF5 data are missing due to a data collection problem.

Lessons Learned

We embarked on our multisensory physiological research in a real flight environment knowing that we needed a very robustly developed data collection apparatus (see Chapter 7). Both the NRC and the OPL had prior experience in flight testing and aircraft instrumentation, and this collective expertise was leveraged in the cognitive avionics flight tests. Sorties on flight test platforms are very expensive, and the research equipment must be integrated such that it can reliably operate in the demanding environment of real flight. Moreover, the equipment must be designed and implemented according to stringent airworthiness certification rules. Most clinical data collection devices need to be specifically flight-hardened for use in the flight test environment, and the installation should comply with standards and best practices applicable for avionics equipment. We intend to develop our apparatus toward an integrated avionics–grade system, such as our CPH concept, that will integrate all neurophysiological and physiological sensors, as well as signal-processing microelectronics on an aviator helmet. The CPH concept will enable us to use pilot state–monitoring sensors without preparation and setup time.

Except for airworthiness certification requirements, we believe that the requirements for, and lessons learned from, flight testing translate very well to other neurophysiological field research. For example, OPL operates an instrument-equipped car that uses exactly the same electronics boxes that are used in the aircraft, and the same general data collection principles are applied with great success. We believe that our technology and methodology would be well suited for other crew stations, such as uninhabited air vehicle ground control stations, tank and automotive crew stations, and shipboard crew stations. We also believe that with wireless transmitters and miniaturized computers, our architecture could be deployed relatively easily on dismounted warfighters.

We found that the aircraft-in-the-loop simulation technique (see Figure 3.7) was an invaluable step in refining the integration of the equipment, in developing the test cards, in EP training, and in determining differences between simulation and flight.

Review

Synchronized data collection is essential, and it is best to perform the synchronization through a local area network in real time during data collection. Multisensory data should be synchronized properly in real time during data collection to avoid a huge burden on the analyst during data postprocessing. Also, data that are not accurately synchronized will introduce inaccuracies between the occurrence of performance events and the neurophysiological measures. This, of course, is not acceptable for any real-time performance-monitoring systems.

The CATS architecture was specifically designed to synchronize multisensory data and provide the ability to tag events exactly when they happen. CATS is also able to incorporate video and audio data in the same unified data structure. This makes it possible to record voice narrations that can be useful during after-action review.

Also, for reasons of quick and safe egress, we needed to devise a methodology that would not entangle the EP in a mess of wires; rather, we had to design the physiological data collection system such that it can be rapidly disconnected in one place. To further reduce intrusiveness, we made use of nontouch sensors wherever possible.

For the reader's convenience, we have developed a lessons learned list, shown in Tables 3.1 and 3.2 (pp. 70–71).

Table 3.1. Lessons Learned Associated With CATS

Lesson Learned	Why Is it Relevant?
Research equipment must be hardened.	Scrubbing a sortie because of malfunctions is expensive. You want the equipment to be a start-and-forget item as much as possible.
Aircraft-in-the-loop simulation simplifies integration, test card development, and training.	Saves cost. The concept involves all aircraft systems except the engine, and systems interactions can be debugged. Facilitates test card development and participant training.
Data should be synchronized at collection time not post-flight.	Sorties can be hectic, and there is no time to keep track of different files while in flight. Also, postflight synchronization is much more work.
Data should be marked (tagged) during data collection, not post-flight.	Taking notes while in flight can make the FTE really miserable (motion sickness). Better to provide an environment in which the FTE can mark from preselected events. Greatly facilitates data analysis and discarding of nonrelevant data.
Collect sufficient data to document the conditions of the experimental situation.	After the sortie, it is important to have synchronized multichannel audio/video, and eye tracking, to name a few, to be able to explain events and effects found in the data.
Don't collect bad data.	Make sure all sensors are working when collecting the data. Take the time to make the sensors work properly. Scrub the sortie if the quality of the data is compromised. Quality is much more important than quantity.
Automated data analysis tools help to avoid computational mistakes that can occur during tedious manual analysis.	Manual labor is too prone to errors. Use tools to simplify analysis as much as possible.
Focus on a small set of participants, but study them in great detail.	Evaluation pilots will know the drill of a sortie. This saves time and is safer. Also, physiological models tend to validate well within a participant. Caution: Some journal reviewers may disagree; they may think a large N is better.

Best Practices

The most important practice in flight testing is safety. If safety is compromised, there will be no flight test. Safety and risks need to be properly managed using methodologies such as Operational Risk Management (ORM). The entire research organization involved in the flight test needs to adopt the safety first philosophy. Organizational safeguards must be in place so that project pressures and managerial influences cannot compromise safety. The flight test needs to be carefully planned under consideration of the following minimal requirements:

1. Program and project management
    a. Flight test goals and objectives
    b. Organizational structure
    c. Contractual requirements
    d. Interaction with other agencies
2. Aircraft
    a. Modifications needed for study, airworthiness certificates
    b. Maintenance programs
    c. Data collection apparatus
3. Research background
4. Flight test operations
    a. Location, routes, hard-deck, etc.
    b. Flight test schedule
    c. Procedures, maneuvers
    d. Flight envelope, limits
    e. Communication with controlling facilities
    f. Meteorological requirements
    g. Weight and balance
    h. Equipment go/no-go
5. Hazard analysis
    a. Define hazards
    b. Develop controls to mitigate hazard

A focus on safety is essential for augmented cognition applications. In aviation, this means that no flight test should ever be conducted without a safety review that involves an impartial panel of experts who review the proposed flight test in light of flight safety. Flight maneuvers that are documented on the flight cards need to be gradually built up in incremental steps to ensure compliance with the safety plan and flight test envelope.

The SP needs to undergo recurrent training to remain focused on the actual mission: namely, the overall safety of the flight test. This may involve aircraft ground handling, startup, taxi, positive handover of controls, early recognition for trends that may lead to breach of flight envelope, traffic avoidance, communication, approach, landing, taxi, and shutdown. Flight test facilities are specialized to design flight test sorties, maintain the aircraft and crews, develop the test cards, conduct safety reviews, and assist in data collection as a service to the research community.

Table 3.2. Best Practices Associated With Flight Testing

Best Practice	Comments
Be safety conscious. Safety is job #1.	Manage and minimize risk with a professional safety program.
Seek input and help.	Get support from individuals or groups who have professional expertise with your operational environment (e.g., flight testing). They are glad to help. This will make the operation safer and more efficient. Safety first.
Best Practice	**Comments**
Generate good data collection techniques (e.g., flight cards).	Good data collection techniques, such as flight cards, will provide you with data that will allow testing of your hypothesis. Such techniques should be simple. It may be better to test a complex hypothesis in several smaller tests. Be aware of participant fatigue, learning curves, and sequence effects.
Keep refining the data collection plan (e.g., flight test).	Cutting-edge research may not lead to an immediate breakthrough. It may be necessary to program the project with several test plans (i.e., several flight tests) that allow for a buildup in expertise and robustness in the equipment.

Design Guidelines

Our experience has been that augmented cognition applications should focus on measurement and classification (see Chapters 1 and 2) before they should focus on mitigation (see Chapters 5 and 6). The classification needs to be robust in the environment for which it is designed. Frequent false classifications that lead to incorrect triggering of a mitigation may cause frustration and mistrust in the user community (Chapter 7).

Furthermore, as designers, we should always ask if the mitigation really depends on an augmented cognition component (see Figure 3.5). To that extent, it may be a good idea to include a condition in the evaluation in which the mitigation is active at all times and is treated like another user interface format that may have a benefit in its own right.

Designers should also specifically include test conditions to ensure that augmented cognition applications cannot lead to a feed-forward deterioration of operator state through frustration with incorrect classifications. That is, we want to be sure that in no situation would an operator or crew receive a mitigation that causes an undesirable physiological reaction that would trigger the mitigation over and over again, snaring the operator in a deepening spiral of sensing, incorrect classification, and mitigation. For the time being, it may be beneficial to consider installing an easy-to-reach "off" switch for the mitigation.

Finally, in dealing with the relatively conservative nature of the operational environment (e.g., pilot, avionics design, certification community), we need to be very concerned with the intended function of the augmented cognition application. We also need to be sure that the application will not provide any hazardous misleading information to the operator (e.g., pilot and crew). All stakeholders, including the certification body, should be brought to the table early to define the intended function of the application.

Parting Message

Flight testing of cognitive avionics systems is very challenging and can be very rewarding as the technology application is tested in an actual target environment. Avionics is the major growth factor in new airframes. Our parting message is to put safety first. "Keep them flying."

Appendix: Explanation of Special Terms

The field of flight testing uses a well-established approach and procedure with the primary goal of safety of flight while ensuring accomplishment of the test objectives. Several specialized terms deserve a brief discussion. The following is by no means a complete list of functions and terms used in flight testing.

- *Sortie:* A test flight from briefing to debriefing.

- *Test card:* A graphical and textual description of the maneuver that is to be flown for a particular test point. Several test points may be accomplished in one sortie by using several test cards.

- *Test director (TD):* The person who is in charge of the science component of the flight.

- *Flight test engineer (FTE):* A person who is in charge of the technical apparatus that accomplishes the science goal of the sortie. The FTE is often the person who obtains IRB consent from participants, holds briefings, makes sure that there is enough disk space for the data, extracts the data after the sortie, and so on.

- *Hard deck:* A specified altitude above ground level (AGL) above which test points will be conducted. The hard deck will be penetrated downward only for returning to base (RTB) or in case of an emergency.

- *Evaluation pilot (EP):* The participant in an airborne experiment. Usually the EP is a pilot who is qualified to operate the aircraft or who is even rated for the type of aircraft being used, although there are exceptions to that rule. The EP will receive full or limited control authority from the safety pilot only for the conduct of the test point.

- *Safety pilot (SP):* The SP is the pilot in command (PIC) and, by definition, the ultimate onboard authority. The SP is responsible for safety of flight and is usually the sole manipulator of flight controls (unless the aircraft requires a crew) when operating below the hard deck. The SP plans the technical portions of the sortie and obtains all necessary briefings from regulating authorities to safely conduct the flight. The SP usually makes the ultimate go/no-go decision prior to a sortie, but other individuals may share in that responsibility. The SP repositions and configures the aircraft for each test point and will hand over control to the EP using an agreed-upon positive hand-over protocol. The SP monitors the performance of the EP during a test point with regard to trends that may lead to exceedance of an aircraft limitation and will take control when the trends indicate that it is necessary to do so. The SP also guards the controls sufficiently to prevent the EP from generating any control inputs that could cause a flight hazard. Typically, the SP has flight instructor and/or test pilot credentials.

References

Gubbels, A. W., Carignan, S., & Ellis, K. (2000). Bell 412 ASRA safety system assessment. *FRL Laboratory Technical Report LTR-FR-162*, National Research Council, Ottawa, Canada.

Gubbels, A. W., Morgan, M., & Baillie, S. W. (1997). Modifications to the NRC Bell 205 Airborne Simulator Safety System in response to a recent incident. In *Proceedings of the American Helicopter Society's 53rd Annual Forum* (pp. 1495-1502). American Helicopter Society, Alexandria, VA.

Kobayashi, A., Kikukawa, A., & Onozawa, A. (2002). Effect of muscle tensing on cerebral oxygen status during sustained high +Gz. *Aviation, Space, and Environmental Medicine, 73*, 597–600.

Kobayashi, A., Tong, A., & Kikukawa, A. (2002). Pilot cerebral oxygen status during air-to-air combat maneuvering. *Aviation, Space, and Environmental Medicine, 73*, 919–924.

Samel, A., Wegmann, H. H., Vejvoda, M., & Wittiber, K. (1996). Stress and fatigue in long distance 2-man cockpit crew. *Wiener Medizinische Wochenschrift, 146*(13–14), 272–276.

Schmorrow, D., Stanney, K. M., Wilson, G., & Young, P. (2006). Augmented cognition in human-system interaction. In G. Salvendy (Ed.), *Handbook of human factors and ergonomics* (3rd ed., chapter 52). New York: Wiley.

Taylor, R. M., Howells, H., & Watson, D. (2000). The cognitive cockpit: Operational requirement and technical challenge. In P. T. McCabe, M. A. Hanson, & S. A. Robertson (Eds.), *Contemporary Ergonomics 2000* (pp. 55–59). London: Taylor & Francis.

Wilson, G. F., & Russell, C. A. (1999). Operator functional state classification using neural networks with combined physiological and performance features. In *Proceedings of the Human Factors and Ergonomics Society 43rd Annual Meeting* (pp. 1099–1102). Santa Monica, CA: Human Factors and Ergonomics Society.

Wilson, G. F., & Russell, C. A. (2003). Real-time assessment of mental workload using psychophysiological measures and artificial neural networks. *Human Factors, 45*, 635–643.

Wilson, G. F. (2005). Operator functional state assessment in aviation environments using psychophysiological measures. *Cognitive Systems: Human Cognitive Models in System Design Workshop*. Santa Fe, NM.

Chapter 4

Cognitive State Estimation in Mobile Environments

Michael C. Dorneich, Santosh Mathan, Patricia May Ververs, and Stephen D. Whitlow
Honeywell Laboratories

This chapter presents the unique challenges encountered, difficult trade-offs required, and promising techniques employed when moving cognitive state assessment from the laboratory to a mobile field environment.

Introduction

Work in the field of augmented cognition began by classifying aspects of cognitive processing (attention, working memory, executive function, and sensory processing) with well-defined, well-understood laboratory tasks (often referred to informally as "Psych 101" tasks). As researchers have moved from the laboratory environment to the field environment, they have introduced the artifacts (motion, electrical, networking traffic, and disconnect) and stressors (information overload, physical load, competition, and threat of pain) inherent in some operational environments to which augmented cognition systems would be transitioned. **The move from the laboratory to mobile field environments brings a number of unique challenges that must be addressed if cognitive state assessment is to be used successfully in task domains that require the operator to be mobile.** Tough sacrifices need to be made, with limitations on the sensors to be used, processing power, and knowledge of the task environment. Therefore, unique techniques must be developed to enable this technology to move beyond sedentary operator domains.

Adaptive Automation
Adaptive automation, in which the automation adapts to the current task environment during task execution, either can make a certain component of a task simpler or can aid with adaptive task allocation, shifting a task from a larger multitask context to automa-

tion (Parasuraman, Mouloua, & Hilburn, 1999). Adaptive systems must make timely decisions on how best to use varying levels of automation to provide support to humans. For an adaptive system to decide when to intervene, it must have some model of the context of operations, be it a functional model of system performance or possibly a model of the operator's functional state.

Many adaptive systems derive their inferences about the cognitive state of the operator from mental models, performance on the task, or external factors related directly to the task environment (Wickens & Hollands, 2000). For example, those working in the 1980s in the field of associate systems developed adaptive information and automation management technologies that depended on a common understanding (between the automation and the human operator) of the mission, the current state of the world, the platform, and the state of the operator him- or herself. Associate systems then used that shared knowledge to plan and suggest courses of action and to adapt information displays and the behavior of automation to better serve the inferred operator intent and needs (Miller & Dorneich, 2006). Associate systems were developed for numerous domains, including single-seat fighter aircraft (Banks & Lizza, 1991), attack/scout helicopter operations (Robertson, 2000), petrochemical plants (Cochran, Miller & Bullemer, 1996), and in-home monitoring and caregiving for the elderly (Miller, Wu, Kirchbaum, & Kiff, 2004). In contrast to augmented cognition adaptive systems, the primary means used in associate systems to infer operator intent was logical deduction based on knowledge of the mission plan and the functional capabilities of the platform (Geddes, 1985).

As Scerbo et al. (2001) noted in their comparison of various adaptive automation techniques, task performance and operator modeling have advantages and disadvantages when used to drive adaptive systems. Although measurement of performance on the task has the advantage of being an online technique that can respond to unpredictable changes in the cognitive state of the operator, the method is only as good as the ability to measure performance. The use of behavioral responses to track cognitive function requires regular or periodic performance assessments to keep an updated assessment of performance capabilities. Few systems provide opportunities to track overt responses for monitoring operator performance (Parasuraman, 2003). Additionally, diagnosing cognitive state degradation via human performance degradation occurs after the fact, and thus an opportunity to proactively adapt the system to maintain performance is limited.

Modeling techniques have the advantages of offline implementation and ease of incorporation into rule-based expert systems. However, these techniques are only as good as the underlying models, and they are susceptible to model *brittleness*. Brittleness occurs because the systems model is necessarily an incomplete representation of the world (i.e., the system model does not account for all possible scenarios), and therefore the system could produce a dramatically incorrect solution when an important, but unmodeled, feature of the problem affects the choice of optimal solution (Smith, McCoy, & Layton, 1997). The more complex the task, the greater the likelihood is that the model will not anticipate all aspects of human operator performance.

Valuable Information
Augmented cognition technologies drive system adaptations by using physiological responses and brain activity to infer the availability of cognitive resources to cope with mission-relevant task demands. The goal is to enhance human performance when task-related demands surpass the human operator's assessed current cognitive capacity, which fluctuates subject to fatigue, stress, overload, or boredom. Neurophysiologically and physiologically triggered adaptive automation offers many advantages over the more traditional model-based approaches to automation by basing estimates of operator state directly on sensed data. These systems hold the promise of leveraging the strengths of humans and machines, augmenting human performance with automation specifically when assessed human cognitive capacity falls short of the demands imposed by task environments. With more refined estimates of the operator's cognitive state, measured in real time, adaptive automation also offers the opportunity to provide aid before the operator even knows he or she needs it.

The potential applications of augmented cognition cover a wide range of human-computer joint cognitive systems. One such application would be a closed-loop adaptive system to help optimize the performance of a stationary operator. Such systems may include operators who interact with information displays, such as an unmanned air vehicle ground control station operator (Snow, Barker, O'Neill, Offer, & Edwards, 2006), or an operator of a weapon control system such as the Tactical Tomahawk (Tremoulet et al. 2006).

Augmented cognition technologies can also be used for studying skill acquisition during training. Krebs et al. (1998) used positron emission tomography (PET) scans to study the learning progression of a novice operating a telerobotic arm. Cognitive state assessments during training could also be used to diagnose student difficulties in real time and provide appropriate context-specific assistance (Mathan & Dorneich, 2005). However, moving these technologies to mobile contexts remains a challenge if they are to be used in operational environments.

Knowledge of instantaneous cognitive state can be used to drive adaptive systems in many mobile contexts. Examples include pilots, dismounted soldiers, and ground vehicle operators (Dorneich, Ververs, Mathan, & Whitlow, 2005; Schnell et al., 2006, Snow et al., 2006). A truly adaptive system that manages information flow will require the ability to operate in a dynamic operational situation with a high degree of fidelity in cognitive state assessment and temporal resolution.

Physiological Measures of Cognitive State
Neurophysiologically and physiologically based assessments of cognitive state have been captured in several different ways, including electrocardiogram (ECG), electroencephalogram (EEG), and functional near-infrared (fNIR) imaging (see Chapters 1 and 2). ECG measures include heart-rate variability (HRV) in the time domain to assess mental load (Kalsbeek & Ettema, 1963), tonic heart rate to evaluate the impact of continuous information processing (Wildervanck, Mulder, & Michon, 1978), variability in the spectral domain as an index of cognitive workload (Wilson & Eggemeier, 1991), and T-wave amplitude during math interruption task performance (Heslegrave & Furedy, 1979). As discussed in Chapter 2, with fNIR spectroscopy one can conduct

functional brain studies using wavelengths of light introduced at the scalp to measure cognition-related hemodynamic changes and to assess cognitive state (Izzetoglu, Bunce, Onaral, Pourrezaei, & Chance, 2004).

Other physiological measures used to assess cognitive state are galvanic skin response (Verwey & Veltman, 1996), eyelid movement (Neumann, 2002; Stern, Boyer, & Schroeder, 1994; Veltman & Gaillard, 1998; Yamada, 1998), pupil response (Beatty, 1982; Partala & Surakka, 2003), and respiratory patterns (Backs & Seljos, 1994; Boiten 1998; Porges & Byrne, 1992; Veltman & Gaillard, 1998; Wientjes, 1992). For a more complete review of physiological measures of mental workload, see Chapters 1 and 2, as well as Kramer (1991).

The suitability of a particular sensor to measure cognitive state depends on many factors, including ability to detect the underlying cognitive state of interest, temporal resolution needed to effectively drive mitigations, and fieldability issues in the context of use (e.g., sensor intrusiveness, processing power required, degree of operator motion).

As the "gold standard" for providing high-resolution temporal indices of cognitive activity, EEG has been used in the context of adaptive systems. Research has shown that EEG activity can be used to assess a variety of cognitive states that affect complex task performance. These include working memory (Gevins & Smith, 2000), alertness (Makeig & Jung, 1995), executive control (Garavan, Ross, Li, & Stein, 2000), and visual information processing (Thorpe, Fize, & Marlot, 1996). These findings point to the potential for using EEG measurements as the basis for driving adaptive systems that demonstrate a high degree of sensitivity and adaptability to human operators in complex task environments.

For instance, researchers have used the engagement index, developed by NASA researchers, in the context of mixed-initiative control of an automated system (Pope, Bogart, & Bartolome, 1995). This method uses a ratio of power in common frequency bands (beta / [alpha + theta]), where cognitively alert and focused are represented in beta, wakeful and relaxed in alpha, and a daydream state in theta. Higher engagement index values indicate increased levels of task engagement.

The efficacy of the engagement index as the basis for adaptive task allocation has been experimentally established. For instance, under manipulations of vigilance levels (Mikulka, Hadley, Freeman, & Scerbo, 1999) and workload (Prinzel, Freeman, Scerbo, Mikulka, & Pope, 2000), an adaptive system effectively detected states in which human performance was likely to fall and took steps to allocate tasks in a manner that would raise overall task performance. In a different domain, adaptive scheduling of communications based on cognitive state assessment of the readiness to process information resulted in a twofold increase in message comprehension and situation awareness (Dorneich et al., 2005). These results highlight the potential benefits of a neurophysiologically triggered adaptive automation.

Challenges Inherent in Mobile Cognitive State Classification

The effectiveness of neurophysiologically triggered adaptive systems hinges on reliable and effective signal processing and cognitive state classification (see Chapter 3).

Although these are difficult technical challenges in any context, they are particularly pronounced in a system designed for mobile contexts. Assessment of an operator's state can be notably more difficult if the operator is permitted to move freely to perform cognitive tasks in conjunction with physical tasks. What is already a difficult problem—gathering clean and robust signals on which to classify cognitive state—is further complicated by signal artifacts induced by motion.

Given the potential usefulness of augmented cognition systems in mobile contexts, methods have been developed to classify cognitive state in ambulatory contexts. In this chapter, we describe the challenges inherent in mobile cognitive state classification, including the ability to (a) collect robust and clean signals, (b) create a mobile computing and data-processing infrastructure, (c) reliably classify cognitive state, and (d) experimentally assess the accuracy and specificity of the algorithms in a mobile operational setting.

The work described in this chapter was developed for the dismounted soldier—potentially one of the harshest, most mobile application domains for cognitive state estimation. Any cognitive state classification solution in this domain must be portable, efficient, and robust to extremes of conditions and motion. A robust solution that meets the challenges of this domain would result in techniques that are applicable to almost any other domain in which motion is a key component of the operator's work environment. We describe the approaches outlined in this chapter in the context of a field evaluation that tested the ability to classify cognitive workload level in an unconstrained, free-play operation with soldiers executing missions in an urban environment.

Scenario

This section presents the mobile dismounted soldier domain and the operational task context in which augmented cognition technologies were applied.

Dismounted Soldier Domain

Dismounted soldiers experience many stressors that are inherent in the operational environment, and these stressors have a direct impact on overall cognitive capabilities. For instance, physical exertion is one of the primary stressors a soldier faces on the battlefield. Simply moving to a rally point in a mission is made difficult when it requires the soldier to carry a load weighing 80 to 120 lb. (Girolamo, 2005). Other common stressors that can diminish cognition include heat (Buller, Hoyt, Ames, Latzka, & Freund, 2005; Steinman, 1987), cold, limited food and water (Buller et al., 2005; Montain, Sawka, & Wenger, 2001), fear, and sleep deprivation. Stress affects all aspects of information processing, including general arousal, selective attention, speed and accuracy of performance, and working memory (Hockey, 1986). The degradation in cognitive performance that often results from the effects of stress can have catastrophic outcomes (Balles, 1988; U.S. Navy, 1988).

In addition to the physical stressors inherent in military operations, netcentric capabilities impose cognitive stressors on dismounted soldiers. The highly dynamic, information-rich environment of the dismounted soldier motivated the development of a toolkit for mobile classification of cognitive state. The next-generation dismounted soldier relies on netted communications to build situation awareness—the kind of situational understanding that drives decisive actions. Information exchange requirements are being pushed to the lowest levels, with the goal of enhancing the capabilities of a squad (9–10 soldiers) so that it can cover the battlefield in the same way that a platoon (16–44 soldiers) now does. The network will be characterized by a network of humans collaborating through a system of C4ISR (command, control, communications, computers, intelligence, surveillance, and reconnaissance) technologies. Small netted units will have robust team communication, state-of-the-art distributed and fused (thermal and image intensification) sensors, organic (i.e., belonging to the unit) tactical intelligence/collection assets (e.g., unmanned aerial vehicles, unmanned ground vehicles, unmanned ground sensors), and linkage to other assets to enhance situational understanding and on-the-move planning (*Future Force Warrior*, 2004). Mission success will depend on the individual soldier's ability to sort through the vast array of continuous information flow afforded by a full range of netted communications. This situation will demand the most from leaders in the field.

The increase in information flow does not come without a cost, however. Effective use of information sources is constrained by the limitations of the human cognitive system. Real-time, dynamic exchange of information in a C4ISR environment can be expected to increase the likelihood of information overload, such that postulated information superiority becomes a profound liability. Potential data overload, coupled with the efficiency of information flow required in executing military doctrine, places an overreliance on the individual soldier. One way to ensure that soldiers are supported appropriately is to develop adaptive information management systems to promote superior situation awareness on the battlefield by assessing the soldier's readiness to receive and process information. The efficacy of such systems is contingent on reliable and timely cognitive assessment.

Domain Challenges

Soldiers are subject to extremes of motion, multiple physical and mental stressors, and a wide range of cognitive activities (long periods of vigilance punctuated by extreme periods of activity). Thus, any approach to the real-time assessment of cognitive state has to be robust to motion and noise artifacts. In addition, any cognitive state classification approach has to be robust to the potentially wide range of cognitive tasks that soldiers perform, which include simple tasks such as sentry duty or defending a position and highly complex tasks such as coordinating medical evacuations or the movements of several squads or replanning tactical moves.

General Approach and Associated Toolkit

The toolkit described here consists of four principal techniques:

1. signal collection and processing of neurophysiological and physiological signals in a mobile context;
2. classification algorithms that address individualization, bias, and generalizability;
3. a computational and experimental infrastructure to support assessment; and
4. the design of experiments to assess classification in a domain-relevant operational setting with soldiers.

Signal Collection in a Mobile Environment

Inferring cognitive state from noninvasive physiological sensors is a challenging task even in pristine laboratory environments. The signal is subject to artifacts—sensor activity that obscures or distorts information associated with the cognitive activity of interest. There is a wide range of neurophysiological and psychophysiological measures, such as EEG, ECG, PET, fNIR, functional magnetic resolution imaging (fMRI), and pupilometry, to name a few (see Chapters 1 and 2). Such measures can be used to detect and determine the cognitive state of a human user. However, only a small subset of these measures is uniquely suited for a mobile environment. Because the PET and fMRI imaging scanners are not portable, this equipment is ill-suited for mobile data collection.

The use of EEG as the basis for cognitive state assessment is motivated by characteristics such as good temporal resolution, low invasiveness, low cost, and portability. The use of ECG is motivated by the strength of the signal and maturity of ECG detection sensors. However, techniques that use EEG and ECG, as well as fNIR sensors, need to focus on ruggedizing the sensor suites and performing advanced signal processing on the data collected in order to reduce the effect of artifacts inherent with mobile participants. If the classification output is desired in real time, then designers must consider ways to efficiently perform data calculations within the time, memory, and power constraints of mobile processors. Although real-time signal processing and classification of physiological signals have been implemented previously (Berka et al., 2004; Gevins & Smith, 2003; see also Chapter 3), they have not been realized in a truly mobile, ambulatory environment.

Artifact detection and reduction, necessary to create a "clean" signal that can be classified, is driven by a consideration of the characteristics of the noise artifacts themselves. How noise artifacts are handled depends on where the noise lies in the frequency band in relation to the signal (see Figure 4.1).

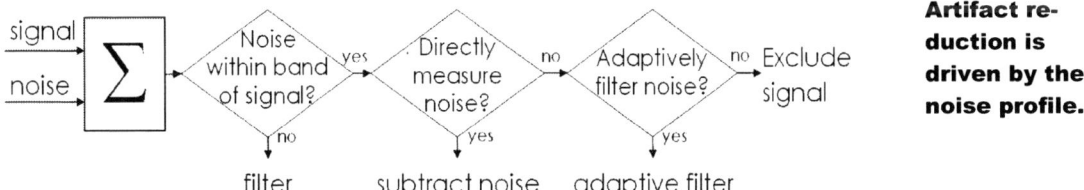

Artifact reduction is driven by the noise profile.

Figure 4.1. Frequency characteristics of the noise in relation to the signal of interest.

Noise signals that lie out of the band of the signal of interest can be removed with filtering. Out-of-band artifacts, such as DC drift and 60-Hz line noise, typically have well-known characteristics and can be filtered out easily. Noise artifacts that lie within the same band as the signal require more sophisticated artifact detection and reduction. If the noise can be measured (e.g., eyeblinks measured via a dedicated ocular sensor), the sensor data can be subtracted to decontaminate the signal. However, when the noise cannot be directly measured, adaptive filtering can be applied to estimate the noise. When adaptive filtering is not feasible, the noise should be detected and the resultant data rejected, to avoid compromising downstream classification. Signal-processing approaches are specific to the signal type and dependent on factors such as signal-to-noise ratio and how specific artifacts affect the signal of interest. Two examples are discussed below: EEG and ECG.

The EEG signal is particularly subject to artifacts because of low power in the underlying signal. High-amplitude artifacts can easily mask the lower-amplitude electrical signals associated with cognitive functions. In addition to the typical sources of signal contamination, mobile applications must consider the effects of artifacts induced by shock, cable movement, and gross muscle movement. Artifacts related to participant motion include high-frequency muscle activity, verbal communication, and ocular artifacts consisting of eye movements and blinks. Artifacts related to the operational environment include electrical noise that creates interference with the EEG signal (cf. Kramer, 1991). These concerns drive an effort to reduce the number of EEG sensors to a minimum. The minimum number of channels is dictated by the spatial resolution and underlying cognitive function of interest. More detail on these techniques is provided later in the chapter.

The challenges inherent in signal processing of ECG signals are different than those associated with EEG. Although heart rate is a very strong signal compared with an EEG signal, it is heart-rate variability that is of interest. HRV is sensitive to task demands (Aasman, Mulder, & Mulder, 1987; Beh, 1990; Porges & Raskin, 1969); thus, it is important to detect ECG peaks with a high degree of accuracy in order to identify small changes in the interbeat interval (IBI) between heart beats. In addition, it is important to account for missed peaks correctly and ignore spurious peaks, as they can have a large detrimental effect on HRV calculation.

Cognitive State Classification
Once signal processing has been applied to create a "clean" signal, the classification stage can commence. In this section we discuss how to assess classification approaches and provide some examples of classification approaches, but this is by no means exhaustive (see also Chapter 3). Considerations are presented that help frame decisions on how to select, assess, and optimize classification.

Classification Assessment Approaches
Effective cognitive state classification approaches need to discriminate between two or more classes on a moment-to-moment basis. For discussion purposes, consider a classification algorithm that should discriminate between low and high cognitive workload—workload being an example of a cognitive state of interest. Typically, in the course of evaluating such a cognitive state classification approach, one would create a task environment containing distinct periods of high and low workload. Often, statistical tests are then conducted on the resultant data to determine if the means were statistically different in the two conditions. However, **statistical significance does not**

suffice when determining if the classification approach is useful for cognitive state assessment. Tests of statistical significance, by definition, look at averages over an entire data set to create two distributions from the data and then determine if the means are statistically different.

Figure 4.2 shows three notional boxplot distributions between two classes. All three are statically significant given enough data points and show that the classification algorithm is tracking workload *on average*. However, it is important to know how effectively the classification approach differentiated between high and low workload on a moment-to-moment basis. The rightmost (third) distribution, though statistically different, shows considerable overlap in the distributions between the two classes. Thus, given a cognitive state classification value, it is impossible to say with confidence to which distribution it belongs.

Although differences may show up in averages, real-time classification requires an approach whereby index values in low and high conditions have minimal overlap, as in the first two distributions. all three plots may be statistically significant, but only the two at left show enough discrimination to be useful in classification between two states.

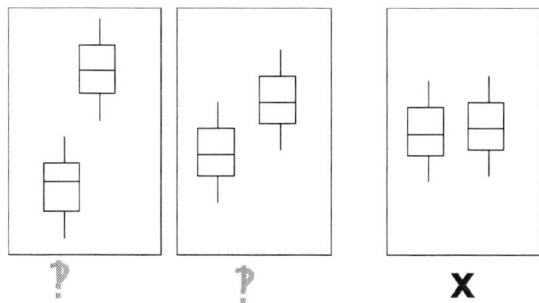

> It is important to know how effectively a classification approach can differentiate between classes on a moment-to-moment basis.

Figure 4.2. Boxplot distributions.

One approach to accomplishing this is to create indices that classify workload based on each individual's unique pattern of electrophysiological activity in response to task demands. In this section, we introduce two broad approaches to classification: (a) generative classifiers that model the distribution of features in each class (examples include probability density estimation techniques such as K-nearest neighbor, Parzen windows, and Gaussian mixture models) and (b) discriminative classifiers that model a mapping function between a set of features and class labels (examples include neural nets, support vector machines, and logistic regression).

Four potential classification approaches are introduced next. The descriptions are by no means exhaustive, as many techniques have been employed, but they represent examples of both generative and discriminative approaches.

Valuable Information
Determining a suitable classifier for a given problem is an art. Unfortunately, no single classifier approach works best on all given problems. Considerations for choosing an appropriate classifier include real-time performance, training time, and sensitivity to parameter setting. Constraints on real-time performance and training time may be dictated by the operational context. Parameter setting is important to tune a classifier to data characteristics.

Generative Classifier Approaches

Generative classification approaches have been used successfully in a mobile but constrained test task environment in which estimates of spectral power formed the input features to a pattern classification system (Mathan et al., 2005). In this example, classification systems used parametric and nonparametric techniques to assess likely cognitive state on the basis of spectral features; that is, estimate *p(cognitive state | spectral features)*. The classification process relied on probability density estimates derived from a set of spectral samples. **It is important to note that when using a pattern recognition process to train the classifier, the feature set should be gathered from tasks that most closely represent the target task environment.** Three examples of generative classification approaches are briefly introduced next: K-nearest neighbor, Parzen windows, and Gaussian mixture models.

K-nearest neighbor (KNN). The K-nearest neighbor approach is one of the simplest machine learning algorithms. It is a nonparametric technique that makes no assumption about the form of the probability densities underlying a particular set of data. Given a particular sample x, the classification process identifies k samples whose features come closest (as assessed by Euclidian or Mahalanobis [1936] distance metrics) to the features represented in x. The sample x is assigned the modal class of the nearest k neighbors.

For example, consider the data point represented by the question mark in Figure 4.3. Based on k = 5, it would be assigned the label associated with the most common class category of its five nearest neighbors (i.e., Class 1).

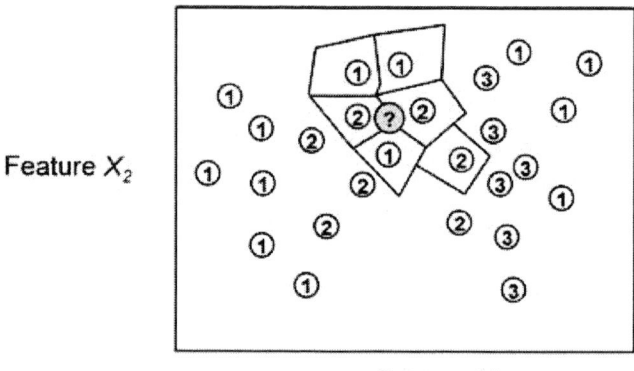

Gaussian kernels placed over each data point are used to estimate the distribution of features in each class.

Figure 4.3. K-nearest neighbor.

Parzen windows. Parzen windows (Parzen, 1967) are a generalization of the K-nearest neighbor technique. Instead of choosing the nearest neighbors and assigning a sample x with the label associated with the modal class of its neighbors, each vote is weighed by using a kernel function. With Gaussian kernels, the weight decreases exponentially with the square of the distance. As a consequence, faraway points become insignificant.

Kernel volumes constrain the region within which neighbors are considered. Consequently, Parzen windows are a better choice when there are large differences in the variability associated with each class. The data point (?) shown in Figure 4.4 is assigned to the dominant class in its immediate vicinity (i.e., class category 2).

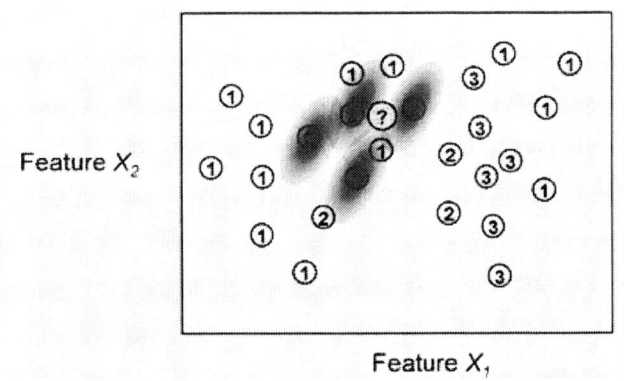

A given feature vector is assigned the class label associated with the modal class of the k samples that are the most similar to it.

Figure 4.4. Parzen windows.

Gaussian mixture models (GMM). Gaussian mixture models provide a way to model the probability density functions of spectral features associated with each cognitive state. This can be accomplished using a superposition of Gaussian kernels (see Figure 4.5). The unknown probability density associated with each class or cognitive state can be approximated by the weighted linear combination of Gaussian density components. Given an appropriate number of Gaussian components and appropriately chosen component parameters (mean and covariance matrix associated with each component), a Gaussian mixture model can model any probability density to an arbitrary degree of precision. For more details, see Dempster, Laird, and Rubin (1977).

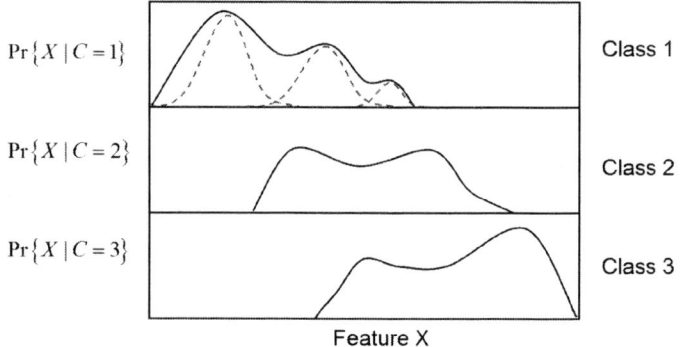

Small numbers of Gaussian kernels (dotted lines) are used to approximate the distribution of features in each class.

Figure 4.5. Gaussian mixture models.

These statistical classification techniques have an advantage over multilayer neural networks because they require minimal training time. KNN and Parzen windows require no training, whereas the GMM converges relatively quickly. KNN and Parzen window approaches require all patterns to be held in memory. Every new feature vector has to be compared with each of these patterns. However, despite the computational cost of these comparisons at run time, such systems have been shown to output classification decisions well within real-time constraints (Erdogmus, Adami, Pavel, Lan, et al. 2005).

Discriminative Classifier Approach

A discriminant function analyses approach has been employed in a fully operational mobile task evaluation, described later in the chapter. This approach used a support vector machine to discriminate between periods of low and high workload (Mathan, Dorneich, Whitlow, & Ververs, 2007).

Support vector machine (SVM). Support vector machines are linear classifiers that use a quadratic optimization procedure to find an optimal orientation and location for a discriminating hyperplane between classes. The optimization procedure finds a location and orientation for the hyperplane that lies as far as possible from examples in each class that are likely to be confused with each other (Figure 4.6).

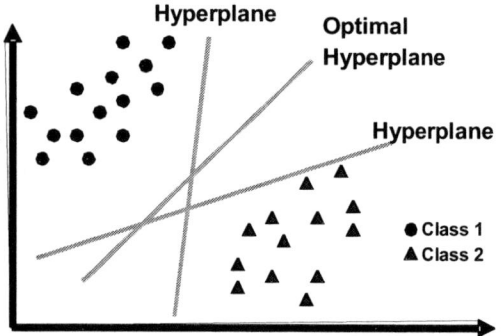

Separating hyperplanes that are identified using the SVM procedure has been shown to maximize generalization performance.

Figure 4.6. Optimal hyperplane orientation could lead to better generalization.
Adapted from Takahashi (2006).

Separating hyperplanes that are identified using the SVM procedure has been shown to maximize generalization performance (Vapnick, 1999). Although they are linear classifiers, SVMs can be used to solve nonlinear problems by means of the so-called kernel trick. Data that may not be linearly separable in the original feature space can be projected into a high dimensional space where the data may be linearly separable (Figure 4.7).

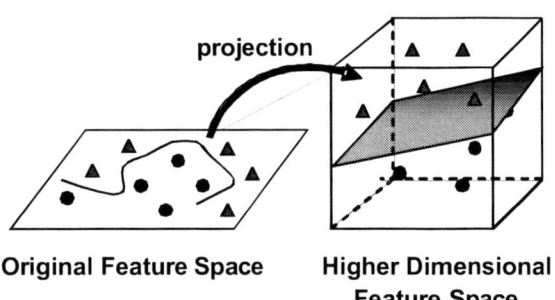

Quadratic optimization procedures find an optimal orientation and location for a discriminating hyperplane between two classes.

Figure 4.7. Transforms to higher dimensional space may result in separable data. Adapted from Takahashi (2006).

Fusion and Composite Techniques

Often it is possible to employ more than one sensor, or to employ more than one classification approach. There are two approaches, outlined next, that depend on whether you combine the input (i.e., sensor data) into the classifier or combine the output of the classifiers.

Sensor fusion. Sensor fusion uses multiple sources of sensor data to create the fusion at the sensor level before the discriminate features are calculated. This strategy for robust classification in noisy field environments integrates information from multiple sensor sources (assuming time synchronization) into a common feature vector that serves as input into a single classifier (see Chapter 7). Such an approach exploits the joint strengths of different data sources while minimizing their individual weaknesses.

Composite classifier fusion. Unlike sensor fusion (in which the fusion happens at the sensor input stage), composite classifiers fuse the output of multiple classifiers to create a final determination. A composite classification system (see Figure 4.8) has been developed that uses this technique. It employs three distinct classification approaches (K-nearest neighbor, Parzen windows, and Gaussian mixture models) and then fuses their outputs to make a final determination of cognitive state (Mathan et. al., 2005).

Actual Implementation: Classification

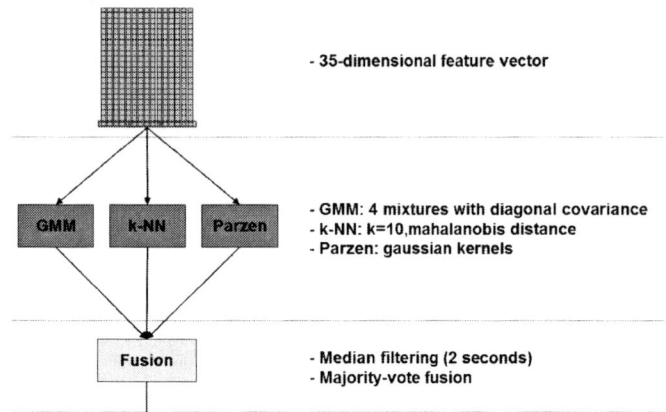

Classification system uses a composite of three distinct classification approaches: K-nearest neighbor (KNN), Parzen windows, and Gaussian mixture models (GMM).

Figure 4.8. Composite classification system.

The composite classification system regards the output from each classifier as a vote for the likely cognitive state. The majority vote of the three component classifiers forms the output of the composite classifier. Fusing the outputs of multiple classifiers using a voting scheme is a widely used strategy to increase the robustness of a classification system. The equal weighting of different classifiers implicit in the voting scheme reflects the fact that no single classifier produces consistently superior results across participants and tasks.

Simple, vote-based fusion has been shown to improve the overall performance of classification systems (Kittler, Hatef, Duin, & Matas, 1998). There are a variety of alternative options for combining diverse classifiers. Exploring these options is an objective of future research.

Considerations in Evaluating Cognitive State Classification in Mobile Environments

Cognitive state classification can be achieved with a variety of methods. Each approach discussed earlier uses statistical pattern recognition techniques to define and, later, recognize unique classes of interest. The effectiveness of a particular cognitive state classification approach in mobile environments is framed by the following research issues:

- *Bias, variance, and temporal smoothing*
  - How well can the classifier fit and discriminate between workload classes in an inherently noisy and dynamic environment?
  - How well does the classifier generalize to unseen data over spans of tens of minutes—when task characteristics remain the same?
  - Can classification accuracy improve as the output of the classifier is integrated over time?

- *Discriminating features*: What aspects of signal serve to discriminate between high and low workload?

- *Fusion:* Can overall classification accuracy be improved by integrating additional sensor sources?

- *Sensor density:* How many channels (of EEG, for instance) are required for accurate classification?

- *Long-term generalization:* How well is the classifier likely to generalize over time spans of days as the task context and patterns of general physiological activity change (e.g., sleep, stimulants)?

Applied Exercise: Classification Design Considerations
Choice of classification approach is driven by multiple considerations. First and foremost, what is practical given the context? How many sensors need to be deployed? How much individualization is necessary, and how much training of classifiers will be required? Which classification approach is best able to discriminate between the cognitive states of interest? Within any one approach, how much tuning of parameters is needed to achieve good performance? Finally, how can the classification approach be meaningfully assessed to ensure that the resultant algorithms will allow for moment-to-moment classification? Design a classification approach that addresses these considerations.

Computational and Experimental Infrastructure

This section briefly outlines the computational and experimental infrastructure needed to classify cognitive state in an operational setting where mobility is a major challenge. In general, a system constructed for a mobile application environment to assess cognitive state classification algorithms consists of the following:

- *Sensors:* a variety of sensors to collect raw physiological and neuro-physiological data.

- *Mobile processing:* mobile semirugged computer platforms to process the raw sensor data into cognitive state classification assessments.

- *Wireless data network:* a wireless data infrastructure to send the classification assessment to automation to close the loop, or to convey open-loop feedback of subordinates' state to human leaders.

- *Experimenter's base station:* a computing infrastructure and base station to control the IT component of the experiment and to troubleshoot any unexpected problems.

Design of Experiments

The evaluation of cognitive state classification algorithms in a mobile setting is fraught with several unique challenges not found in laboratory settings. First, it is much more difficult to design a task environment that reliably produces the cognitive state of interest when moving from a constrained task environment to a "free-play" operational environment. Second, because the task environment is not subject to the normal level of experimental control, it is much more difficult to know *ground truth* (i.e., the actual cognitive states experienced by the participant). Finally, the metrics used to assess the viability of the classification algorithms to distinguish the cognitive states of interest must take into account moment-to-moment discriminability, as opposed to averages of means over time (as discussed earlier).

Manipulating the Cognitive State of Interest

In addition to the practical and system configuration challenges faced when moving from the laboratory to field studies, there are issues of experimental control and the characterization of cognitive state in less constrained environments. It is essential to select tasks both that are operationally relevant and that afford reasonable adaptations that improve performance. In the laboratory it is possible to develop simple tasks in which the cognitive state of interest (e.g., cognitive workload) is manipulated precisely and consistently. Additionally, a user's performance can be collected and evaluated accurately. This makes it relatively easy to establish ground truth about a user's likely workload, for instance. However, when developing operationally relevant tasks in a field environment, it becomes substantially harder to manipulate workload precisely and to interpret and assess a user's performance without compromising operational realism.

In many operational settings, it is not always possible to vary workload directly. Instead, one must vary task load to induce cognitive workload. Furthermore, the amount of cognitive workload induced in a participant is a function of factors such as stress, fatigue, training, experience, and individual differences in capabilities. Thus, methods must be devised to correlate task load directly to workload in a systematic way in order to derive ground truth.

Ground Truth

In order to calculate the accuracy of a classification approach, classifier results are compared with ground truth. The output of the classifier at any moment is then compared with ground truth to determine the accuracy of the classifier.

The principal issue in scenario and task design is to create detectable and sustained periods (5–10 min) of high or low workload multiple times within any single data collection session. Definable periods of high and low workload sustained by participants are difficult to obtain directly from task characteristics, for the reasons discussed earlier. Thus, indirect methods must be used.

There are several classes of indirect methods: observation, secondary task performance, and participant self-reporting. Often, human experts can observe the experiment and determine ground truth based on their knowledge of task demands and the demonstrated behavior of the participant. When possible, secondary tasks can be introduced as discrete probes (in which metrics include performance and response latency) or as continuous tasks in which performance declines on the secondary tasks when the primary tasks induce higher workload. Participants can do a postscenario cognitive walkthrough, often with time-stamped videos, and report their self-assessment of their level of cognitive workload. These methods can be used individually or together to produce ground truth.

There are two important considerations when using some or all of these techniques. First, both the ground truth data and the classification data must be on a common (and accurate) time-stamp system, to allow for a moment-to-moment determination of classification accuracy. Second, for participants doing self-assessments, the terms *high workload* and *low workload* should be defined in an understandable manner.

- Operationally, *low workload* can be defined as times when the participant would have been able to take on additional cognitive tasks.

- *High workload* can be defined as times when it was not possible for the participant to take on any additional tasks and/or was not able to handle the current task load.

Classification Metrics

As discussed earlier, it is important to know how effectively a classification approach can differentiate between classes on a moment-to-moment basis. A metric used to evaluate classification performance is the Area under the Receiver Operating Characteristic (ROC) curve (see Duda, Hart, & Stork, 2001; see Chapter 6). ROC curves plot true positives (on the *y* axis) against false positives (on the *x* axis) as a threshold for discriminating between targets and distractors.

The ROC curve provides a way to assess the degree of overlap between two univariate distributions. It is widely used to evaluate human and machine signal detection capabilities. In addition, the ROC curve provides a way to assess the degree of overlap between the output of a classifier for two classes of data. Perfect classification produces an area under the curve value (Az) of 1.0, and chance performance produces an Az value of 0.5.

Test Your Knowledge
What is the motivation for using an ROC curve versus a simple statistical test for evaluating a significant difference between the means of distributions (classes)?

Tuning Classification Parameters

A major concern in the environments in which dismounted soldiers function is that noise from myriad sources could completely mask features that could be used to discriminate between high and low workload. Thus, a classifier may fail to discriminate adequately between workload classes. The capacity of a classifier to overfit training data is known as the *bias of the classifier*. These noise characteristics can also change dramatically over time—so that even if a classifier is able to effectively discriminate between workload classes over a short temporal window, it may fail to generalize adequately to unseen data collected a few seconds or minutes beyond the duration of the data used to train the classifier. The capacity of a classifier to generalize is referred to as the *variance of the classifier*.

One way to explore the bias and variance of a classifier is through a process called *n-fold cross validation*. This procedure entails splitting the data into *n* subsets. At each iteration of the validation procedure, one of these subsets (n_i) is used for testing the classifier, and the remaining $1 - 1/n$ sets are used for training the classifier. A typical choice of *n* is 10. Estimates of bias and variance get more conservative as the size of *n* decreases—the classifier has to be trained with less of the data and is assessed by generalizing to a larger subset of unseen data.

Advanced Warfighting Experiment

The general toolkit described in the previous section was realized in a mobile system that facilitated the evaluation of cognitive state algorithms (Dorneich, Mathan, Ververs, & Whitlow, 2007). The work discussed in the remainder of this chapter is grounded in an experiment conducted in an outdoor field environment. The Advanced Warfighting Experiment (AWE) was an evaluation of a MOUT (Mobile Operations in Urban Terrain) exercise at the U.S. Army Aberdeen Proving Ground. The overall objective of the AWE was to evaluate the effectiveness of the toolkit's sensor-driven cognitive state assessment technologies in a realistic, operational, mobile environment.

Scenario Description

The AWE used a full Army platoon as participants. Of the 32 soldiers, four key leaders were instrumented with an augmented cognition system. There were two principal phases of the 12-day training session: part-mission training and full-mission execution. In part-mission training, the tasks changed each day, starting from simple entry techniques (e.g., door and wall breaching, upper-level entry, use of suppression devices), progressing to clearing techniques (e.g., room, hall, and stairwell entry and clearing; reflexive fire techniques), then on to defensive techniques (e.g., hasty defense of an urban area, security, protection, fields of fire), and finally to battle drills. Soldiers mastered a technique before moving to the next, as each technique built upon what was learned previously.

Battle drills were a culmination of all the training that soldiers received and enabled them to establish their own standard operating procedures. Examples included conducting a platoon attack, entering and clearing a building, reacting to an ambush, and securing at a halt. These tasks were not performed until the individual teams and squads demonstrated proficiency in all basic skills.

The second phase of the AWE was a full-mission evaluation involving a 24-hour training exercise. For this exercise, soldiers used techniques and skills learned during the part-mission training. The 24-hour period was divided into three 8-h phases:

1. Conduct dismounted movement along the lines of communication to an objective to ensure routes are free of mines and obstacles.

2. Conduct a cordon and search of the objective to kill, capture, or expel opposition forces operating in an urban area.

3. Prepare to defend the objective for an extended period and report any enemy activity in and around this key terrain.

This evaluation focused primarily on a platoon leader (PL), platoon sergeant (PSG), and two squad leaders (SL1 and SL2); however, the activities of their subordinates and responses from senior leaders had a direct impact on stress levels experienced by the leaders. The MOUT training facility, known as Mulberry Point, contained preassault staging and assault areas. It was a compound with several single- and multistory buildings with windows, doors, and hallways. This site served as a close combat training area.

The test site was equipped for data collection, including cameras in and around buildings. These data were used by experts in determining the workload ground truth. There were several stressors that the platoon-level training exercise introduced in the MOUT facility. These stressors were used to ensure that participants were placed in the cognitive states of interest (low and high workload). Each is summarized in Table 4.1.

In the remainder of this section, we address how the four principal challenges discussed in the "General Approach and Associated Toolkit" section (i.e., signal processing, classification, computational and experimental infrastructure, and design of experiments) were addressed in the AWE.

Table 4.1. Stressors Encountered by Soldiers in a MOUT Environment

Category	Example Stressors
Distributed operations	Distributed squads, loss of sight, reliance on faulty radios
Fatigue	Extended operational period (e.g., 24 h of operation): lengthy march followed by assault and then lengthy occupation of site in defensive posture
Realistic threats	Use of human OPFOR to prevent assault on urban facility, and to "hit" the friendly forces at different times; use of simunitions (soap bullets)
Evaluation stress	Evaluation of performance by commanders, Army trainers
Surprise / confusion	Unexpected elements imposed that affect plan, conditions, and mission; loss of communications and assets
Severe weather	Periods of high heat and humidity; intense rainfall
Information Gaps	Information flow variations from subordinates and commander

Signal Processing

Signal processing on the EEG signal was performed with a system that supported an independent signal-processing stream. Six channels were sampled at 256 samples/s with a bandpass from 0.5 Hz and 65 Hz (at 3 dB attenuation) obtained digitally with Sigma-Delta A/D converters. Quantification of the EEG in real time was achieved using signal analysis techniques that identified and decontaminated eyeblinks and identified and rejected data points contaminated with electromyographic (EMG), amplifier saturation, and/or excursions attributable to movement artifacts (see Berka et al., 2004, for a detailed description of the artifact decontamination procedures).

Decontaminated EEG was then segmented into overlapping 256 data-point windows called *overlays*. An epoch (the temporal window of analysis) consisted of three consecutive overlays. Fast-Fourier Transform (FFT) was applied to each overlay of the decontaminated EEG signal, multiplied by the Kaiser window ($\alpha = 6.0$), to compute the power spectral density (PSD). The PSDs were adjusted to take into account zero values inserted for artifact contaminated data points. The PSDs between 70 and 128 Hz were used to detect EMG artifact. Overlays with excessive EMG artifacts or with fewer than 128 data points were rejected. The remaining overlays were then averaged to derive a PSD for each epoch with a 50% overlapping window. Epochs with two or more overlays with EMG or missing data were classified as invalid. For each channel, PSD values were derived for each 1-Hz bin from 3 to 40 Hz and the total PSD from 3 to 40 Hz. Relative power variables were also computed for each channel and bin using the formula (total band power/total bin power).

Signal processing on the ECG signal focused on the importance of detecting the ECG peaks with a high degree of accuracy, accounting for missed peaks, and ignoring spurious peaks. Without appropriate correction, missing a single valid beat or adding a single spurious beat can lead to questionable estimates of spectral HRV measures. Heartbeat peaks within areas of the signal are identified with high-frequency content. False peaks are eliminated based on statistical comparisons to expected QRS morphology (QRS waves are related to the contraction of the left and right ventricles).

Scerbo et al. (2001) found that that the efficacy of HRV measures has mostly been limited to lab contexts, whereas IBI is a better measure in operational contexts. The IBI was computed as the time between successive peaks. The IBI was resampled at a rate of 256 Hz using a cubic spline interpolation. The resampled IBI was used as the workload indicator. Spectral methods, though not as effective, were also employed. The IBI signal was spectrally decomposed and the power in the band between .05 and .15 was used as an additional workload indicator.

Classification Approach
Cognitive state estimation was based on a support vector machine approach. The support vector machine used in this effort employed a radial basis function kernel with a kernel parameter of 1 and a slack parameter of .05.

The AWE evaluated the effectiveness of the classification algorithms to detect the user's cognitive state by correlating classification output to performance in various task load conditions. Experimentally, the principal hypothesis that was tested in the AWE was as follows:

> *The cognitive state classification algorithms would be able to differentiate periods of high and low cognitive workload using a combination of physiological (ECG) and neurophysiological (EEG) sensors.*

Computational and Experimental Infrastructure
Cognitive state classification was based on two sensor sources: EEG and ECG. An elaborate experimental infrastructure was developed to meet some of the challenges of collecting data in a harsh, mobile environment.

Sensors

EEG data were collected from the Advanced Brain Monitoring (ABM) EEG sensor headset (see Figures 4.9a and 4.9b). The sensor headset acquired six channels of EEG using a bipolar montage. Differential EEG were sampled from bipolar channels CzPOz, FzPOz, F3Cz, F3F4, FzC3, and C3C4 obtained digitally. Data were transmitted across a Bluetooth RF link to the collection laptop via an RS232 interface.

The Sensor Headset was developed by ABM as a portable system to record EEG signals. The headset fit snugly on the head and housed EEG sensors. It was important to minimize the movement of sensors to reduce signal artifacts. Snug-fitting caps reduced signal noise caused by sheering of the sensors against the head and scalp. Physiological recordings were made with an experimental eight-channel digital physiological recorder with low-powered EEG and EOG amplifiers designed specifically for ambulatory recordings. Amplification at the electrode site was important to boost the signal-to-noise ratio.

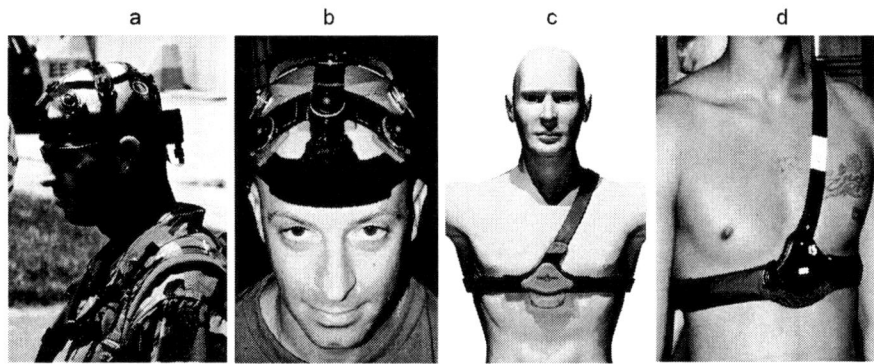

Figure 4.9. (a & b) ABM's Wireless EEG Sensor Headset, (c & d) Hidalgo VSDS for ECG.

The Hidalgo Vital Signs Detection System (VSDS) measured heart rate, respiration rate, and body motion and position (see Figures 4.9c and 4.9d). Both waveform and summary data were transmitted across a Bluetooth communications link. The AWE used the ECG waveform (two views, sampled at 256 Hz) and three-axis accelerometry waveform (sampled at 25.6 Hz) signals.

Mobile Processing and Data Collection Platform

Each of the four primary soldier participants (PL, PSG, SL1, and SL2) was followed by a member of the experimental personnel in the role of shadower. Each shadower remained within a 30-m range of his/her participant to ensure Bluetooth® connectivity. Each shadower carried a specially designed backpack (based on the MOdular Lightweight Load-carrying Equipment [MOLLE] system), which contained a Panasonic Toughbook® CF-51 computer equipped to receive Bluetooth communication from the participant's EEG and ECG sensors and audio from a wireless microphone. In addition to logging data, shadowers processed raw sensor data on their computers using Honeywell's Cognitive State Classification algorithms to produce a real-time assessment of their participants' cognitive state. That cognitive state assessment was then transmitted to the base station via a wireless data network (see next section).

The use of shadowers allowed sufficient computing power to process the raw data from the sensors without interfering with the mobility of the participant, as well as protecting the equipment from being broken during the more physical aspects of the missions. Ideally, small mobile, ruggedized systems configured with sufficient computing power worn on the body by the soldier would be used once parameters are downselected through the research process. Additionally, the shadower wore a webcam and logged video to the computer for later review by experts to enable determination of the workload ground truth. The participant wore a wireless microphone, and the resultant audio stream was multiplexed into the webcam video.

The base station received data from the four shadowers' computers via the wireless data network. The base station was the test team's command and control center for the devices and facilitated the diagnosis of problems, resetting of systems, and monitoring of system status. The base station performed several functions, including remote control of the four shadower computers (ability to stop/start processes), monitoring of processes on four shadower computers, running of the master radio, remote troubleshooting of the shadower computers, data collection, and the shutting down of processes at the end of a trial.

Wireless Network

The AWE employed a 900-MHz radio modem system to create a wireless data network connecting the four shadowers' computers to the base station. The ABM EEG and the Hidalgo VSDS communicated to the shadower computer via Bluetooth. Figure 4.10 illustrates the final data collection system and experimental infrastructure configuration.

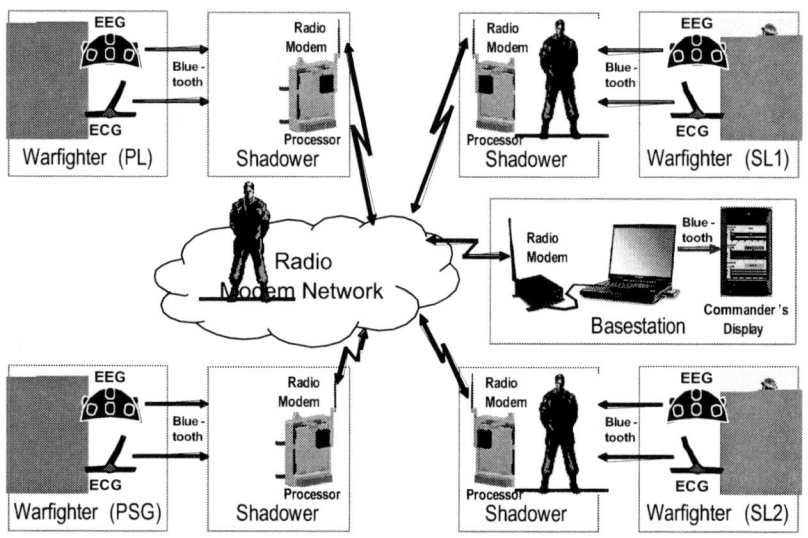

Creating a stable and robust experimental infrastructure is key to successfully field-testing classification approaches.

Figure 4.10. Final data collection system and experimental infrastructure.

Several practical challenges were encountered during the AWE. First and foremost, the pace of the training was subject to each soldier's progress through a predefined set of tasks, drills, and procedures. Soldiers were trained to performance on battle drills. The use of "simunitions" (soap bullets) implied that all hardware—including potentially sensitive equipment such as EEG sensors—had to be hardened to withstand a direct hit of a simunition round. During the experiment, the ABM EEG system sustained a direct hit but was not damaged.

The weather was another challenge: During two of the days of training, 12 inches of rain fell. The AWE required that wireless connectivity be maintained over two networks, the Bluetooth connections (between sensors and shadowers) and the 900-MHz radio modem network. Power consumption of the mobile equipment was always a challenge, and battery management was key to ensuring that all devices continued to function despite inevitable delays and schedule changes.

Finally, EEG sensor integration with soldier's standard equipment was a challenge that required special modifications to the padding and padding configuration under the soldiers' helmets.

Design of Experiments
The independent variable in the AWE was workload (for all phases). The experimental scenarios were manipulated to ensure definable periods of high and low cognitive workload. Low-workload periods were characterized by engagement in a single task that was well within the current cognitive capability of the soldier, usually under little or no stress or time pressure. Periods of low workload included completing initial paperwork, reporting activities, preplanning, establishing a hasty defense position (e.g., foxhole), consolidation/transition, after-action reviews, and periods of low activity during missions.

High-workload periods were characterized by multiple-task performance, often under time pressure and fatigue. Examples of high workload included replanning caused by changing circumstance (e.g., enemy location, available squads, loss of communication), directing squad movements during preassault, assault, managing multiple communications (i.e., responding to commanders, squad leaders, or other platoon leaders), or calling for fire. Stressors that contributed to high workload included frustration, loss of communication, lack of asset availability, and loss of situation awareness of squad locations and activities.

Ground Truth
During the AWE, multiple streams of data were collected with the objective of providing experts with enough insight to make a determination of ground truth levels of workload for each participant in each scenario. Data included video from a roaming camcorder (focused on the platoon-level action), video from the webcam of the shadower (focused on the participant), notes from an observer at a central (video) monitoring site, annotations radioed in from the shadower and entered at the base station into the time-stamped data stream via an annotator's interface, postscenario cognitive walkthroughs with the participants as they reviewed (with an experimenter) the video of

the day's events, postscenario NASA TLX surveys (Hart & Staveland, 1988), and questionnaires.

Not all data were collected for every part-mission and full-mission scenario, but some combination of data streams was available for expert review. The notes, annotations, and cognitive walkthrough feedback data streams were merged (by time stamp) into a spreadsheet. An expert then reviewed the video streams, taking into account the various data sources, to make a moment-to-moment assessment of the cognitive workload being experienced by the participant at any given time stamp. The result was a time-stamped series of blocks of low, medium, or high cognitive workload. Physical load was also assessed by the experts.

Two experts independently performed the ground truth analysis described earlier. Their respective results were then compared to gain a measure of interrater reliability on the cognitive workload assessments of ground truth. For the data sets analyzed, agreement between the raters was high: Agreement in the rating of physical load was 94.9%, and agreement in the rating of cognitive workload was 87.9%.

A final, canonical, assessment of ground truth was created by reconciling the two individual expert's assessments. Periods of disagreement were flagged. The two experts then jointly reviewed the video and other data streams to make a final assessment of the workload in the disputed block. In cases in which no consensus was reached, a third rater was available to resolve the disagreement; however, this option was never needed. Reconciled ground truth tables were used to calculate the accuracy metric of the classification algorithms.

Test Your Knowledge
The AWE was an evaluation in a mobile, operational context using participants (i.e., soldiers) performing their natural, domain-specific tasks. The elaborate setup described was designed to meet four challenges, which were made particularly difficult in a mobile setting (i.e., signal processing, classification, computational and experimental infrastructure, and design of experiments). The objective of the evaluation was to assess workload classification techniques during multiple operational tasks requiring different levels of cognitive and physical engagement. How were these challenges overcome? Which strategies were the most effective in overcoming these challenges?

Lessons Learned

The AWE data analysis forms a good example of the specific lessons learned when evaluating the general approach outlined earlier in the chapter. For a more extensive description of classification results, see Mathan, Whitlow, Dorneich, and Ververs (2007). The lessons derived from the results reviewed here fall into three principal categories: (a) choosing parameters of the classification approaches to improve performance in mobile environments, (b) creating a set of features that adequately captures the cognitive state of interest, and (c) improving classification while minimizing computational demands.

Tuning Classification Parameters

In noisy operational environments, EEG and other electrophysiological sensors could be compromised by noise over short temporal windows. One strategy for dealing with momentary fluctuations in classification accuracy is to median-filter the output of the classifier over different time windows. One consequence of such temporal smoothing of classifier output is it may introduce a lag in the decision process. The analysis must consider the trade-off in accuracy as the temporal window of output smoothing is varied.

The classification approach was assessed with two individuals, the platoon leader and the platoon sergeant, both employing the widely used tenfold cross-validation approach and the more conservative twofold cross-validation procedure.

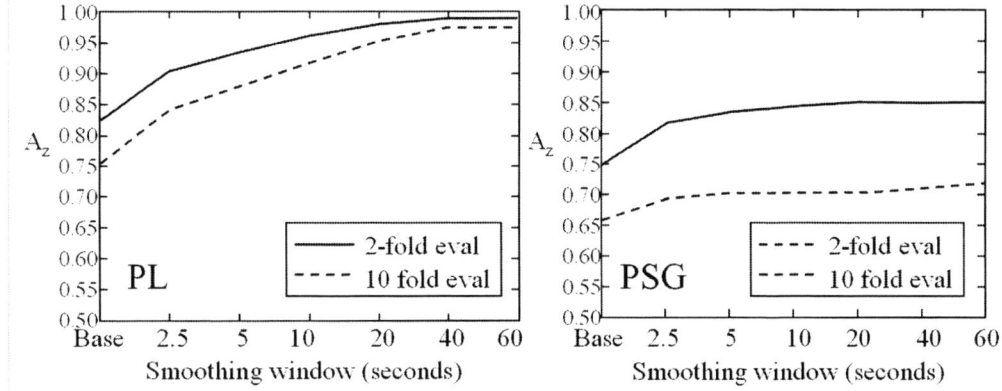

Figure 4.11. EEG-based classification accuracy for the PL (left) and PSG (right) as a function of validation technique and temporal smoothing window.

As Figure 4.11 illustrates, base EEG classification accuracy for the PL ranged from 0.76 (using twofold cross validation) to 0.83 (using tenfold cross validation). Base results for the PSG ranged from 0.66 (using twofold cross validation) to 0.75 (using tenfold cross validation), as seen in Figure 4.11 (right). Accuracy for both soldiers rose monotonically up to a 1-min-long temporal smoothing window. However, the rate at which temporal smoothing benefited accuracy diminished beyond approximately 2 to 3 s of smoothing. This analysis confirms the lesson learned that the single-point analysis of classifier accuracy does not convey the bias, variance, and generalizability of a classifier approach.

The discrepancy between the more conservative twofold validation and more optimistic tenfold cross validation was more pronounced for the PSG than it was for the PL. This could indicate some change in the features that serve to discriminate between high and low workload over time; these changes could stem from changes in task, strategy, artifacts, or a variety of physiological factors.

Discriminating Features

The analysis included a qualitative examination of the spectral features that serve to discriminate between high and low workload. Figure 4.12 (p. 101) depicts the power spectral densities (PSD) for high and low workload across six channels of EEG for the PL (left) and the PSG (right).

Each graph in Figure 4.12 represents a channel. The x axis in each graph represents frequency, and the y axis represents amplitude. One line in each graph represents average spectral power in the high-workload condition; another line represents average spectral power in the low-workload condition. Finally, the center line in each graph corresponds to the mean spectral power across both high- and low-workload conditions. Qualitatively, the key distinction is the separation (if any) between the high and low spectral power lines.

An analysis of the graphs for both participants suggests that power in the beta (12 to 30 Hz) and gamma (30 to 40 Hz) bands is the most discriminative feature for both participants. However, this pattern is most pronounced for PL and may account for the superior classification results observed relative to PSG. This discrepancy across individuals also points to the lesson of the importance of an individualized approach to classification, rather than an approach that relies on group norms.

Review: Techniques for Improving Classification in Mobile Environments
Mobile environments require that the number of sensors and the computational demands of the classification algorithms are minimized while accurate classification is maintained in the presence of motion-induced noise artifacts. Robust classification is addressed by sensor fusion techniques. Minimizing computational demand requires that the minimum number of sensors be identified.

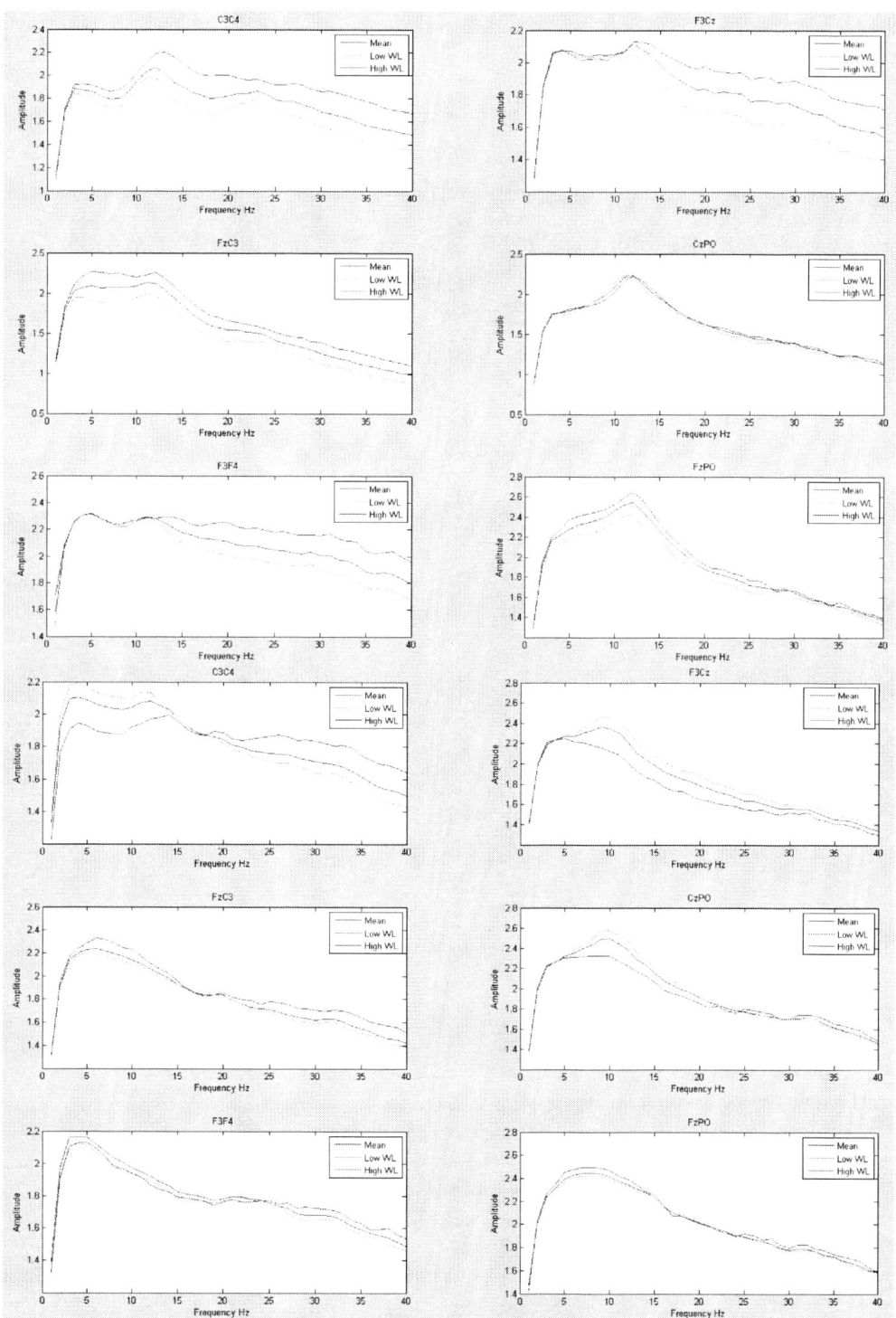

Figure 4.12. Power spectral densities in each band for the PL (left) and PSG (right).

Sensor Fusion

One lesson learned in the work reported here was the utility of fusing data from multiple sensor sources to improve classification in noisy field environments. Such an approach exploits the joint strengths of different data sources while minimizing their individual weaknesses. Fusing multiple sensor sources into a common feature vector allows a classifier to find an optimal weighting for each feature based on training data.

We assessed the effect of including IBI estimates as a feature for classification. The fusion of cardiac data provided a substantial boost to overall classification performance—these improvements were most pronounced for PSG, as seen in Figure 4.13. Base classification for PL went up from 0.76–0.83 to 0.87–0.95, and base classification for PSG went up from 0.66–0.75 to 0.83–0.86.

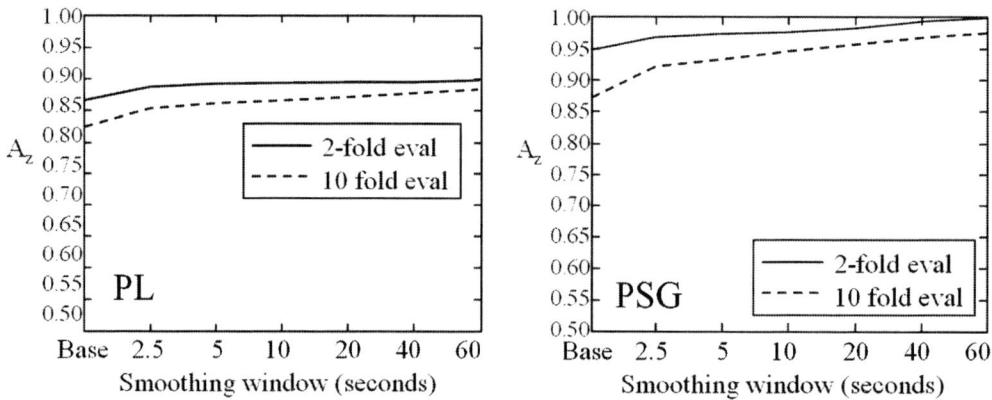

Figure 4.13. Classification accuracy for the fused sensor data for the PL (left) and PSG (right).

Sensor Density

The EEG system used in the field evaluation consisted of a six-channel system. In a mobile setting, such as dismounted soldier operations, it is important to reduce the number of sensors to the minimum required to capture the underlying cognitive state of interest. An analysis was conducted to identify a subset of the six EEG channels that could match or exceed the performance of all channels together. With each iteration of a backward elimination-ranking algorithm, each channel of the current set was sequentially eliminated from consideration. The channel whose exclusion led to the best performance results was eliminated from further consideration.

The ranking assigned to each channel corresponded to the order in which it was eliminated. The first channel to be eliminated was ranked as being last in importance, whereas the last channel to remain was regarded as being of the highest importance. Performance of each feature subset was assessed using tenfold cross validation. The performance metric used was the area under the Receiver Operating Curve (A_z; see Chapter 6). The channel-ranking procedure produced the channel ranks shown in Figure 4.14 (PL left and PSG right), which plots classification accuracy as a function of the top n channels.

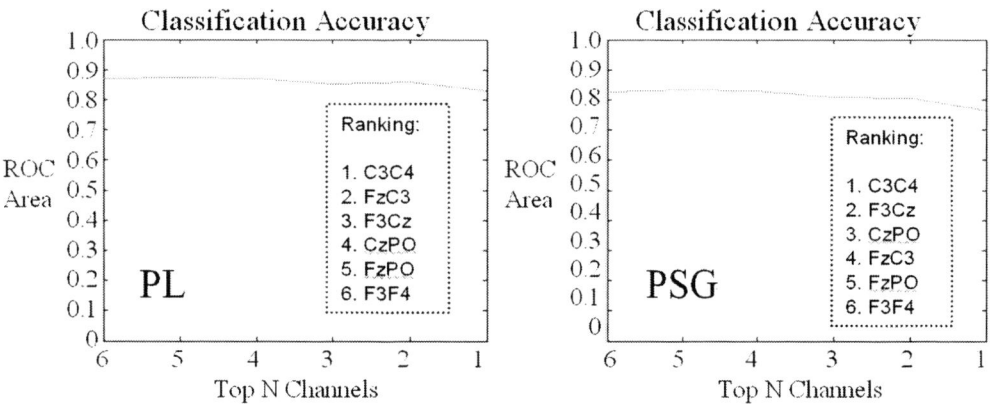

Figure 4.14. Classification accuracy as a function of the top *n* channels.

The channel-ranking procedure yielded a consistent set of features for both participants. Classification performance suffered little with the exclusion of all but the two most salient channels. The top channels were identical for both participants (C3C4). These channels, which were located at the apex of the skull, are likely to have been least affected by helmet-related artifacts because of good clearance between the sensors and the helmet at this location. The lesson learned was the importance of stable electrode sensor sites, which is possible even in noisy conditions.

Although these results require further validation, the lesson learned was that accurate workload classification may be feasible with as few as one or two sensors. This has compelling implications for the design of practical EEG systems that could be integrated easily within helmets and find broad user acceptance.

Best Practices

The best practices derived from this work stem from the approaches taken to overcome four principal challenges: (a) collecting and processing EEG signals in a mobile context; (b) developing classification algorithms that address individualization, bias, and generalizability; (c) designing experiments to assess classification in a domain-relevant operational setting with soldiers; and (d) building a computational and experimental infrastructure to support assessment. For each challenge, a table of best practices and practical recommendations is given.

Meeting the Mobility Challenge

Table 4.2 outlines some of the best practices, guidelines, and recommendations in several areas when meeting the challenge of collecting and processing EEG signals in a mobile context.

Table 4.2. Best Practices, Guidelines, and Recommendations Addressing Mobility Challenges

Area	Best Practices	Guidelines / Recommendations
Equipment	Develop the capability to collect data in the actual environment.	Select an EEG system that preamplifies the signal at the electrode site to enable low noise measurements. Cabling between the sensor and data collection equipment must be secured to avoid cable sway-induced noise artifacts.
Signal processing	Understand the noise artifacts, and understand the signal of interest.	Understand the noise artifacts by running a series of experiments in which you control (in turn) every type of mobility condition (e.g., stationary, standing, walking, running, head motion). Understand the signal of interest by first collecting data in pristine environments in order to be able to identify the signal later amid all the noise artifacts introduced by the operational environment.
Signal processing	Develop stability controls to improve adaptive filtering (see Chapter 7). When faced with the extreme artifacts in a mobile environment, most adaptive filters would become unstable and unusable.	If the benefits of adaptive filter algorithms are to be obtained in a mobile environment, the algorithms must be stabilized during high-amplitude spikes. See, for instance, Mathan, Dorneich, and Whitlow (2007).
Signal processing	Findings from prior research were quickly identified as inadequate for identifying relevant EEG sites for use in applied operational domains.	Run pilot studies in the operational environment that use the same or a similar task to identify the cognitive states of interest. Start with many EEG sites and run sensor density analysis to rank the channel contributions.
Signal processing	Collect sufficient data to determine how much training data are required to provide good classification performance.	Use pilot studies to determine how much training data are needed. The amount of data needed varies depending on the nature of the task environment, signal-to-noise ratio, and classification techniques used.

Meeting the Classification Challenge

Table 4.3 outlines some of the best practices, guidelines, and recommendations in several areas when meeting the challenge of developing classification algorithms that address individualization, bias, and generalizability.

Table 4.3. Best Practices, Guidelines, and Recommendations Addressing Classification Challenges

Area	Best Practices	Guidelines / Recommendations
Classification	Fit the approach to the constraints of the environment.	Determine the spatial density of EEG sensor arrays based on an understanding of the nature of the tasks, pace of task switching, and specific types of cognitive processing involved. Consider the constraints imposed by sensor density, computational efficiency, precise task adaptation needs, and the desire for a high degree of classification accuracy during ongoing research studies.
Classification	Explore multiple temporal windows.	Temporal smoothing should be employed to stabilize classifier output. Classifier update rates need only satisfy the requirements of the pace of adaptation switching.
Sensor density	Determine the ideal number of sensors by considering processing demands, the operational environment, and the generalizability of classification across multiple situations.	Once the classifier approach goes beyond the most informative features (site by frequency band) the classifier begins to overfit to noise and degrade classification performance, much as does adding unnecessary parameters to a regression model.
Sensor location	Choose the sensor location based on the cognitive state of interest.	Generic workload assessment can employ a low-density array over the frontal central and parietal lobes. Data from sensors located in the occipital area tend to be noisier and do not provide discretionary information, given that most tasks involve visual processing.
Sensor type	Choose the sensor type based on the cognitive state of interest.	If you need immediate feedback on specific events, use a time-locked EEG measure such as evoked response potentials (ERPs). If you need general task loading over extended periods, use oscillatory EEG measures such as PSDs.
Fusion	Utilize complementary measures of cognitive state where appropriate.	ECG provides information on tonic states (i.e., slowly changing), whereas EEG provides high temporal fidelity (i.e., moment-to-moment). Together, the two have been shown to improve classification accuracy.

Meeting the Infrastructure Challenge

Table 4.4 outlines some of the best practices, guidelines, and recommendations in several areas when meeting the challenge of building a computational and experimental infrastructure to support assessment.

Table 4.4. Best Practices, Guidelines, and Recommendations Addressing Infrastructure Challenges

Area	Best Practices	Guidelines / Recommendations
System integration	Ruggedize the equipment for testing in a field environment.	Use ruggedized laptops that come with shock-mounted hard drives to protect your data, and include effective thermal management.
System integration	Ruggedize all connections.	Secure all cable connections. For instance, typical USB connectors were not designed to maintain a connection under mobile conditions.

Meeting the Assessment Challenge

Table 4.5 outlines some of the best practices, guidelines, and recommendations in several areas when meeting the challenge of designing experiments to assess classification in a domain-relevant operational setting with soldiers.

Table 4.5. Best Practices, Guidelines and Recommendations Addressing Assessment Challenges

Area	Best Practices	Guidelines / Recommendations
Task definition	Consult domain experts.	Even if it is not possible to perform the actual task in early experiments, developing representative tasks lends confidence that the findings will be transferable to the actual domain. Not only does designing tasks with input from domain experts save considerable time, but results will be better received because of their ecological validity.
Task definition	Baseline tasks early and often to ensure that representative participants perform and perceive different task loads as low and high.	Task load does not produce the same workload in different participants, or even in the same participant over time. Maximize discrepant task loading for good binary classification.
Experimental control	Evaluation of techniques in operational environments often results in a loss of experimental control as evaluations move from the lab to the field.	Free-play evaluations with high ecological validity are very effective in loading leaders with varying levels of workload. Even in a free-play evaluation, an operator or controller can manipulate the workload of participants by changing the scenario, introducing unexpected events, and controlling the pace of operations.
Experimental design	Whenever possible, simplify the experimental design to reduce the complexity of conducting field studies.	Inevitably, the system integration phase will take three times longer than expected. Limit the number of research questions of interest and avoid rolling up everything into a single study.
Risk management	Consider an experimental design that includes segments with severable benefits.	Ensure that data analysis is possible on the cumulative data collected (i.e., each day's data), so if data collection becomes impossible, the experiment can still produce results on whatever data were collected thus far.
Ground truth	Explicitly design the data collection plan for ground truth.	If you are videoing the participant, make sure that a microphone channel is included, as it is often difficult to decipher the state of the participant from video alone. Reviewing video (and audio) with the participant immediately after the experimental trial provides the single best data source for insight into the participant's cognitive loading at any given moment. Make sure that all data streams share a common time stamp.

Design Guidelines

Design guidelines take the form of overarching considerations that become important when cognitive state classification work is matured outside the laboratory and is used in real-world, mobile, operational contexts. Table 4.6 captures a principal guideline in each of the four challenge areas.

Table 4.6. Design Guidelines

Area	Design Guideline
Mobility	Thorough advanced signal-processing algorithms are essential to ensure a clean signal for cognitive state classifiers. It is particularly important to remove or identify noise artifacts in harsh operational environments.
Classification	There is not a one-size-fits-all approach to cognitive state classification. Individualized measurements are necessary for each individual. In addition, because of changes in physiological data over time, regularly scheduled baselines will need to be captured to maintain a high level of classification accuracy.
Infrastructure	There is a need to further ruggedize physiological and neurophysiological sensors and sensor systems to enable the deployment of this capability.
Assessment	The assessment of classification effectiveness will always require an evaluation to capture the context of the mission and task and incorporate user feedback as a basis of ground truth information. In addition to a complete understanding of the target environment, thorough interviews with participants and multiple raters of ground truth classification will help to minimize errors in cognitive state classification caused by poor insight into the cognitive loading requirements of the task environment.

Parting Message

The evaluation of cognitive state classification techniques outside the laboratory is wrought with challenges apart from the classification techniques themselves. Successful assessment will depend on the ability to collect valid signals robustly in a noisy environment, and require a computing and experimental infrastructure that can enable realistic experiments in the domain of use. The design of these experiments is itself a major challenge, as one no longer has the benefit of well-defined, well-understood laboratory tasks that engage the cognitive state of interest. Failure to address any of these challenges severely compromises the ability to draw meaningful conclusions about the use of cognitive state classification algorithms in the target operational domain.

References

Aasman, J., Mulder, G., & Mulder, L. J. M. (1987). Operator effort and the measurement of heart rate variability. *Human Factors, 29*, 161–170.

Backs, R. W., & Seljos, K. A. (1994). Metabolic and cardiorespiratory measures of mental effort: The effects of level of difficulty in a working-memory task. *International Journal of Psychophysiology, 16*(1), 57–68.

Balles, J. (1988, December). Vincennes. Findings could have helped avert tragedy: Scientists tell Hill panel. *APA Monitor*, pp. 10–11.

Banks, S., & Lizza, C. (1991, June). Pilot's associate: A cooperative, knowledge-based system application. *IEEE Expert*, pp. 18–29.

Beatty, J. (1982). Task-evoked pupillary responses, processing load, and the structure of processing resources. *Psychological Bulletin, 91*(2), 276–292.

Beh, H. C. (1990). Achievement motivation, performance and cardiovascular activity. *International Journal of Psychophysiology, 10,* 39–45.

Berka, C., Levendowski, D., Cvetinovic, M. M., Petrovic, M. M., Davis, G., Lumicao, M. N., Zivkovic, V. T., Popovic, M. V., & Olmstead, R. (2004). Real-time analysis of EEG indices of alertness, cognition, and memory acquired with a wireless EEG headset. *International Journal of Human-Computer Interaction, 17*(2), 151–170.

Boiten, F. A. (1998). The effects of emotional behavior on components of the respiratory cycle. *Biological Psychology, 49*(1–2), 29–51.

Buller, M. J., Hoyt, R. W., Ames, J., Latzka, W., & Freund, B. (2005). Enhancing warfighter readiness through physiologic situational awareness—The warfighter physiological status monitoring—Initial capability. *Proceedings of the 1st International Conference on Augmented Cognition* (pp. 335–344). Mahwah, NJ: Erlbaum.

Cochran, E., Miller, C., & Bullemer, P. (1996). Abnormal situation management in petrochemical plants: Can a pilot's associate crack crude? In *Proceedings of the NAECON, 1996* (pp. 806–813). Houston, TX.

Dempster, A. P., Laird, N. M., & Rubin, D. B. (1977). Maximum likelihood from incomplete data via the EM algorithm. *Journal of the Royal Statistical Society, 39,* 1–38.

Dorneich, M. C., Mathan, S., Ververs, P. M., & Whitlow, S. D. (2007). An evaluation of real-time cognitive state classification in a harsh operational environment. In *Proceedings of the Human Factors and Ergonomics Society 51st Annual Meeting* (pp. 146–150). Santa Monica, CA: Human Factors and Ergonomics Society.

Dorneich, M. C., Ververs, P. M., Mathan, S., & Whitlow, S. D. (2005, October). A joint human-automation cognitive system to support rapid decision-making in hostile environments. Presented at the International Conference on Systems, Man and Cybernetics, Waikoloa, HI.

Duda, R. O., Hart, R. E., & Stork, D. G. (2001). *Pattern classification* (2nd ed.). New York: Wiley.

Erdogmus, D., Adami, A., Pavel, M., Lan, T., Mathan, S., Whitlow, S., & Dorneich, M. (2005, March). *Cognitive state estimation based on EEG for augmented cognition.* Presented at the 2nd IEEE EMBS International Conference on Neural Engineering, Arlington VA.

Future Force Warrior. (2004). Retrieved August 11, 2008, from http://nsrdec.natick.army.mil/about/techprog/index.htm

Garavan, H., Ross, T. J., Li, S.-J., & Stein, E. A. (2000). A parametric manipulation of central executive functioning using fMRI. *Cerebral Cortex, 10,* 585–592.

Geddes, N. D. (1985, September). Intent inferencing using scripts and plans. In *Proceedings of the First Annual Aerospace Application of Artificial Intelligence Conference* (pp. 160–172), Dayton, OH.

Gevins, A., & Smith, M. E. (2000). Neurophysiological measures of working memory and individual differences in cognitive ability and cognitive style. *Cerebral Cortex, 10,* 829–839.

Gevins, A., & Smith, M. (2003). Neurophysiological measure of cognitive workload during human-computer interaction. *Theoretical Issues in Ergonomics Science, 4*(1–2), 113–132.

Girolamo, H. J. (2005). Augmented cognition for warfighters: A beta test for future applications. *Proceedings of the 1st International Conference on Augmented Cognition.* Mahwah, NJ: Erlbaum.

Hart, S. G., & Staveland, L. E. (1988). Development of a multi-dimensional workload rating scale: Results of empirical and theoretical research. In P. Hancock & N. Meshkati (Eds.), *Human mental workload* (pp. 138–183). Amsterdam: Elsevier.

Heslegrave, R. J., & Furedy, J. J. (1979). Sensitivities of HR and T-wave amplitude for detecting cognitive and anticipatory stress. *Physiology & Behavior, 22*(1), 17–23.

Hockey, G. R. J. (1986). Changes in operator efficiency as a function of environmental stress, fatigue, and circadian rhythms. In K. R. Boff, L. Kaufman, & J. P. Thomas (Eds.), *Handbook of perception and human performance* (Vol. II, pp. 44/1–44). New York: Wiley.

Izzetoglu, K., S. Bunce, S., Onaral, B., Pourrezaei, K., & Chance, B. (2004). Functional optical brain imaging using near-infrared during cognitive tasks. *International Journal of Human-Computer Interaction, 17*(2), 211–227.

Kalsbeek, J. W. H., & Ettema, J. H. (1963). Scored irregularity of the heart pattern and measurement of perceptual or mental load. *Ergonomics, 6*, 306–307.

Kittler, M., Hatef, R., Duin, R. P. W., & Matas, J. (1998). On combining classifiers. *IEEE Transactions on Pattern Analysis and Machine Intelligence, 20*(3), 226–239.

Kramer, A. (1991). Physiological metrics of mental workload: A review of recent progress. In D. Damos (Ed.), *Multiple task performance* (pp. 279–328). London: Taylor & Francis.

Krebs, H., Brashers-Krug, T., Rauch, S., Savage, C. R., Hogan, N., Rubin, R. H., Fischman, A. J., & Alpert, N. M. (1998). Robot-aided functional imaging: Application to a motor learning study. *Human Brain Mapping, 6*, 59–72.

Mahalanobis, P. C. (1936). On the generalized distance in statistics. *Proceedings of the National Institute of Science of India, 12*, 49–55.

Makeig, S., & Jung, T -P. (1995). Changes in alertness are a principal component of variance in the EEG spectrum. *NeuroReport, 7*(1), 213–216.

Mathan, S., & Dorneich, M. (2005, July). *Augmented tutoring: Improving military training through model tracing and real-time neurophysiological sensing.* Presented at the 11th International Conference on Human-Computer Interaction (Augmented Cognition International), Las Vegas, NV.

Mathan, S., Dorneich, M. C., & Whitlow, S. D. (2007). *Statistical control of adaptive ocular filter stability* (U.S. patent application #20070239814). Washington, DC: U.S. Patent and Trademark Office.

Mathan, S., Mazaeva, N., Whitlow, S., Adami, A., Erdogmus, D., Lan, T., & Pavel M. (2005, July). Sensor-based cognitive state assessment in a mobile environment. Presented at the 11th International Conference on Human-Computer Interaction (Augmented Cognition International), Las Vegas, NV.

Mathan, S., Whitlow, S., Dorneich, M., & Ververs, T. (2007). Neurophysiological estimation of interruptibility: Demonstrating feasibility in a field context. *Proceedings of the 4th International Conference of the Augmented Cognition Society*. Baltimore, MD.

Mikulka, P., Hadley, G., Freeman, F., & Scerbo, M. (1999). The effects of a biocybernetic system on vigilance decrement. In *Proceedings of the Human Factors and Ergonomics Society 43rd Annual Meeting* (p. 1410). Santa Monica, CA: Human Factors and Ergonomics Society.

Miller, C. A., & Dorneich, M. C. (2006). From associate systems to augmented cognition: 25 years of user adaptation in high criticality systems. In *Proceedings of the 2nd Augmented Cognition International*. San Francisco, CA.

Miller, C., Wu, P., Kirchbaum, K., & Kiff, L. (2004). Automated elder home care: Long term adaptive aiding and support we can live with. In *Proceedings of the AAAI Spring Symposium on Interaction Between Humans and Autonomous Systems over Extended Operation*, Stanford, CA.

Montain, S. J., Sawka, M. N., & Wenger, C. B. (2001). Hyponatremia associated with exercise: Risk and pathogenesis. *Exercise Sports Science Review, 29*, 113–117.

Neumann, D. L. (2002). Effect of varying levels of mental workload on startle eyeblink modulation. *Ergonomics, 45*, 583–602.

Parasuraman, R. (2003). Neuroergonomics: Research and practice. *Theoretical Issues in Ergonomics Science, 4*(1–2), 5–20.

Parasuraman, R., Mouloua, M., & Hilburn, B. (1999). Adaptive aiding and adaptive task allocation enhance human-machine interaction. In M. W. Scerbo & M. Mouloua (Eds.), *Auto-

mation technology and human performance: Current research and trends (pp. 129–133). Mahwah, NJ: Erlbaum.

Partala, T., & Surakka, V. (2003). Pupil size variation as an indication of affective processing. *International Journal of Human-Computer Studies, 59*(1–2), 185–198.

Parzen, E. (1967). On estimation of a probability density function and mode. *Time Series Analysis Papers*. San Diego, CA: Holden-Day.

Pope, A. T., Bogart, E. H., & Bartolome, D. (1995). Biocybernetic system evaluates indices of operator engagement. *Biological Psychology, 40*, 187–196.

Porges, S. W., & Byrne, E. A. (1992). Research methods for measurement of heart-rate and respiration. *Biological Psychology, 34*(2–3), 93–130.

Porges, S. W., & Raskin, D. C. (1969). Respiratory and heart rate components of attention. *Journal of Experimental Psychology, 81*, 497–503.

Prinzel, L. J., Freeman, F. G., Scerbo, M. W., Mikulka, P. J., & Pope, A. T. (2000). A closed-loop system for examining psychophysiological measures for adaptive automation. *International Journal of Aviation Psychology, 10*, 393–410.

Robertson, G. (2000). Flight demonstration of an associate system—A rotorcraft pilot's associate example. In *Proceedings of the AHS International, Annual Forum* (Vol, 56, pp. 431–445), Virginia Beach, VA.

Scerbo, M. W., Freeman, F. G., Mikulka, P. J., Parasuraman, R., DiNocero, F., & Prinzell, L. J. III. (2001). *The efficacy of physophusiological measures for implementing adaptive technology* (NASA Tech. Report NASA/TP-2001-211018). Hampton, VA: NASA Langley Research Center.

Schnell, T., Macuda, T., Poolman, P., Craig, G., Erdos, R., Carignan, S., Allison, R., Lenert, A., Jennings, S., Swail, C., Ellis, K., & Gubbels, A. W. (2006). Toward the "cognitive cockpit": Flight test platforms and methods for monitoring pilot mental state. In D. Schmorrow, K. Stanney, & L. Reeves (Eds.), *Foundations of augmented cognition* (2nd ed.). Arlington, VA: Strategic Analysis, Inc.

Smith, P. J., McCoy, E. C., & Layton, C. (1997, May). Brittleness in the design of cooperative problem-solving systems: The effects of user performance. *IEEE Transaction on Systems, Man, and Cybernetics—Part A, 27*(3), 360–371.

Snow, M. P., Barker, R. A., O'Neill, K. R., Offer, B. W., & Edwards, R. E. (2006). Augmented cognition in a prototype uninhabited combat air vehicle operator console. In D. Schmorrow, K. Stanney, & L. Reeves (Eds.), *Foundations of augmented cognition* (2nd ed., pp. 279–288). Arlington, VA: Strategic Analysis, Inc.

Steinman, A. M. (1987). Adverse effects of heat and cold on military operations: History and current solutions. *Military Medicine, 152*, 389–392.

Stern, J. A., Boyer, D., & Schroeder, D. (1994). Blink rate: A possible measure of fatigue. *Human Factors, 36*, 285–297.

Takahashi, N. (2006). Efficient learning algorithms for support vector machines. Retrieved from http://www-kairo.csce.kyushu-u.ac.jp/~norikazu/research.en.html

Thorpe S., Fize D., & Marlot C. (1996). Speed of processing in the human visual system. *Nature, 381*, 520–522.

Tremoulet, P., Barton, J., Craven, R., Gifford, A., Morizio, N., Belov, N., Stibler, K., Regli, S. H., & Thomas, M. (2006). Augmented cognition for tactical Tomahawk weapons control system operators. In D. Schmorrow, K. Stanney, & L. Reeves (Eds.), *Foundations of augmented cognition* (2nd ed., pp. 313–318). Arlington, VA: Strategic Analysis, Inc.

U.S. Navy. (1988). *Investigation report: Formal investigation into the circumstances surrounding the downing of Iran Air Flight 655 on 3 July 1988* (Department of Defense Investigation Report). Washington, DC: Author.

Vapnik, V. (1999). *The nature of statistical learning theory*. New York: Springer-Verlag.

Veltman, J. A., & Gaillard, A. W. K. (1998). Physiological workload reactions to increasing levels of task difficulty. *Ergonomics, 41*, 656–669.

Verwey, W. B., & Veltman, H. A. (1996). Detecting short periods of elevated workload: A comparison of nine workload assessment techniques. *Journal of Experimental Psychology-Applied, 2*(3), 270–285.

Wickens, C. D., & Hollands, J. (2000). Engineering psychology and human performance (3rd ed.). Upper Saddle River, NJ: Prentice Hall.

Wientjes, C. J. E. (1992). Respiration in psychophysiology: Methods and applications. *Biological Psychology, 34*(2–3), 179–203.

Wildervanck, C., Mulder, G., & Michon, J. A. (1978). Mapping mental load in car driving. *Ergonomics, 21*, 225–229.

Wilson, G. F., & Eggemeier, F. T. (1991). Physiological measures of workload in multi-task environments. In D. Damos (Ed.), *Multiple-task performance* (pp. 329–360). London: Taylor & Francis.

Yamada, F. (1998). Frontal midline theta rhythm and eyeblinking activity during a VDT task and a video game: Useful tools for psychophysiology in ergonomics. *Ergonomics, 41*, 678–688.

Chapter 5

A Mitigation Framework for Enhancing Situation Awareness

Sven Fuchs[1], Kelly S. Hale[1], Chris Berka[2], and Joseph Juhnke[3]
[1]Design Interactive, Inc.
[2]Advanced Brain Monitoring, Inc.
[3]Tanagram Partners, Inc.

Augmented cognition can be applied to mitigate problems with situation awareness by targeting specific perception events.

Introduction

To date, many systems have been developed that use physiological measures to enhance operator performance based on real-time assessment of general cognitive processes. These systems succeeded in quantifying cognitive parameters and the identification of cognitive breakdowns with regard to directly detectable cognitive state indicators, such as workload or arousal, which triggered predetermined mitigation strategies when cognitive gauges surpassed a threshold. But augmented cognition systems to date have been limited in the number and variety of mitigation strategies, in part because of limited knowledge regarding context. When addressing more complex cognitive constructs, such as situation awareness (SA), direct measures of particular cognitive activity do not suffice in driving real-time mitigation of cognitive problems—the cognitive variable must instead encompass a multitude of parameters, including comprehensive information about the task or system status that caused the problem.

To account for these demands, in this chapter, we introduce **a different approach to augmented cognition: event-based cognitive assessment to drive dynamic real-time system mitigation**. The method described here provides a more prescriptive way to evaluate cognitive processes in real time through the selection of key events as good/bad performance indicators, the measurement of the physiological reaction to these specific events, and the configuration of mitigation strategies on the fly, depending on the system and operator state. This chapter focuses on SA—specifically, **how to measure perception of events and mitigate associated cognitive breakdowns** in a dynamic, complex environment.

Many prior implementations of augmented cognition systems had a problem with *inconsiderate augmentation* (a term coined by Stanney & Reeves, 2005), in which SA was sometimes compromised for performance benefits. Inconsiderate mitigation describes system adaptation that is based on physiological parameters alone, with little or no consideration of the task or system context or the cognitive impact of the evoked mitigation strategies. When one of the intended benefits of the system is to increase SA, such problems must be minimized or eliminated to avoid a vicious cycle in which the mitigation, triggered to increase SA, has detrimental effects on the same variable.

The **considerate mitigation** approach offered here targets the specific tasks that are driving decrements in SA and mitigates shortcomings while **accounting for individual task parameters and the cognitive impact of the overall task context**. We outline lessons learned from an implementation effort—including advantages, disadvantages, and generalizability—to assist practitioners in adapting the event-based augmented cognition approach to suit their needs. In addition, a scoring approach is offered that can be implemented in real time for the evaluation of accomplished SA benefits.

Examples provided in this chapter focus on enhancing SA in a simulated Naval Command air-monitoring task. Specific events of interest are related to perception of key events, and electroencephalography event-related potentials (EEG/ERP) serve as cognitive indicators. The closed-loop system developed for this event-based approach includes a simulated command and control environment in which participants monitor and identify air tracks within a given area of responsibility, respond appropriately based on a given set of rules of engagement, and interact with the system (i.e., simulated team members) via chat and auditory communications.

Scenario

This scenario illustrates a day in the service of Flag Tactical Action Officer (FTAO) Phil Stevenson and highlights how an event-based augmented cognition system aids in optimizing Phil's SA throughout his daily tasks. (Please note that this scenario serves as an example only and does not reflect actual practices or rules of engagement.)

"Another day, another tussle." Those words were becoming Phil Stevenson's mantra lately. It seemed like this conflict was never going to end. Phil and his crew were ranked among the highest-performing teams in the fleet, a feat Phil knew had much to do with his augmented cognition equipment. Phil's crew had been outfitted with some new, just-being-tested, gear about six months ago. He smirks when he recalls the joke he and his crew made about the hassle it was going to be. But that is all water under the bridge; he is a convert now.

His watch begins like any other day. He puts on a baseball cap–like device that houses EEG sensors and starts reviewing the last watch summary on his augmented Personal Viewing System (aPVS) while he sips his daily glass of orange juice. He is feeling pretty sharp today and is validated when the aPVS briefs the preceding watch to him in a matter of minutes. He smiles, remembering old briefings that used to take an hour. It is amazing how hours of planning, execution, decision making, and performance evaluation can be communicated using summary graphics, animations, text logs, and audio, among other things. Phil also enjoys the fact that the system knows when he has understood a point and moves on. Phil can now start his watch with a complete understanding of mission objectives he has yet to achieve.

"Let's see what we've got today!" Phil announces over group comms. Projection screens are already displaying summaries of team activities and status. As usual, his task is to monitor the air space over the carrier group, which is close to a sector that is heavily trafficked by commercial planes. Phil recalls an intelligence alert that terrorists may attempt an attack with hijacked commercial airliners.

Dynamically assigned their areas of responsibility (AOR), Phil's team begins scanning and validating traffic. The five-man team is strong—except for Rudy, who is still pretty green and not aligned with the predictive workload models as well as the others. Things are just a bit harder for him than the system demands. Phil notes that the detail level on Rudy's AOR display has already been reduced; the augmented cognition system doesn't usually declutter a display unless events have been missed despite less intrusive system intervention. This concerns him, but the buzzing of the vibrating actuators integrated into his vest reminds him that he has bigger things to worry about right now.

He looks back at his aPVS; it is already providing blinking cues of the most important situational details: A 737 has deviated from the commercial air lane, and there is no response to ID requests. That is when things get interesting. Sam, one of Phil's brightest, exclaims, "Oh dear, those are dots I didn't want to connect!" A radio transmission comes in from the Nemesis, but Phil doesn't listen because his attention is focused rightly on Sam's concerns. "If she's concerned, I'm concerned," he thinks, as the system automatically translates the unattended radio channel to visual text on the aPVS display. Thanks to 3-D headphones, the remaining radio transmissions have been redistributed around his head. He recalls the tech guy calling it "spatialized audio"; he is a big fan, as it allows him to monitor and distinguish several voice channels simultaneously.

Phil focuses on Sam's analysis of historical patterns, airport reports, and traffic logs. The same summarizing tools used in his morning briefing allow him to clearly understand that the maybe 30 regular tracks in the AOR have been reliably traveling the same route over the past 20 days. Today, however, the traffic log history indicates a 10-mile route diversion for one of them. On top of that, the crew cannot be reached for a status report—a sign of trouble?

"Rudy, what's your status?"

A MITIGATION FRAMEWORK

"Looks like business as usual, sir," Rudy responds.

"You kidding? Look what I got here." Phil points at the big screen where his assessment is shared. "We need ID and a status update from these guys. And get a bird ready—just in case!"

A high-pitched tone reminds Rudy to redirect his attention to the display. Phil likes this considerate adaptation. Their first augmented cognition system used to interrupt him often, beeping and blinking when he was in the middle of something important. The new system is much better. When it detects that his eyes are busy, it often switches to presenting auditory information in an opportune moment.

"We need to find out whether these folks are baddies. And check if there are more," Phil commands. Rudy immediately knows what he is talking about because confirmed civil planes have been reduced to mere dots on his display, whereas the track and data for the plane with suspicious deviations are plotted in bright colors. Luckily, it soon becomes evident that it is the only threat, but it will be entering the no-fly zone of the carrier group in 8 min.

Phil orders the weapons teams on alert and cues the rest of the crews on watch of his team's findings. Communications jump exponentially. The entire fleet is hot now. With increasing importance of tasking, low-priority information is removed from his aPVS, and the constant, distracting chatter of the radio comms has been converted to sequenced text messages with response buttons where appropriate. Skimming these text messages, Phil can process a dozen or more in as many seconds. During a few points of very intense, high-priority activity, Phil notices the communications seem to slow down. However, he knows it is the system holding back comms from lower-priority channels until more important information is attended to.

After a short while, the conflict resolves. It was nothing but a commercial aircraft with electronics gone bad. Without instruments and autopilot, the pilots could not stay on route, and a dead radio kept them from communicating their status to flight control. But now things are back to normal. As communication and activity slow down, the various system augmentations are also faded out, returning displays and comm channel format back to normal.

Test Your Knowledge
What are the benefits that Phil Stevenson experienced during use of his new augmented cognition system? What strategies were used to obtain these benefits?

General Approach and Associated Toolkit

> *"Of 32 SA errors identified . . ., twenty-three (72%) were attributed to problems with Level 1 SA, a failure to correctly perceive some pieces of information in the situation."*
>
> *Endsley (1999), p. 3*

The preceding scenario presents an impression of how an augmented cognition system can provide seamless mitigation of perceptual shortcomings by augmenting missed system events in a multimodal fashion (e.g., triggering auditory alerts when Phil did not respond to the visual cues on his display).

Creation of an event-based augmented cognition system requires three major tasks, each with several subtasks: (a) a task analysis during which the augmented cognition practitioner must gather information about the system to be augmented and identify key events that represent both good and bad SA, (b) extraction of unique cognitive signatures that indicate good and bad performance for each of the key events identified in the first step, and (c) implementation of a real-time mitigation management framework in which system and user events are monitored and detected problems are mitigated by display adaptations. All necessary steps are discussed in detail within this section. Later, based on the immediately evaluated event perception, we introduce a scoring method that makes possible the real-time quantification of Level 1 SA, which represents a tool for the validation of accomplished mitigation benefits (see Chapter 6).

Valuable Information: Situation Awareness (SA)
With the rapid evolution of new technologies, warfighters often become overwhelmed by the multitude of data available or presented to them without consideration of the current status of those data. Since this problem was recognized, SA has become an important concept within defense-related research, because it constitutes a key aspect within netcentric warfare decision making in which operators consolidate data from high-tech sensors into an understanding of the "big picture" that dynamically updates as the situation changes. Although there are many definitions of SA, the one most commonly used is from Endsley (1995): the perception and comprehension of events within a given environment into a mental model of the current situation, and the prediction of future events based on current understanding.

Numerous approaches have been developed since 1995 to measure and increase SA; however, this has proven difficult, as SA is a multidimensional concept that is influenced by a variety of parameters and thus cannot be assessed unitarily or measured directly with a single variable (Endsley, 2000). Traditional methods to measure SA include subjective metrics of SA reported at the end of a scenario (e.g., the Situation Awareness Rating Technique [SART; Taylor, 1990] and the Mission Awareness Rating Scale [MARS; Matthews & Beal, 2002] questionnaires), but these require interruption of the task and are thus not suited for use in an operational environment or for real-time evaluation such as that needed for augmented cognition systems.

Using other approaches, researchers have attempted to create derivate measures in which cognitive processes (e.g., workload or performance metrics) are identified that supposedly correlate with SA (Endsley & Garland, 2000). This is in line with Endsley's assumption that SA cannot be measured directly, but imposes problems: Measured processes can be strongly influenced by other parameters than SA. For example, if an operator capitulates because of information overload, the cognitive load score goes down; however, this does not mean that SA has increased. Furthermore, metrics frequently employed capture only certain aspects of SA, and one must therefore be careful that one aspect of SA is not gained by decreasing SA

> on another, equally important, aspect (Endsley, 2000). For example, a cue
> may draw attention to one task, thereby interrupting a second task.
>
> Finally, an ideal system would mitigate bad SA before it occurs. However,
> when indirect indicators such as cognitive performance breakdowns are detected, it is already too late for prevention and the system can only assist
> with recovery. Hence, problems with SA should be detected at an early
> stage, before overall task performance is affected.

In Endsley's (1999) review of SA-related accidents on aircraft carriers, problems with the first SA level (i.e., perception) accounted for over 70% of errors. Jones and Endsley (1996) provided a similar figure. If a system could detect problems at this level, second- and third-level errors (where information is not, or not fully, comprehended, wrong predictions are made about the future state, and erroneous actions are taken) could potentially be avoided. This is the underlying idea of our approach to augmented cognition.

The goal of this chapter is to provide guidance on how physiological sensors can be implemented to monitor perception of events that are critical for SA, detect problems, and, when appropriate, trigger mitigation strategies that adapt the interface to support perception when and where necessary while considering the current cognitive status of the operator. The last-mentioned component is particularly important because mitigation strategies come at a potential cost (e.g., imposed by task switching or the cognitive impact of the information display adaptation), which itself can have negative consequences for SA.

Framework for an Event-Based Augmented Cognition System

In this section, we detail the three major tasks required to create an event-based augmented cognition system for the detection and mitigation of problems with Level 1 SA. These include identifying key events that represent both good and bad SA, extracting unique cognitive signatures that indicate good and bad performance, and implementing a real-time mitigation management framework.

1. Identification of Key Events

The first steps in an event-based augmented cognition approach are to identify, define, prioritize, and assign predictive resource consumption values to key events that will be monitored and used to determine if and when mitigation is required and help define what mitigations may be most appropriate for the given situation. Details of each of the four steps are listed next.

1.A Identify events with a high impact on SA. Working from a comprehensive task analysis (which may include document review, subject matter expert interviews and surveys, and task observations), key tasks related to the augmented cognition system's goals need to be identified. The system objective discussed in this chapter is enhancement of Level 1 SA (perception); thus, key tasks related to perception of new information within a dynamic task environment are of interest. Figure 5.1 provides an impression of the simulated tactical Naval Command task used for Design Interactive's augmented cognition endeavor (Hale et al., 2006). In this exercise, participants were required to perceive, monitor, identify, and respond appropriately to air tracks in their area of responsibility as well as monitor auditory and chat communications. A new track on the display, a deviation in route, or new chat messages are examples of events that affect SA.

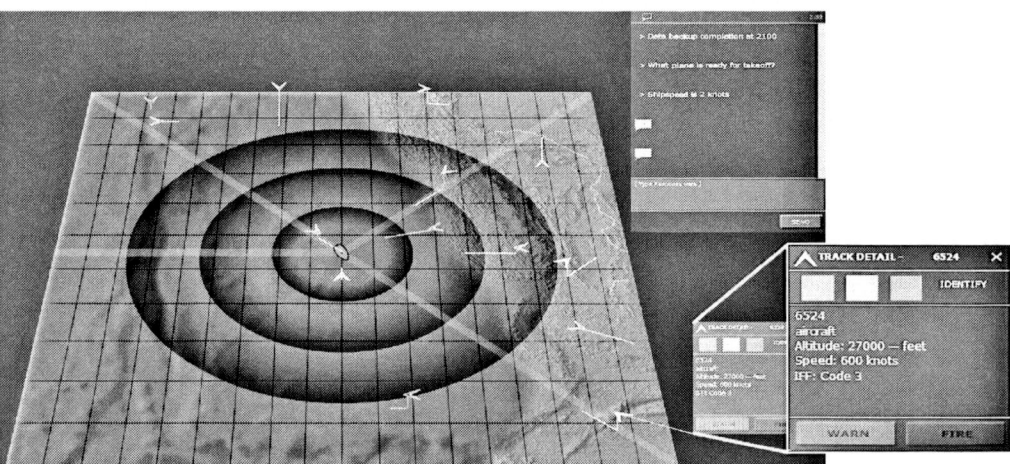

Figure 5.1. Screen shot of the simulated naval command task.

Figure 5.2 (p. 119) shows the associated task flow. Lines marked with a P indicate events related to the perception of information. Within this Naval Command task, participants were required to perceive new tracks on the screen, when a track entered weapon range (WR), when sufficient time had passed after warning an unknown track in WR (i.e., time to fire), when audio messages were played, when chat information appeared, and when an incoming chat message required a response. If participants failed to perceive any of these events when deemed critical to the task at hand (i.e., missed events were of highest importance at that point in time), mitigation may have been required to enhance their perception.

1.B Define events. Once tasks related to the cognitive state of interest (in our case, perception) are identified, specific start and stop triggers for each event must be defined and used to categorize events. As shown in Table 5.1, each perceptual event from the Naval Command task had an associated starting and stopping point that was represented by either a system event or a behavioral response. Note that for some tasks, there may have been more than one stop trigger. For example, when an unidentified (i.e., new) track appeared outside WR, this track could be selected either to view details (which would change its status) or enter into WR without operator activity (which changes its status to "unidentified track in WR"). With start and stop triggers defined for each event, these could be used to determine when key events are ongoing during operations and serve as a point of alignment for cognitive state metrics—the corresponding "perception" signatures—that indicate good or bad Level 1 SA.

Given that this was a dynamic task environment with multiple events potentially occurring simultaneously, it was reasonable to assume that not all events would be perceived immediately, and thus cognitive signatures might indicate a "not perceived" response. By also defining stop triggers for events that were based on either system events (e.g., track enters WR) or behavioral responses (e.g., track selected), these could be used to define when an event was appropriately acted upon in a specified period, providing a backup system for events that were not perceived (as evident with physiological sensors) at the time of stimulus presentation.

1.C Prioritize events relative to other key events in the system. As mentioned in the previous section, multiple tasks may occur simultaneously, and operators are often limited in the amount of events to which they can respond at one time. While high-priority tasks are being attended to, lower-priority events may go unperceived. Thus, priority values for each perception task from the Naval Command environment were required for each key task based on event criticality and the impact of missed perception on the overall task goal. These were used to organize ongoing events in the dynamic task environment.

Table 5.1 shows exemplary priority values for key perceptual events from the Naval Command environment. The values were derived from a task analysis and converted to a linear scale ranging from 0 to 100.

Table 5.1. Definition of Key Perceptual Events for Naval Command Task

Start Trigger	Stop Trigger(s)	Priority
Unidentified track appears outside WR	Track identified Track enters WR	10
Unidentified track appears inside WR	Track identified	60
Unidentified track enters WR	Track identified	60
Track ID as hostile in WR	Track fired	100
Track ID as unknown in WR	Track warned	90
Hostile track enters WR	Track fired	95
Unknown track enters WR	Track warned	55
Unknown track in WR ready to be fired	Track fired	90
Incoming chat information	Envelope selected	30
Chat question appears in message box	Response box selected	35
Audio message plays	N/A	30

AUGMENTED COGNITION: A PRACTITIONER'S GUIDE

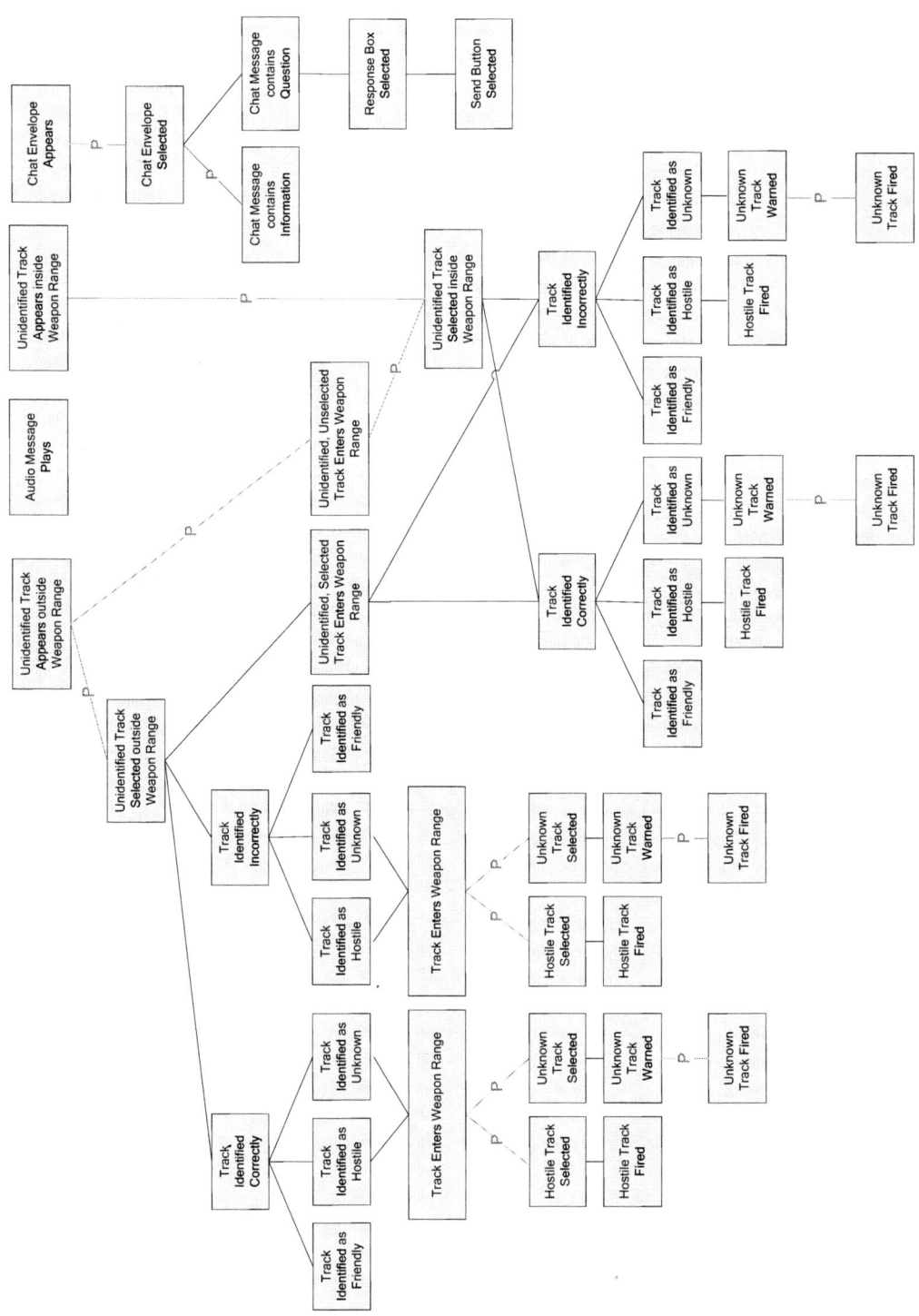

Figure 5.2: Task flow of the naval command task.
Note: Lines with a "P" indicate where perception of new information is required.

1.D Assign resource consumption values to events. Whereas the foregoing three processes can define key events and assist in determining when and what to mitigate, further information regarding how to mitigate may be provided by assigning predicted resource consumption values to each key event. Based on active key events, a multimodal approach to mitigation can be implemented whereby resource consumption values can predict the amount of load placed by each event on various sensory systems (e.g., visual, auditory, haptic), cognitive systems (e.g., verbal and spatial working memory), and response systems (e.g., motor, speech), and thus they can be used to determine which resources have capacity available and ideally should be used to mitigate the current situation to enhance perception of a missed event (Fuchs, Hale, Berka, Levendowski, & Juhnke, 2006).

The Workload Index (W/INDEX) predictive equation (North & Riley, 1988) has been used in various systems (e.g., in the Improved Performance Research Integration Tool [IMPRINT; US Army Research Laboratory, 2005] and the Multimodal Information Design Support [MIDS; Hale, Reeves, Samman, Axelsson, & Stanney, 2005]) to predict cognitive resource consumption (see Chapter 6). Values from tools such as these are appropriate to define which presentation modality is ideal for additional information at any point in time, depending on the current task status.

Test Your Knowledge: Determining the Optimal Adaptive Strategy

Table 5.2 presents a list of active events at a given time in the order of their priority. Values of cognitive resource consumption have been added to each event. Imagine that Rudy, the operator from the scenario, missed the top-priority event, and a mitigation strategy was to be triggered to support perception of a hostile track in WR. What would be a good strategy from a cognitive resource perspective?

Note that several events already impose visual load (ViL) on Rudy. In addition, an audio message is playing (Event 3), occupying much of his auditory processing capacity. The haptic channel, however, is currently unused. Using a tactile cue, it may thus be possible to cue his attention without affecting his perception of other ongoing events.

Table 5.2. Analyzing Values of Cognitive Resource Consumption

	Event	Priority	Response Event	ViL	AuL	HaL	SpL	VeL
1	Track in weapon range identified as hostile	90	Fire at track	4.0	0	0	0	4.6
2	Unknown track enters weapon range	80	Warn track	4.0	0	0	4.6	0
3	An audio message plays	40	Listen to audio msg.	0	5.9	0	0	3.7
4	Unidentified track enters area of responsibility	10	Identify track	3.0	0	0	3.7	6.8
5	Friendly track enters weapon range	0	[no response required]	3.0	0	0	3.7	0
			Total Cognitive Load	14.0	5.9	0	12.0	15.1

ViL = visual load; AuL = auditory load; HaL = haptic load; SpL = spatial load; VeL = verbal load.

2. Extraction of Event-Related Signatures

Once a task model is in place, it is necessary to create at least two distinct indicators for each event identified in the task analysis: one for positive assessment of perception and another for negative assessment (i.e., event not perceived or response too late). There are various methods to evaluate the operator's reaction to events. The simplest approach is monitoring system and user activity (e.g., Van Orden, Viirre, & Kobus, 2007), whereby system events and associated user activity are used to determine whether the event was appropriately processed. Using physiological sensors, however, one can also evaluate the perception of events when no behavioral response is present. For example, a multitasking situation may not allow the operator to react immediately, although the event was perceived. Other events, such as status updates, may not require a behavioral response at all, even though they contribute substantially to SA.

By definition, the concept of augmented cognition mandates the use of physiological sensors. At least two distinct physiological signatures must be created for each event identified in the task analysis: one for positive physiological assessment of perception, and another for negative physiological assessment (i.e., event not perceived or response too late). Various techniques are available for measuring cognitive activity (e.g., eye movements, pupil diameter, galvanic skin response, EEG, and functional magnetic resonance imaging [fMRI]). Although it should be possible to implement effective systems based on relatively simple measures—for example, by determining the attentional focus using ocular gaze data—our effort embraced the identification of event-related cognitive signatures of SA using electroencephalography methods that produce distinct neurophysiological signatures (event-related potentials, or ERPs) in response to cognitive processing. To date, EEG has been shown to best reflect subtle shifts in alertness, attention, and workload in real time and with high temporal resolution (Berka, Levendowski, Ramsey, et al., 2005). Accurate detection of these subtle changes and system-sensor synchronization at a millisecond level are necessary to obtain reliable signatures that are unique for each event.

To obtain cognitive signatures that can be used to evaluate perceptual SA, participants can undergo user testing with scenarios that present significant events first in isolation and then in context. For a first round of data generation, it is important to isolate events so that a clean signal can be captured without interference from other events. Responses are classified into successful and unsuccessful responses. Event-related signatures are then averaged across participants and compared between successful and unsuccessful trials, with the goal of finding indicators that are significantly different for both outcomes so they can later help in evaluating whether perception was successful. In further participant trials, context should be introduced to initially verify the robustness of obtained signatures under operational conditions (see Chapter 7).

Review: Cognitive Signature from Event-Related Potentials (ERPs)

For developing physiological signatures in the described testbed environment, we used Advanced Brain Monitoring's 9-sensor system (Figure 5.3). Power spectral analysis was used to compute mean power spectra time-locked to a specific stimulus presentation or to a specific user response in the testbed environment. In our case, the EEG analysis window (between

> 250 and 2000 ms) was positioned to align either with a specific stimulus presentation event or over a response event to calculate the specific EEG power spectra associated with processing of the stimulus or with generation of the response, respectively.

Although there are many ways in which event-related signatures could be extracted, the details of our EEG approach will now be discussed; the general techniques described should be generalizable to other extraction approaches.

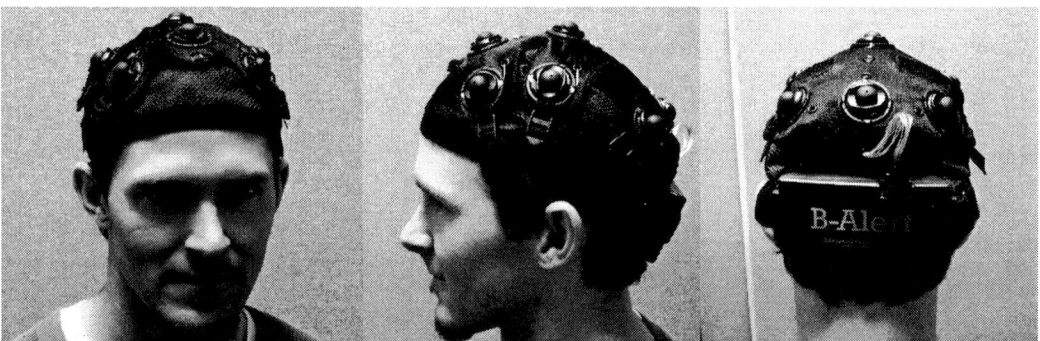

Figure 5.3. The ABM 9-Channel EEG system.

In order to obtain event-specific signatures, we used algorithms to compute mean power spectra to quantify the EEG within discrete intervals. Signal analysis techniques were used to identify and decontaminate eye blinks and to identify and reject data points that were contaminated with excessive muscle activity, amplifier saturation, and/or excursions attributed to movement artifacts (see Berka et al., 2004, for a detailed description of the artifact decontamination procedures). Fast-Fourier transform was then applied to each overlay of the decontaminated EEG signal to compute power spectral density (PSD) values derived for each 1-Hz bin from 1 Hz to 40 Hz. The EEG signal was decomposed using a wavelets transformation in order to compute additional variables for use in developing the neural signature templates.

Figure 5.4 presents ERP templates and data from a single participant during the encoding period of a Standard Image Learning and Recognition (i.e., imagery analysis) task using just one EEG sensor positioned over the parietal-occipital region. The mean power spectral values during the encoding period were computed for only 10 of the images that the participant correctly identified during recognition ("Correctly Encoded") and for 9 that were incorrectly identified during recognition ("Images Incorrect"). There is a clear and distinctive difference in the power, suggesting a dominant peak of theta activity in the EEG associated with correct encoding, which in turn suggests that this pattern—if consistent across participants—could be used as a distinct signature template to evaluate recognition.

The next step was to determine whether the single trial ERP could be used to accurately classify correctly encoded images in comparison with those that were incorrect. In this example, using a simple threshold based on the total power in the theta band (3 Hz–7 Hz) allowed a fast, accurate real-time classification of 9 of the 10 correctly encoded images and 8 of the 9 incorrectly encoded images (overall classification accuracy = 89.5%). In an operational environment, pattern-matching algorithms would compare the real-time ERP over an event-locked time window with the templates.

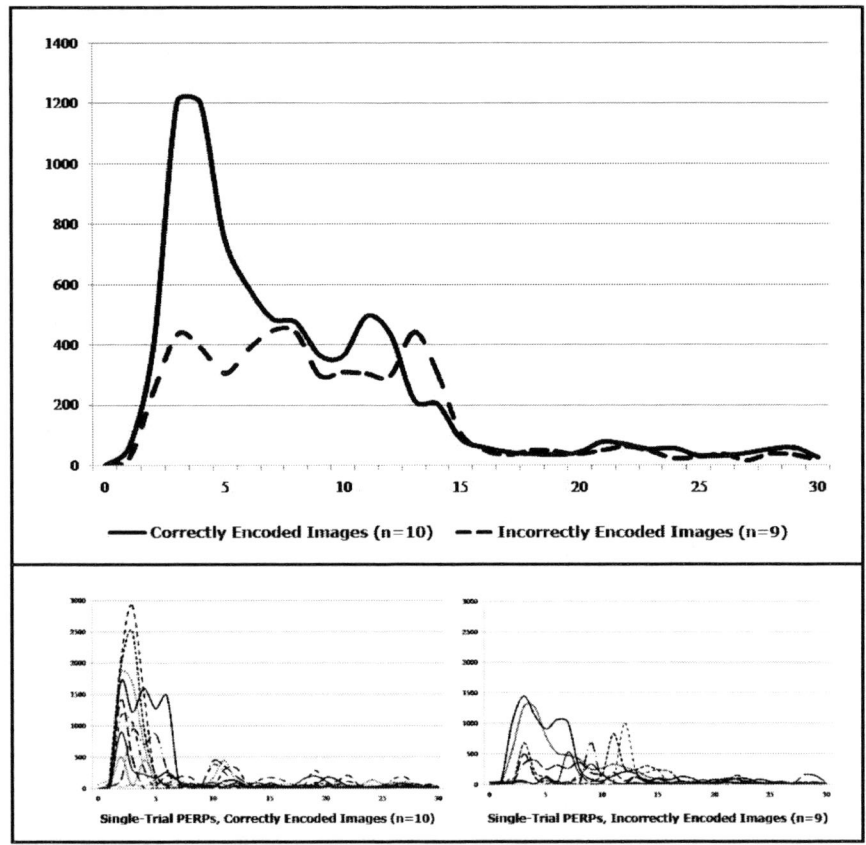

Figure 5.4. ERP data from the encoding stage of the standard image learning and recognition task.
Top: Mean event-related EEG power spectra for correctly encoded images and incorrectly encoded images during a recognition task show distinct differences that can serve as a cognitive signature template with which real-time data are then compared. **Bottom left:** Single-trial event-related power spectra for correctly encoded images. **Bottom right:** Single-trial event-related power spectra for incorrectly encoded images (data from one participant).

3. Mitigation Management Framework

A mitigation management framework directs when, what, and how to mitigate (see the conceptual model in Figure 5.5). Referring to Figure 5.5, Step A depicts an event manager that determines which events are currently active and whether an appropriate physiological or behavioral response was detected for each event. Step B encompasses a mitigation selector, which applies a decision matrix to determine if and for what event a mitigation strategy should be triggered. Step C represents the core of a **considerate mitigation** method, whereby the most appropriate strategy is selected based on the current task and cognitive status. This section outlines our implementation of such a mitigation management framework, which constitutes a real-time system based on a task model and event details, as well as the associated event-related cognitive signatures obtained in the first two steps.

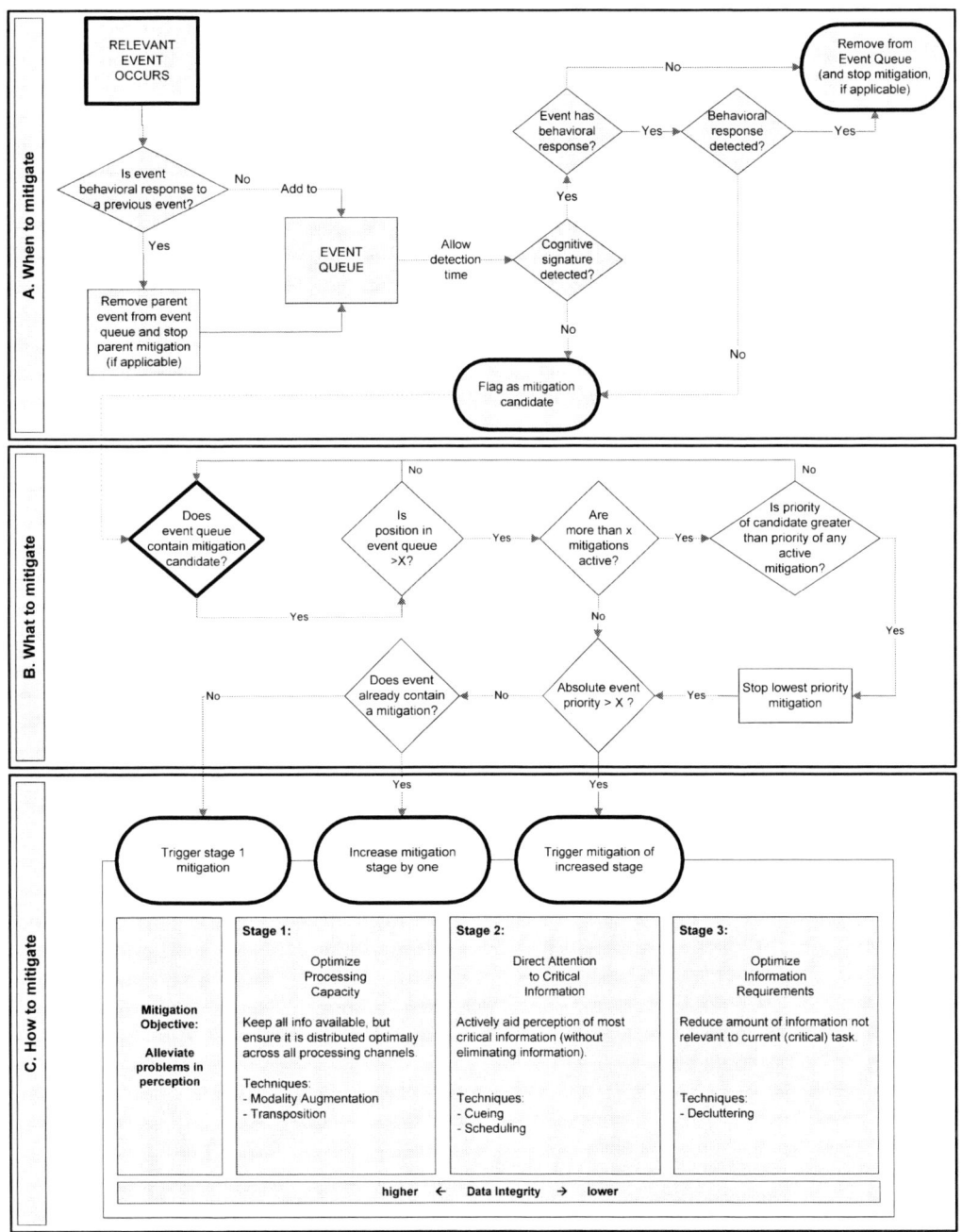

Figure 5.5. Event-based mitigation framework.
A: Event Manager, B: Mitigation Selector, C: Stages of Mitigation (from Fuchs, Hale, Stanney, Juhnke, & Schmorrow, 2007).

3.A The event manager—What to mitigate. The event manager component is concerned with the evaluation of the current task status and associated responses. For this matter, an event queue is maintained that holds all currently active events in a prioritized list. Upon occurrence of a new event initiated by the user or the system, it is determined whether this new event is in response to a currently active event (e.g., selection of a track would be an appropriate response to a "new track occurred" event). If it was a response, the pending parent event that was responded to is removed from the event queue prior to adding the new event. If not, it is directly added to the event queue.

Upon occurrence of an event and its addition to the queue, the event manager monitors physiological activity through pattern-matching algorithms that compare the cognitive signature template of the event (event perceived or event not perceived) with the real-time sensor data stream in a certain time window around the event. Additionally, possible behavioral response events (if appropriate) are monitored as a means of backup. For example, if the event was not immediately perceived, and physiological data indicate a missed perceptual event, an associated behavioral event later in time may be used to indicate that the event has been addressed. If an appropriate response occurred, it can be assumed that the event has been perceived and, hence, can be removed from the queue of pending events. If no response was detected, it must be assumed that the event was not perceived appropriately. In this case, the event is flagged as a candidate for mitigation.

Review

Consider the command and control scenario described at the beginning of this chapter. Little is going on, and Rudy, the rookie operator, is attending to an ongoing conversation in the chat system. Suddenly, a track identified as hostile occurs in his area of responsibility but outside weapon range, along with an additional unknown track that has already entered WR. Both are added to the systems event queue. The hostile track is distant and therefore does not pose an immediate threat. However, the unknown track is a potential threat and is moving toward the ship.

Now assume Rudy notices the track in WR (indicated by streams of physiological or behavioral data). That event is then removed from the event queue, leaving the distant hostile track as the most critical current event. Because the most critical event has not been perceived, it becomes a mitigation candidate if Rudy resumes attention to the chat system instead of addressing the hostile track on display.

3.B The mitigation selector—When to mitigate. As described earlier, problems with missed events are not immediately mitigated but, rather, are flagged as candidates, for the following reason: In complex multitasking systems, such as air traffic control or command and control environments, many events may happen at the same time. Immediate, simultaneous display adaptation for all concurrent events may have a significant impact on cognitive load and may result in cognitive breakdown, operator confusion, and detrimental effects on SA. To prevent this from happening, only the most important problems should be considered for mitigation based on the priority value assigned during the identification of key tasks—Phase I of this process—and only a limited number of mitigations should be activated simultaneously. Other candidates should remain in queue until events with the highest priority have been responded to. However, in the case in which a high-priority event is missed, it may be mitigated regardless of the number of active mitigations. In this case, the active mitigation(s) with the lowest priority could be deactivated in favor of the one with higher importance.

To further avoid unnecessary interruption or task switching, additional conditions can be implemented as appropriate to limit the number of mitigations triggered. In the case of our implementation, a criterion for absolute priority was included so that time-critical events can be directly triggered with the most intrusive strategy.

Review
The hostile track that occurred outside WR in the previous example has no immediate impact on the security of the ship. Accordingly, this track has a low priority assigned to it, as operators are not yet able to address the threat. Because Rudy did not perceive the track's appearance, it was flagged as a mitigation candidate, but an incoming chat message had a higher priority assigned (it could potentially include important information). Because Rudy attended to the chat message instead of the hostile track, he maintained focus on the most critical task at that time. Thus, no mitigation was triggered for the missed track in order to keep attention focused on the higher-priority event.

The situation changes as the track changes course and starts moving toward the ship. A follow-up event is triggered when the track enters WR, replacing the original (low-priority) event. Because the track is now a potential threat to the ship, the follow-up event has a high priority. However, with removal of the original event, the "mitigation candidate" status is revoked to give the operator a chance of detecting the change on his own. If still missed after the newly allotted response time, candidate status will be assigned to the new event—with the now higher priority—and mitigation may be triggered.

3.C Mitigation stages—How to mitigate. To best support human performance, mitigations should be enacted in stages. Note that mitigation strategies have been found to come with an associated cost (see Fuchs et al., 2007) and should thus be evoked only when a need is indicated (in the case of augmented cognition, an indication of "need" is provided by physiological sensors). This makes mitigation strategies different from design improvements, which provide a constant benefit when present. Thus, when instantiating a mitigation strategy, one should consider available cognitive resources and make an effort to ensure that negative effects of mitigation are kept at a minimum.

To accomplish the latter, we compiled a number of potential strategies suitable for attention redirection and assigned them to three incremental stages based on their negative effects; the stages are outlined next. Note that these stages were developed specifically for mitigation of perception of missed events during our implementation of the Naval Command task and may require adjustments for use in other task environments to optimize the cognitive construct of interest and, ultimately, human performance. A higher or lower number of stages may be adequate, and many different approaches exist for each of the introduced strategies, some of which may or may not be well suited for specific systems.

Fuchs et al. (2007) provided a comprehensive overview, in which further strategies, appropriate use cases, potential benefits, and associated costs are laid out as a reference for the practitioner. A summary of findings is included in the Design Guidelines section of this chapter.

Stage 1: Optimize Processing Capacity. The initial mitigation stage attempts to reroute information to available resources. Because all currently ongoing tasks are listed in an event queue and values of predicted resource consumption have been assigned during identification of key events (see Section 1.D), it is possible to determine which sensory and cognitive resources are available by summing up the channel load across all tasks in queue. Assessment by channel can reveal unused resource capacity. Although not all tasks in queue are simultaneously attended to, summing up all possible components creates a worst-case scenario for cognitive load for the given situation.

According to Multiple Resource Theory (MRT; e.g., Wickens, 2002), additional information ideally should be presented in available resource channels to increase the probability of adequate cognitive processing. Modality augmentation/transposition, whereby the sensory (i.e., augmentation) or cognitive (i.e., transposition) channel is changed or additional channels are used, are appropriate strategies to make use of available cognitive resources (see Table 5.3). Because these resources are not fully tapped, it is more likely that adapted information will be noticed without actively redirecting attention. Therefore, these techniques are expected to have minimal detrimental effects on the user and should not affect data integrity.

Table 5.3. Mitigation Strategies for Stage 1 Mitigation

Modality augmentation/switching	Utilize the individual processing capabilities of separate sensory and cognitive resources (Wickens, 2002) to increase information throughput by optimally distributing information to available sensory-processing channels.
Transposition	Change the way displayed information will be processed to the user from verbal to spatial or vice versa (Dufresne, Martial, Ramstein, & Mabilleau, 1996).

Stage 2: Direct Attention to Critical Information. Stage 2 mitigations actively redirect the operator's attention by making the missed information stand out from other information on the display (see Table 5.4). Appropriate strategies here include cueing, whereby the salience of an item is increased by implementing attention grabbers in the current modality (e.g., highlight, blink for visual) or another modality (e.g., play auditory cue or provide a tactile cue), and scheduling, in which the order of task presentation is changed, resulting in the most critical information being displayed first with less multitasking. As in Stage 1, noted earlier, the predicted cognitive resource consumption can be used to optimally configure Stage 2 mitigation strategies. For example, if visual cognitive resources are fully tapped, new incoming visual information of lower priority could be held back temporarily. With this approach, SA may be slightly decreased as attention is unnaturally directed, and task-switching costs may occur.

A MITIGATION FRAMEWORK

Table 5.4. Mitigation Strategies for Stage 2 Mitigation

Cueing	Capture user's attention by enhancing information on display (Posner, 1980).
Sequencing/scheduling	Defer or escalate tasks on timeline (Dorneich et al., 2004).

Stage 3: Optimize Information Requirements. More intrusive means of mitigation beyond Stage 1 and Stage 2 may be required to enhance operators' perception of missed high-relevance information. These mitigation approaches may compromise data integrity in favor of easier detection. For example, decluttering techniques can be implemented to make lower-priority items less salient, reduce level of detail, or remove them completely, so that the complexity of presented information is reduced (see Table 5.5). Again, predicted cognitive load can be used to determine which information to declutter: For example, if only the visual channel is overloaded, it makes sense to trim the amount of visual information.

Although decluttering increases the likelihood of detection of the remaining information, it bears problematic implications for SA, because decluttered information may be harder to find or not accessible at all while the mitigation is active. If a quick status change occurs, task-switching issues may have a high impact. In addition, the operator must "catch up" after the issue has been resolved and the display resumes normal mode. Thus, higher-stage mitigation should be used rarely and with caution.

Table 5.5. Mitigation Strategies for Stage 3 Mitigation

Decluttering	Reduce the amount or complexity of information displayed (Kroft & Wickens, 2001).

Review
Consider the previous example. Although the hostile track has now moved into WR, Rudy is still busy with the chat system. Mitigation is therefore initiated at Stage 2 (Stage 1 is skipped because of the high priority and time-criticality of the event). Rudy's current cognitive load is determined to be visual-verbal (reading chat messages), so the system predicts that cognitive capacity is available in the auditory channel and triggers an auditory cue to direct attention to the track on the visual radar screen. If, for example, ambient noise kept Rudy from detecting this auditory cue as well, the system could increase the intrusion level, initiating a Stage 3 mitigation: To declutter the display and force the user's attention to the radar screen, the chat box on the user's display could be minimized and/or faded.

Event-Based Situation Awareness Metric

In order to demonstrate obtained SA improvements, an SA scoring metric was developed—building on the event-based approach presented in this chapter—by quantifying so-called good SA versus bad SA relative to system events across time. Compared with traditional, subjective metrics of SA that are reported at the end of a scenario (e.g., SART, Taylor, 1990; Mission Awareness Rating Scale [MARS], Matthews & Beal, 2002), the method presented here is more prescriptive, as dynamic changes in SA over time can been measured, and specific events that led to increased or decreased SA can be identified. For demonstration purposes, a minute-by-minute analysis based on behavioral responses of the Naval Command task is presented here; however, the methodology can be expanded to a second-by-second analysis that considers both physiological indicators of SA (e.g., perception) as well as behavioral responses. In addition, the inclusion of comprehension events would further enhance the accuracy and analytic capability of this SA scoring method.

To complete a minute-by-minute analysis of SA based on the Naval Command task, we evaluated performance outcomes within each minute for SA-relevant events and categorized them into (a) system events, (b) good-SA events, or (c) bad-SA events. The SA score considers all three categories of events as well as event importance using a weighted scoring method:

$$\text{SA Score} = (\text{Good-SA Score} - \text{Bad-SA Score}) / \text{System Events Score} * 100,$$

where Good-SA Score = Σ (number of occurrences$_x$ * priority$_x$); Bad-SA Score = Σ (number of occurrences$_y$ * priority$_y$); System Events Score = Σ (number of occurrences$_n$ * priority$_n$); x represents tasks associated with Good-SA; y represents tasks associated with Bad-SA; and n represents system event tasks.

The score was baselined, so that ideal SA is reflected by a score of 100 and poor SA is reflected by a score of -100. Categorization of events as indications of good versus bad SA was done based on input from subject matter experts.

Table 5.6 (p. 132) displays an example of calculated values for a single participant in the first 5 min of a scenario from the simulated Naval Command task. As can be seen from the overall SA score, SA was ideal during Minutes 1 and 3 (100%), when all system events were responded to with Good-SA actions. SA was lower during Minutes 2 and 4 (65% and 53%, respectively), when one track's status was incorrectly identified during each minute. Note that during Minute 5, no incorrect actions were recorded that would indicate Bad SA; however, not all system events were responded to, so overall SA was lower than ideal (80.85%).

A MITIGATION FRAMEWORK

Review

Figure 5.6 shows SA scores for three participants during a moderately difficult scenario. As can be seen, individuals fluctuate between good and bad SA throughout a scenario. In addition, substantial individual differences are seen. This fluctuation within as well as between individuals provides additional support for the adoption of an event-based augmented cognition approach, whereby system mitigations are optimized for the individual operator, as opposed to implementing global design changes that do not specifically address changes in SA over time. Further, the scoring method presented here may be used to provide a more prescriptive SA metric in real time compared with currently available SA metrics.

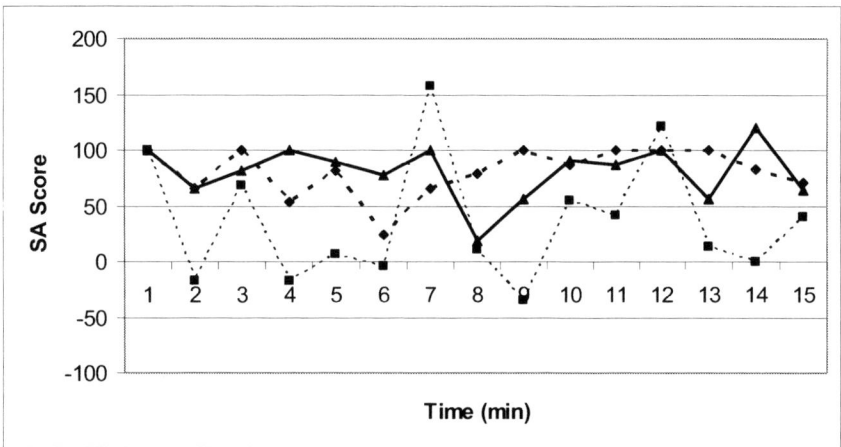

Figure 5.6. Global SA score for three participants based on behavioral responses to system events.

Table 5.6 follows

Table 5.6. Example of Event-Based SA Metric Calculations

Event	Task Priority	# of Task Occurrences Within Each 1-Min Window				
		Min 1	Min 2	Min 3	Min 4	Min 5
SYSTEM EVENTS						
Track appears	40	3	3	4	3	5
Track enters WR—unidentified track	60	0	0	0	0	0
Track enters WR—unknown track	50	0	1	1	1	0
Track enters WR—hostile track	90	1	0	0	0	2
Chat envelope appears (question and information)	30	1	2	3	0	3
SYSTEM EVENT SCORE		240	230	300	170	470
ACTIONS THAT REFLECT GOOD SA						
Number of tracks identified correctly	40	3	2*	4	2*	5
Number of unidentified tracks responded to within 6 s of entering WR	60	0	0	0	0	0
Number of unknown tracks responded to within 6 s of entering WR (i.e., track selected within 6 s) correctly	50	0	1	1	1	0
Number of hostile tracks responded to within 6 s of entering WR (i.e., track selected within 6 s) correctly	90	1	0	0	0	1*
Percentage of chat envelopes selected within 6 s	30	1	2	3	0	3
GOOD SA SCORE		240	190	300	130	380
ACTIONS THAT REFLECT BAD SA						
Number of tracks identified incorrectly	40	0	1*	0	1*	0
Number of unidentified tracks not responded to within 6 s after entering WR	60	0	0	0	0	0
Number of unknown tracks not responded to within 6 s after entering WR	50	0	0	0	0	0
Incorrect warnings within 6 s from entering WR	50	0	0	0	0	0
Number of hostile tracks not responded to within 6 s after entering WR	90	0	0	0	0	0
Incorrect fires within the first 6 s	90	0	0	0	0	0
Number of chat envelopes not selected within 6 s	30	0	0	0	0	0
BAD SA SCORE		0	40	0	40	0
SA SCORE = (GOOD SA - BAD SA) / SYSTEM EVENTS * 100		100	65.21	100	52.94	80.85

Note: Numbers with asterisks (*) indicate nonoptimal SA.

Lessons Learned

Lessons Learned for the Analysis of Environment

Working from a detailed task analysis in conjunction with subject matter experts, it is critical to clearly define key tasks that affect the cognitive parameter of interest. Specifically, each task must be identified by start and end points that can be used to identify a task independent from other, concurrent tasks, and must have a defined time (often either stimulus onset or behavioral response) that will define the alignment of the cognitive signature window.

In addition to clearly defining tasks, assigning priority values is a critical step in optimizing the event-based augmented cognition approach. Priority values need to be considered for each task and subtask and assigned in such a way as to ensure that, at any given moment, a critical task that is being attended to will not be interrupted by a second critical task that just entered the event queue.

Review: Priority Assignment Across Tasks
Rudy notices a track of unknown status entering WR and proceeds to warn this track that it has entered a no-fly zone. If the track retreats, Rudy does not need to take any further action. However, if the track continues on its path toward the ship despite the warning, Rudy must take action to protect the fleet.

While Rudy waits the required 3 s to assess whether the track will retreat, a confirmed hostile track enters WR. Ideally, the operator should be allowed to finish the current task by addressing the unknown track prior to being directed to take care of the hostile track that just entered WR. Thus, although of unknown status, a warned track that is already in WR should take priority over a hostile track that has just entered WR, which ensures that task interruption is kept at a minimum. (Note that all examples are of an illustrative nature and do not describe actual rules of engagement.)

Lessons Learned for Cognitive Signature Creation and Assessment

Earlier in this chapter, we listed some advantages of using physiological signatures, such as the ability to monitor cognitive processes that do not require a physical response. To realize these benefits, one must implement event logging with millisecond resolution and accuracy and align it to physiological sensors, so that cognitive signatures can be time-locked to the system or user events. When one creates scenarios that will be used to capture cognitive signature templates, it is advantageous to simplify conditions so events happen in isolation or sequential order as opposed to simultaneously; this creates uncontaminated templates for individual tasks that then can be used in a more realistic task environment to identify when defined events are perceived.

For the purposes of identification and extraction, physiological signatures can be characterized as the key stimulus-driven (i.e., occurrence of signature is expected after event occurrence) or response-locked (i.e., signature occurrence is expected before the response) activities in the task environment. The latter appears to be mostly true for tasks involving a decision-making component, in which the original event is followed by an evaluation process and, once the decision is made, the response (event) is initiated.

When it comes to accuracy and reliability, a challenge is posed by the fact that cognitive signatures are as individual as fingerprints. In fact, they are even more unique because they vary for the same participant from trial to trial. The event manager monitors physiological activity through pattern-matching algorithms that compare the cognitive signature template of the event with the real-time stream in a certain time window around the event. As the templates are constructed from averaged user data obtained in experimental trials, signatures in an operational context may not be reliably detected because of interference with other external stimuli or individual differences.

Although implemented templates should be robust enough to detect the majority of cases (accuracy at or above 90%), we recommend that physiological or behavioral responses be monitored as a backup solution when possible. This allows for more flexibility in the categorization of mitigation candidates in a complex, high-workload environment, in which multiple stimuli may occur simultaneously and/or in close proximity to one another. Often, cognitive signatures will identify a portion of these stimuli that were perceived, yet others may not be perceived at the time of stimulus presentation, thus providing a cognitive signature of *missed perception*. When adequate behavioral response times (we obtained decent results with RT = 3 s) are incorporated as a backup monitor of event perception (i.e., stimulus has been acted upon in appropriate manner), fewer false alarms related to missed perception should enter the system.

Lessons Learned for the Implementation of Mitigation

A review of past augmented cognition programs (Fuchs et al., 2007) revealed that previous implementations of mitigation strategies did not always follow a single naming convention. For example, some researchers use the term *pacing* (Tremoulet et al., 2005) to describe what we and others call *scheduling* (i.e., deferring or escalating information in the timeline of events). Caution is also advised with regard to observed benefits of previously implemented mitigation strategies; some strategies were more design changes than dynamic mitigations (e.g., tactile navigation cueing system; Dorneich et al., 2004), whereby the provided advantage was independent from the cognitive state of the user. Some systems triggered mitigation based on cognitive state, which were then left active, regardless of changes in cognitive state. Though it may be challenging to develop appropriate exit strategies (i.e., return to unmitigated state when the problem is resolved), this status reversal is necessary to avoid potential costs inflicted by the mitigation strategy (e.g., cognitive load or distraction caused by an unnecessarily continued cue) for the remainder of the scenario. Costs of mitigation have also been associated with state transitions from one mitigation to another; participants reported being confused and labeled the information display as "jumping" and inconsistent (Hale et al., 2006).

Table 5.7 summarizes discovered issues with regard to mitigations and possible solutions to consider in future implementations.

Table 5.7. Lessons Learned for Implementation of Mitigation

Issue	Possible Approaches
Turning mitigations on/off instantly leads to task-switching issues. Displays may appear jumpy and inconsistent.	- Develop enter, exit, and transition techniques to minimize yo-yo effects. - Consider recovery intervals between different display stages.
Augmented cognition systems have mitigated cognitive states with minimal regard for system and task context.	An underlying task model helps identify where to apply mitigation strategies. Theoretical models of cognitive resource consumption can identify how to apply mitigation strategies.
Current implementations of mitigation strategies are limited in creativity and dynamic implementation.	- Consider innovative mitigation techniques from domains in which context is frequently changed in a seamless manner, such as gaming, cinematography, theater, animation, and information art. - These could be used to implement a more fluid human-system interaction that makes use of advanced computer graphic techniques, eases transitions between display states, or conveys information in the background.
Mitigation strategies come with associated costs (e.g., task-switching issues or cognitive load; see Fuchs et al., 2007) that should be kept at a minimum.	- Apply appropriate mitigation strategies that address cognitive, system, and context states at opportune moments, and transition out of mitigation state at the earliest possible moment.

Applied Exercise

Below, some facts and techniques are listed as candidate sources for innovative approaches to mitigation strategy design.

- *Aural-spatial representation:* Based on echo and reverberations, people can hear how big a room is.

- *High-key/low-key lighting:* When a scene is evenly lit, it suggests a familiar world containing few surprises or mysteries, whereas strongly contrasted areas of light and shadow create a sense of mystery.

- *Galvanic-vestibular stimulation:* This technique simulates the sensation of movement by applying electromagnetic signals directly into the inner ear.

- *Ambient lighting:* Decreased ambient light focuses attention. Theaters use lighting to focus attention and notify of upcoming events.

- *Shift in a constant tone draws attention.* Experts can qualify sounds or changes in sounds (e.g., mechanics may hear or even diagnose engine problems based on sound).

How could these techniques be implemented as innovative mitigation strategies? How could they be used to improve current ones? What other techniques can you think of?

Enhanced Event-Based Augmented Cognition Approach Using Multiple Sensors

For the event-based approach to work optimally, one needs distinct physiological indicators for each event. One concern is that in operational environments, mere reliance on a single sensor type may not provide sufficient information to detect perception because numerous other events can occur simultaneously and interfere with the signature to be observed. A sensor suite may substantially enhance perception detection and provide more context to drive mitigation than could any single sensor in isolation.

For example, combining EEG and eye-tracking technology (see Chapter 1) may further expand the utility of event-based systems that drive mitigation in real time by providing further context to direct when, what, and how to mitigate. In this setting, eye tracking would be able to evaluate the current focus of visual attention in real time (i.e., determine what event has been perceived, even for events that do not have a behavioral response). EEG/ERP could then be used to evaluate neurophysiological reactions at relevant fixation points to evaluate SA indicators such as successful mental model integration.

Best Practices

Maintain SA by Using Low-Impact Mitigation Strategies

The use of mitigation strategies may, under certain circumstances, result in reduced SA if the costs of the mitigation strategy outweigh the benefits. For example, decluttering an information display by hiding information may be effective in directing attention to a problem area. By decluttering relevant information (i.e., making it less salient), however, an adverse effect on SA may be produced if the situation evolves in a way that suddenly requires the decluttered data.

Yeh and Wickens (2001), studying electronic map displays, found a significant performance decrease when required information was not present or was less salient when needed. Thus, it is important to carefully maintain information integrity and sacrifice it only when absolutely necessary—for example, when lower-impact strategies have failed. This also applies to temporal integrity when deferring information in time in order to escalate other information (e.g., Sequencing/Scheduling; see Table 5.4). Creating a framework that selects and applies dynamic mitigations in a stepwise manner (as described in Section 3.C) minimizes intrusiveness and avoids unnecessary loss of information integrity.

Cognitive Load Assessment

By integrating predictive resource consumption values into an augmented cognition system, designers can be more prescriptive in mitigation strategies, ensuring that additional information (i.e., mitigations) is in a form that will be readily perceivable based on available resources and/or will supersede currently displayed information to ensure perception of most critical events. Various approaches exist to quantify the attentional demands required to complete a task. For example, the W/Index equation, developed by North and Riley (1988), provides a method for predicting attentional demands across various sensory and cognitive channels. Building from this work, the MIDS workload equation (Hale et al., 2005) expands the number of processing channels to more fully account for humans' ability to process multiple sensory inputs across distinct cognitive resources, and incorporates more descriptive resource load values based on Keller (2002).

Generating ERP-Based Cognitive Signatures

EEG signals can be acquired and decoded using sensor arrays ranging from 1 to 256 scalp sites. If a system is being designed for operational deployment, a minimal number of sensors is desirable because each additional sensor adds to the intrusiveness of the headgear, increases the time required for application, and adds to the power consumption of the system. To optimize sensor selection, investigators may utilize a two-step process, first using a multisensor array in a testbed simulation environment and then down-selecting to obtain the minimum number of sensors required to detect desired cognitive states.

In the case of our effort, nine channels, recorded using monopolar configurations (e.g., single sensor sites referenced to linked mastoids), were used to mathematically derive 36 bipolar EEG combination channels. Investigators then determined the optimal subset of these channels for capturing the desired neural signatures, so the sensor array could be optimized to include even fewer electrodes. Limiting the sensors (in our case, to seven) and channels (in our case, six) ensured that the sensor headset could be applied within 10 min.

ERP signature components have been associated with specific regional distributions across the cortex and with predictable timelocking to the onset of the relevant stimuli. In some examples of neural signatures, the distinctive wave shape and magnitude of a specific component such as the P1-N1-P2, the LPC (Late Positive Component; Berka et al., 2004; Berka, Levendowski, Davis et al., 2005), or the N2-P3 may be sufficient to enable accurate detection using a single trial obtained from one scalp site (usually the parietal region). Alternatively, event-based classification may be based on the identification of characteristic complexes of EEG signatures that delineate the events of interest (e.g., correctly identified targets vs. incorrectly identified or missed targets).

Our approach uses EEG power spectral data from EEG epochs associated with events of interest (stimuli or responses) to derive a set of EEG variables that are characteristic of these events. Training data are selected using those event-related EEG signatures that are representative of the event classes of interest. Sample size is important in addressing the problems associated with individual variation in EEG classifiers and the risk of overfitting the model if the training data set is too small. Ideally, a minimum of 20 participants will provide data for the initial training set. If feasible, larger sample sizes of 50 or more provide a better model development set.

Design Guidelines

Mitigation design strategies differ from interface design strategies in that they are adopted into the system under specific conditions, and only as long as these conditions persist, as they have both associated costs and benefits. Although an interface design strategy benefits an environment as a whole and should therefore be implemented permanently, a mitigation design strategy shows only situation-specific advantages and is introduced and removed based on the dynamic situation. Regardless of this distinction, many design principles from human-computer interaction are also applicable to mitigation design and should be considered to optimize their implementation.

Table 5.8 outlines design guidelines for mitigation strategies that influence perception. Guidelines are organized by mitigation stage, as presented in Figure 5.5.

Test Your Knowledge
(a) The following list describes potential system adaptations. Which are mitigation strategies and which are design improvements? Why?

- A haptic direction cue is presented whenever a soldier reaches an intersection.

- Message display is switched from visual to audio when cognitive state indicators register overload. Once switched, the message presentation modality remains auditive for consistency reasons.

- Low-priority, nonanalytic tasks are offloaded to an automated agent when the operator's attention is required. Upon completion of the high-priority tasks, the operator gradually reobtains responsibility for all tasks.

(b) Review your or others' implementation of mitigation strategies. Were they defined appropriately and applied correctly? How could these approaches be improved?

Applied Exercise
Context-sensitive help provides information specific to the condition or mode that an interactive system is in at the time the help is needed (Sukaviriya & Foley, 1990). It may be an appropriate strategy to use when deficits in understanding are detected by cognitive state sensors (e.g., levels of confusion, inconsistent mental model) to support Level 2 SA (comprehension). How could context-sensitive help be implemented as a dynamic mitigation strategy? What other strategies could be implemented to support comprehension?

A MITIGATION FRAMEWORK

Table 5.8. Design Guidelines for Mitigation Strategies That Influence Perception

Stage 1: Optimize Processing Capacity	**Modality Augmentation/Switching**
	If ideal sensory system for information is loaded and perception of event does not occur, switch presentation to alternative modality that has available resources and/or augment information by presenting complementary or redundant information in multiple modalities.
	Visual displays are well suited for specific types of information, such as spatial information, 2-D localization, or abstract or coded data, which require optimized reaction time, represent change over time, represent physical objects, require persistent viewing, or concern absolute quantitative parameters.
	Auditory displays are well suited for specific types of information, such as omnidirectional data, data that must be encoded while in motion, or data that must be dealt with immediately or are simple and short.
	Haptic displays are well suited for temporal and spatial tasks, or for grabbing attention.
	Combining visual, auditory, and haptic information redundantly reduces reaction time and improves task accuracy compared with unimodal information presentation.
	Transposition
	Transpose information to enhance perception when cognitive sensors indicate overload in one processing resource and available capacity in another, and when information to be altered is appropriate for transposition (i.e., information can be changed from verbal to spatial content without loss of context).
	Interrupt ongoing spatial tasks with verbal information (and vice versa) when respective cognitive resources for an ongoing task are loaded.
Stage 2: Direct Attention to Critical Information	**Cueing**
	Present cues in same information format (e.g., spatial) but different modality to facilitate congruent information format-to-modality mappings, while not unnecessarily conflicting with or overloading format/modal resources being used for an ongoing or new task.
	When high intrusiveness is required, visual cues should be placed in foveal vision, whereas low-priority events may be cued by changes in the peripheral or ambient visual environment.
	Add auditory or haptic cues to benefit visual target detection task, especially when gaze shift needed. • Use auditory cues for rapid alerting, indication of direction, location, and movement. • Use haptic cues for alerting and providing a spatial frame of reference.
	Alerting cues should occur in a modality that is most appropriate for the information source type and one that makes the cue dissimilar enough to the previous and current tasks to allow timely detection.
	Sequencing/Scheduling
	Appropriate when multiple time-critical tasks are required; particularly effective when sequenced tasks are similar in nature. Sequencing can involve change of task order or decomposition of tasks and rearrangement of subtasks to optimize use of cognitive capacity across multiple modal channels.
	Scheduling should be determined by priority of task information, with higher-priority task information (ongoing and new) being presented before lower-priority information.
	Intelligent pacing is appropriate during stressful, overloaded, multitasking conditions because users are suboptimal at scheduling when and how long a task should take under such conditions. • Intelligent pacing should not only coordinate timing of task presentation rates during overload but also prioritize and coordinate when and how queued information is later presented to a user.

Table continues on p. 140

Table 5.8 continued

	Decluttering
Stage 3: Optimize Info. Requirements	Particularly effective when task requires user to focus attention within one domain.
	Provide decluttering techniques that do not remove information completely, as decluttered information may be suddenly needed, which may require display reconfiguration resulting in increased response time and total information retrieval times.
	Provide a cue to what information is on hidden screens at all times.

Parting Message

When using augmented cognition systems to improve SA, designers must consider when, what, and how to mitigate. This may be accomplished by following the approach described in this chapter, including identification of key events related to SA, development of distinct cognitive signatures related to each event, and implementation of a mitigation management system to ensure that necessary mitigations are appropriately applied at opportune moments. Given the sheer volume of events in operational settings, however, it may not be possible to develop distinct physiological signatures using one measure for each and every event, because the cognitive processes reflected in the signature may be too similar for certain event types. Practitioners may consider implementation of more than one physiological sensor to gather additional data that can be used to obtain a more complex but, at the same time, more distinct cognitive signature. Evaluating ocular fixations, for example, is a straightforward way to determine attentional focus and differentiate among similar tasks in different areas.

But even if context sensitivity and reliable sensors are implemented, the effectiveness of an event-based augmented cognition system can stand or fall on the design and implementation of mitigation strategies. Looking beyond the previously described costs of mitigation strategies, additional problems may occur with system stability, resulting in user confusion caused by rapid task switching and loss of control (see Chapter 7). Thus, when invoking mitigation strategies, practitioners must consider enter, exit, and transition phases of mitigation strategies and carefully design approaches to eliminate or minimize associated problems. To do so, creative outside-the-box thinking is needed for next-generation augmented cognition systems that may use strategies inspired by areas such as gaming, animation, cinematography, or other domains that rapidly switch context without disrupting SA. The full potential of information displays should be leveraged when augmented cognition is transitioned from the lab to the real world.

Although the methods described in this chapter focused on enhancing Level 1 SA, this event-based augmented cognition approach has potential to evaluate and enhance other cognitive states as well. Some success in evaluating and mitigating Level 2 SA (comprehension; Hale et al., 2006) and target recognition in imagery review (Hale, Fuchs, Axelsson, Baskin, & Jones, 2007) has been reported; however, more research is needed, and more candidate mitigation strategies must be identified that support cognitive integration and interpretation of information.

References

Arrington, C. M., Altmann, E. M., & Carr, T. H. (2003). Tasks of a feather flock together: Similarity effects in task switching. *Memory & Cognition, 31,* 781–789.

Berka, C., Levendowski, D., Davis, G., Lumicao, M., Ramsey, C., Stanney, K., Reeves, L., Harkness, S., & Tremoulet, P. D. (2005, July). *EEG indices distinguish spatial and verbal working memory processing: Implications for real-time monitoring in a closed-loop tactical Tomahawk weapons simulation.* Paper presented at the 1st International Conference on Augmented Cognition. Las Vegas, NV.

Berka, C., Levendowski, D. J., Olmstead, R. E., et al. (2004). Real-time analysis of EEG indices of alertness, cognition and memory with a wireless EEG headset. *International Journal of Human-Computer Interaction, 17*(2), 151–170.

Berka, C., Levendowski, D. J., Ramsey, C. K., Davis, G., Lumicao, M. N., Stanney, K., Reeves, L., Harkness Regli, S., Tremoulet, P. D., & Stibler, K. (2005). Evaluation of an EEG workload model in an Aegis simulation: Biomonitoring for physiological and cognitive performance during military operations. *Proceedings of the International Society for Optical Engineering, 5797,* 90–99.

Dorneich, M., Whitlow, S., Ververs, P. M., Mathan, S., Raj, A., Muth, E., et al. (2004). *DARPA improving warfighter information intake under stress—Augmented cognition: Concept validation experiment (CVE) analysis report for the Honeywell team* (prepared under contract DAAD16-03-C-0054). Arlington, VA: DARPA/IPTO.

Dufresne, A., Martial, O., Ramstein, C., & Mabilleau, P. (1996). Sound, space, and metaphor: Multimodal access to Windows for blind users. In *Proceedings of ICAD '96: 3rd International Conference on Auditory Display* (pp. 51–58). Palo Alto, CA: ICAD.

Endsley, M. R. (1995). Measurement of situation awareness in dynamic systems. *Human Factors, 37*(1), 65–84.

Endsley, M. R. (1999). Situation awareness and human error: Designing to support human performance. In *Proceedings of the High Consequence Systems Surety Conference,* Albuquerque, NM: Sandia National Laboratories.

Endsley, M. R. (2000). Theoretical underpinnings of situation awareness: A critical review. In M. R. Endsley & D. J. Garland (Eds.), *Situation awareness analysis and measurement* (pp. 3–32). Mahwah, NJ: Erlbaum.

Endsley, M. R., & Garland, D. J. (2000). *Situation awareness analysis and measurement.* Mahwah, NJ: Erlbaum.

Fuchs, S., Hale, K. S., Berka, C., Levendowski, D., & Juhnke, J. (2006). Physiological sensors cannot effectively drive system mitigation alone. In D. D. Schmorrow, K. M. Stanney, & L. M. Reeves (Eds.), *Foundations of augmented cognition* (2nd ed., pp. 193–200). Arlington, VA: Strategic Analysis, Inc.

Fuchs, S., Hale, K. S., Stanney, K. M., Juhnke, J., & Schmorrow, D. (2007). Enhancing mitigation in augmented cognition. *Journal of Cognitive Engineering and Decision Making, 1,* 309–326.

Hale, K. S., Fuchs, S., Axelsson, P., Baskin, A., & Jones, D. (2007). Determining gaze parameters to guide EEG/ERP evaluation of imagery analysis. In D. D. Schmorrow, D. M. Nicholson, J. M. Drexler, & L. M. Reeves (Eds.), *Foundations of augmented cognition* (4th ed., pp. 33–40). Arlington, VA: Strategic Analysis, Inc.

Hale, K. S., Fuchs, S., Axelsson, P., Baskin, A., Jones, D., Berka, C., & Juhnke, J. (2006). *Information delivery and display for shared awareness in the net-centric battlespace* (SBIR Phase I Final Report, Contract No. W31P4Q-06-C-0041). Arlington, VA: DARPA/IXO.

Hale, K., Reeves, L., Samman, S., Axelsson, P., & Stanney, K. (2005). Validation of predictive workload component of the multimodal information design support (MIDS) system. In *Proceedings of the Human Factors and Ergonomics Society 49th Annual Meeting* (pp. 1162–1166). Santa Monica, CA: Human Factors and Ergonomics Society.

Jones, D. G., & Endsley, M. R. (1996). Sources of situation awareness errors in aviation. *Aviation, Space and Environmental Medicine, 67*, 507–512.

Keller, J. (2002). Human performance modeling for discrete-event simulation: Workload. In E. Yücesan, C. H. Chen, J. L. Snowdon, & J. M. Charnes (Eds.), *Proceedings of the 2002 Winter Simulation Conference* (pp. 67–75). Piscataway, NJ: Institute of Electrical and Electronics Engineers.

Kroft, P., & Wickens, C. D. (2001). *The display of multiple geographical data bases: Implications of visual attention* (Report No. ARL-01-2/NASA-01-2). Savoy, IL: University of Illinois, Aviation Research Laboratory. Retrieved March 21, 2006, from http://www.humanfactors.uiuc.edu/Reports&PapersPDFs/TechReport/01-02.pdf

Latorella, K. (1996), Investigating interruptions: An example from the flightdeck. In *Proceedings of the Human Factors and Ergonomics Society 40th Annual Meeting* (pp. 249–253). Santa Monica, CA: Human Factors and Ergonomics Society.

Matthews, M. D., & Beal, S. A. (2002). *Assessing situation awareness in field training exercises* (U.S. Army Research Institute for the Behavioral and Social Sciences Research Report 1795. September 2002). Retrieved February 11, 2006, from http://www.hqda.army.mil/ari/pdf/rr1795.pdf

Mejdal, S., McCauley, M. E., & Beringer, D. B. (2001). *Human factors design guidelines for multifunction displays* (Report No. DOT/FAA/AM-01/17). Retrieved January 31, 2007, from http://www.hf.faa.gov/docs/508/docs/cami/0117.pdf

North, R. A., & Riley, V. (1988). W/Index: A predictive model of operator workload. In G. McMillan et al. (Eds.), *Applications of human performance models to systems design* (pp. 81–89). New York: Plenum Press.

Posner, M. I. (1980). Orienting of attention. *Quarterly Journal of Experimental Psychology, 32*, 3–25.

Reeves, L. M. (2007). *Optimizing the design of multimodal user interfaces.* Unpublished dissertation. University of Central Florida.

Rupert, A. (1997, March). Which way is down? *Naval Aviation News*, 16–17.

Stanney, K., & Reeves, L. (2005, March). *Mitigation strategies and performance effects.* White paper outbrief from a working session at Improving Warfighter Information Intake Under Stress, AugCog PI Meeting, Chantilly, VA.

Stanney, K., Samman, S., Reeves, L., Hale, K., Buff, W., Bowers, C., Goldiez, B., Nicholson, D., & Lackey, S. (2004). A paradigm shift in interactive computing: Deriving multimodal design principles from behavioral and neurological foundations. *International Journal of Human-Computer Interaction, 17*(2), 229–257.

Sukaviriya, P., & Foley, J. D. (1990). Coupling a UI-framework with automatic generation of context-sensitive animated help. In *Proceedings of the 3rd Annual ACM SIGGRAPH Symposium on User Interface Software and Technology* (pp. 152–166). New York: ACM Press.

Taylor, R. M. (1990). Situational Awareness Rating Technique (SART): The development of a tool for aircrew systems design. *Situation Awareness in Aerospace Operations* (AGARD-CP-478, pp. 3/1–3/17). Neuilly-sur-Seine, France: NATO-AGARD.

Tremoulet, P., Barton, P., Craven, P., Corrado, C., Mayer, G., Stibler, K., et al. (2005). *DARPA improving warfighter information intake under stress—Augmented cognition phase 3 concept validation experiment (CVE) analysis report for the Lockheed-Martin ATL Team* (Prepared under contract No. NBCH030032). Arlington, VA: DARPA/IPTO.

U.S. Army Research Laboratory. (2005). *Improved Performance Research Integration Tool (IMPRINT), User's guide, version 7.* Boulder, CO: Micro Analysis and Design, Inc.

Van Orden, K. F., Viirre, E., & Kobus, D. A. (2007). Augmenting task-centered design with operator state assessment technologies. In D. D. Schmorrow & L. M. Reeves (Eds.), *Foundations of augmented cognition* (3rd ed., Lecture Notes in Computer Science 4565, pp. 212–219). Berlin: Springer.

Wickens, C. D. (2002). Multiple resources and performance prediction. *Theoretical Issues in Ergonomics Science, 3*(2), 159–177.

Yeh, M., & Wickens, C. D. (2001). Attentional filtering in the design of electronic map displays: A comparison of color coding, intensity coding, and decluttering techniques. *Human Factors, 43*, 543–562.

Chapter 6

Methodology, Methods, and Metrics for Testing and Evaluating Augmented Cognition Systems

Frank L. Greitzer
Pacific Northwest National Laboratories

This chapter reviews methodologies and metrics that may be used to assess the impact of augmented cognition mitigations in enhancing decision making and training effectiveness.

Introduction

The augmented cognition community seeks cognitive neuroscience–based solutions to improve warfighter performance by applying and managing mitigation strategies to reduce workload and improve the throughput and quality of decisions. The focus of augmented cognition mitigation approaches is to define, demonstrate, and exploit neuroscience and behavioral measures that support inferences about the warfighter's cognitive state that prescribe the nature and timing of mitigation.

A challenge to the successful application of augmented cognition is to develop valid evaluation methodologies, metrics, and measures to assess the impact of augmented cognition mitigations. Two considerations are *external validity*, which is the extent to which the results apply to operational contexts, and *internal validity*, which reflects the reliability of performance measures and the conclusions based on analysis of the results. The scientific rigor of the research methodology employed in conducting empirical investigations largely affects the validity of the findings. External validity requirements also compel us to demonstrate **operational significance of mitigations**. Thus it is important to demonstrate the effectiveness of mitigations under specific conditions.

In this chapter, I review cognitive science and methodological considerations in the use of human performance metrics and analysis methods to assess the impact of augmented cognition mitigations.

Scenario

For amusement, and perhaps to help ground this chapter's discussion about our discipline in some historical roots, consider the following quotations from some of the great minds that have contributed directly or indirectly to our field of inquiry (with added narration to tie the asynchronous threads of ideas together):

> And how will you enquire, Socrates, into that which you do not know? What will you put forth as the subject of enquiry? And if you find what you want, how will you ever know that this is the thing which you did not know? (Plato, 380 B.C.)

This question, taken from Plato's dialogues (Plato, 380 BC) over 2,000 years ago, relates to an age-old problem dealing with consciousness, awareness, or what is most generally referred to today as *metacognition*—that is, in our research toward reasoning about or solving real-world problems that are not well specified or for which there is a lack of "ground truth," how can we assess the status of our efforts? Is there a window into awareness of one's cognitive state? In its focus on brain mechanisms of cognition, augmented cognition research aims to shed light on such metacognitive elements, including perhaps (even unconscious) correlates of mental workload/overload, confusion, or error. Interest in the mind, or cognition and cognitive state as they would be called today, was at the forefront of a nascent field of psychology near the end of the 19th century, as illustrated by a quotation from William James:

> Mental phenomena are not only conditioned *a parte ante* by bodily processes; but they lead to them *a parte post*.... Mental states occasion also changes in the caliber of blood-vessels, or alteration in the heart-beats, or processes more subtle still, in glands and viscera.... Our psychology must therefore take account not only of the conditions antecedent to mental states, but of their resultant consequences as well. (James, 1890, p. 5)

Here James anticipates the augmented cognition research aimed at the development of cognitive state assessors. His description of experimental methods—in particular, reaction-time studies ("ascertainment of the time occupied by nervous events," p. 85)—relates to the early work of James's contemporaries (e.g., Donders, 1969; Wundt, 1880), who used reaction times to infer the speed of higher mental processes, and the more recent work of Sternberg (1966, 1969) and Posner (1978). Posner's research related both spatial and temporal characteristics of neural activity with mental processes. William James established the functionalist discipline of psychology that is concerned with the study of the mind—the foundations of cognitive psychology—and was a participant in fostering the emergence of the new field of experimental psychology, with its focus on learning, sensation, and perception and it application of experimental methods, measurement, and tests.

Half a century later, a visionary in an emerging field of computer science introduced new concepts to inspire this new field as well as to suggest scenarios that, at the time, could only be construed as science fiction:

> We know that when the eye sees, all the consequent information is transmitted to the brain by means of electrical vibrations in the channel of the optic nerve. This is an exact analogy with the electrical vibrations which occur in the cable of a television set: they convey the picture from the photocells which see it to the radio transmitter from which it is broadcast. We know further that if we can approach that cable with the proper instruments, we do not need to touch it; we can pick up those vibrations by electrical induction and thus discover and reproduce the scene which is being transmitted, just as a telephone wire may be tapped for its message. (Vannevar Bush, 1945, p. 108)

In this 1945 article, Bush famously described a modern vision of computer systems with associative organization and access to knowledge via his Memex concept. But in the foregoing excerpt he anticipated the notion of developing brain-computer interfaces that remind us of the thrust in the augmented cognition research program aimed at medical implant technologies to restore lost functionality because of neurological trauma, debilitating diseases, or other physical impairment.

Nowhere is the early conceptualization of human-machine symbiosis better articulated than in the 1960 paper by J. C. R. Licklider:

> The hope is that in not too many years, human brains and computing machines will be coupled together very tightly, and that the resulting partnership will think as no human brain has ever thought and process data in a way not approached by the information-handling machines we know today. (Licklider, 1960, p. 5)

In the nearly half century that followed the publication of Licklider's vision, with most researchers pursuing topics in human factors engineering/human-system interaction research for display, control, and interaction design, much progress was made in human factors, user-centered design, and experimental design/research methodology. But the vision of symbiosis was largely left dormant. With major advances in the last decade in the fields of neuroscience and computer science, the augmented cognition community is on the threshold of resurrecting Licklider's vision of symbiosis through the exploration of technologies to acquire, measure, and validate neurological cognitive state sensors that will inform mitigation management systems about when and how to augment performance.

When viewed from a distinct perspective of experimental methodology and cognitive science, the field of augmented cognition presents critical challenges in addressing basic practical questions reflected in or underlying the quotations provided earlier—challenges to provide scientifically valid, operationally significant demonstrations of performance enhancement through the application of augmented cognition technologies that have been developed, tested, and validated within a theoretical and methodological framework that guides the enterprise.

General Approach

Two general augmented cognition R&D areas that require assessments of the impact of mitigations are **decision making** and **training**. (More broadly, we are referring to *information processing* and *decision making,* but we shall use the term *decision making* as a shorthand reference to these cognitive processes, as presumably the end result is a decision.)

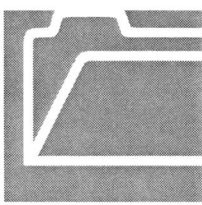

Valuable Information

For decision-making research, the term *mitigation strategies* refers to the *type* of mitigation, when to apply it, and when to withdraw it (see chapter 5). A mitigation that always improves performance does not meet the goals of adaptive mitigation and would more appropriately be considered a design change. The objective is to define and exploit neuroscience and behavioral measures to enable inferences about cognitive state that prescribe mitigation strategies to improve performance.

For research aimed at improving training, strategies are examined for their efficacy in enhancing understanding and speeding the learning process. Here the challenge is to identify neurophysiological signatures that identify propitious opportunities to apply targeted training mitigations to accelerate the transitions to higher learning stages corresponding to increasing levels of expertise.

Augmented cognition applications for both decision-making and training tasks share a common requirement to assess the cognitive state of the operator and to measure the impact of mitigations. In decision making, the goal is to identify opportunities for mitigation to relieve information-processing bottlenecks and improve performance. In training situations, the goal is to identify opportunities to accelerate learning, for instance by applying adaptive learning mitigations at appropriate times.

Underlying each of these challenges is the need to define cognitive state gauges that signal such opportunities (see Chapters 1 and 2). Assessment of cognitive load serves as a basis for such mitigation management decisions. *Cognitive load* refers to the amount of mental effort exerted in performing a task. Much of the research conducted in augmented cognition has focused on correlating hypothesized neurophysiological measures of cognitive load with workload ratings to validate augmented cognition mitigation management strategies.

In addition to establishing an inventory of validated measures, the augmented cognition community must develop and demonstrate effective **mitigation strategies** (see Chapter 5). Here the challenge is to develop cognitive interventions and support functions for information processing, decision making, and training and to **demonstrate significant impact of such mitigations on performance.**

To address these challenges, this chapter is organized as follows: The first topic is a description of approaches for testing and validating measures of cognitive load, which is a prerequisite for any application of mitigation strategies. The next topic concerns methodological issues surrounding the testing and validation of mitigation strategies. This includes some additional considerations about augmented cognition approaches for enhanced training. The chapter concludes with a discussion of lessons learned, best practices, and recommendations or guidelines for advancing the mitigation management goals of augmented cognition.

Cognitive Load Measurement

To effectively apply augmented cognition technology, one must understand the relationships among **measures of workload, measures of physiological processes, and measures of performance.** Appropriate use of these measures in isolation and in concert will enable the development of more robust metrics for managing augmented cognition mitigation (see Chapter 7). Three general approaches to assessing cognitive load are the assignment of workload indices, behavioral performance measures, and neurophysiological measures.

Workload indices. Augmented cognition practitioners have employed NASA's Task Load Index (TLX; Hart & Staveland, 1988) and Design Interactive's Multimodal Information Design Support (MIDS) tool (Hale, Axelsson, Samman, & Stanney, 2005; also see Chapter 5) to obtain measures of workload. The NASA TLX is typically derived from user ratings (subjective measures). The MIDS tool, which is usually applied to the tasks or components of tasks derived from cognitive task analysis, is used to develop workload measures based on *expert* ratings. Based on the W/INDEX predictive workload equation—North and Riley's (1988) and Wickens's (1992) Multiple Resource Theory—the MIDS tool is applied by human factors experts to predict a workload value by summing additive workload effects across attentional channels, adding a penalty attributable to demand conflicts within channels, and adding a penalty arising from demand conflicts between channels. The MIDS measure has been shown to correlate with subjective workload measures (Hale et al., 2005) and is useful in predicting cognitive workload for experimental tasks on a second-by-second basis to correlate and validate neurophysiological workload measures (e.g., Polythress et al., 2006).

Behavioral correlates of workload. Behavioral performance measures may be used to assess cognitive demands or study the effects of varying workloads. One way to study the effects of task demand or difficulty is to attempt to control it as an experimental variable—that is, compare the effects of proposed mitigations under conditions of high workload (or difficulty level) versus low workload. Another method to experimentally study and manipulate workload is the dual-task experiment. This method, which has been used for decades, introduces a secondary task that must be performed concurrently with a primary task. Inspired in part by Kahneman's (1973) central capacity theory, which holds that cognitive resources are limited by the resources or capacity of a central processor that can be used flexibly across numerous activities, dual-task and divided-attention tasks are studied under a variety of conditions, such as task similarity, task difficulty, and amount of practice. Performance of participants on the secondary task is indicative of the amount of available cognitive resources while performing the primary task. Thus, if one compares performance on a secondary task under varying (difficulty) conditions of a primary task, differing levels of performance on the secondary task are inversely related to the difficulty or processing load of the primary task conditions.

Research by Wickens (1984) on dual-task performance showing differential effects of secondary tasks depending on factors such as modality led to his Multiple Resource Theory, which assumes the existence of different pools of resources. If two tasks make use of different pools of resources, then people should be able to perform both tasks without disruption (see Chapter 5).

Both task load/difficulty manipulation and dual-task experiments have been employed in augmented cognition applications. St. John et al. (2004) described the Warship Commander Task (WCT), developed by Pacific Science & Engineering Group, which was used as a foundation for studying and validating cognitive state sensors (see Chapter 9). Task difficulty could be experimentally manipulated by varying the quantity and type of targets with which the participant had to deal. The WCT—which combines perceptual, motor, spatial, auditory, verbal, memory, and decision-making processing—also provides a means of studying and evaluating cognitive-state gauges in the context of dual-task or multiple-task performance.

Dual-task methodologies have been employed in a number of augmented cognition experiments. Berka et al. (2004) described experiments using the WCT and several other secondary tasks to examine the effects of cognitive workload on EEG measures.

Neurophysiological correlates of workload. One reason there is great interest in developing and validating neurophysiological correlates of workload is the potential for such measures to be computed in real time, whereby they could provide a mechanism for managing and directing mitigation options when and where necessary. Research has linked cognitive workload with physiological and neurological activity such as galvanic skin response (GSR), pupil dilation, heart rate (electrocardiogram, or EKG), electrical activity in the brain (electroencephalogram, or EEG), and cortical blood flow (functional near infrared, or fNIR).

These measures are difficult to validate, for a variety of reasons. Electrical activity in the brain, specifically event-related potentials (ERPs), represent signals that are difficult to detect because of their low signal-to-noise ratios: ERPs range in amplitude from 1 to 10 microvolts, whereas background EEG activity ranges from 10 to 100 microvolts (this activity is easily obscured by common events such as eye blinks). Disambiguating these signals under such conditions and in properly controlled settings is a difficult current research topic.

An overview of early results is provided by St. John et al. (2004). Craven et al. (2006) reported on more recent work on real-time classification methods for cognitive workload measures using EEG, heart rate, GSR, and fNIR sensors (see Chapters 1 and 2).

Test Your Knowledge
1. Describe some distinctions between the NASA TLX and the MIDS tool for workload assessment.
2. What is the basic idea behind the use of dual-task performance to assess cognitive load?

Challenges for development and validation of cognitive load metrics. It is often difficult to operationally define task difficulty, especially in experimental research using realistic tasks. Most of the experimental investigations within the augmented cognition field have focused on cognitive behavior in the areas of attention, perceptual processes, and working memory, exemplified by the WCT (see Chapter 9), imagery analysis (see Chapter 5), and simulated tasks such as piloting fixed- or rotary-wing aircraft (see Chapter 3). For such tasks, operational definitions of workload and task difficulty are more readily and unambiguously obtained than for tasks characterized by more complex cognitive processes, such as information analysis, synthesis, and reasoning. Such complex decision making behavior was the topic of the 2005 film *The Future of Augmented Cognition* (Singer, 2005), a film that embraces a vision of the future through a scenario describing the augmentation of intelligence analysis.

Greitzer (2005a) and Greitzer and Griffith (2006) argued for the application of augmented cognition to more complex cognitive tasks, such as intelligence analysis. Here it is more challenging to operationally define task difficulty and therefore also to experimentally control difficulty and cognitive load. More focused research aimed at identifying neural and physiological correlates of cognitive load for these complex decision-making tasks remains a critical need for the field of augmented cognition.

Interpretation of neural correlates of workload is further complicated by possible effects of operator overload on cognitive state measures. Craven et al. (2006) pointed out the potential confusion between neural correlates of low workload, high workload, and "a third state referred to as *overloaded*" (p. 72). Data obtained in digit span tasks, for example, suggested that neurological signals for an overloaded participant might be difficult to distinguish from those obtained from participants experiencing low cognitive effort. As participants become overloaded, they may "give up" at the highest task-difficulty levels, and their neural signatures may be similar to those of participants working on a task requiring a low cognitive effort.

Challenges such as those just described—discriminating cognitive load using neural signatures, controlling cognitive load, and disambiguating neurophysiological signatures of low workload and "overload"—strongly suggest that *convergent* methods of cognitive load measurement are needed that integrate behavioral data, task/context data, and a mix of physiological and neural sensors (Craven et al., 2006; Fuchs et al., 2006; also see Chapter 5).

Review: Cognitive Load Measurement
Three approaches to cognitive load measurement have been described: workload indices, neurophysiological correlates of workload, and behavioral correlates of workload. The recently developed MIDS tool uses current theory and practice to derive useful cognitive load measures that correlate with subjective workload measures and that are useful for predicting cognitive workload for experimental tasks and for validating neurophysiological workload measures. Recent promising neuroscience research has led to real-time classification methods for measuring cognitive workload using EEG, heart rate, GSR, and fNIR sensors. Performance/behavioral-based measures of workload have been developed based on research on attention and

multiple resource theory, with an emphasis in current practice on the use of dual- or multiple-task performance experiments to study cognitive load. Of particular interest to the augmented cognition program is the development of neurophysiological measures of cognitive load that may be applied or related specifically to the development of expertise or stages of learning.

This research is critical to the development of methods to assess the impact of decision aids and to help manage the application of training mitigation strategies. Research challenges for the identification and validation of neural signatures for cognitive load include deriving more fine-tuned distinctions in neural signatures of cognitive load, developing experimental methods that provide more control of cognitive load, and separating effects of training and skill acquisition. A comprehensive approach that seeks to define convergent methods of cognitive load measurement—integrating results from studies involving workload indices, performance/behavioral-based measures, and neurophysiological data monitoring—appears to be the most fruitful plan for advancing the science and practice in augmented cognition cognitive state assessment.

Testing and Validation of Mitigation Strategies

As cognitive state gauges are validated, more attention will focus on developing effective schemes for mitigation management based on cognitive principles and proving that augmented cognition mitigations yield the hypothesized performance improvement. The challenge in effective mitigation management is to determine when mitigation is needed and to select the most appropriate form of mitigation to produce positive results (see Chapter 5). As stated in the introduction to this chapter, a device or manipulation that always improves performance may be more appropriately termed a design change than a mitigation strategy. Augmented cognition practitioners must exploit neuroscience and behavioral measures of cognitive state to reach decisions about when mitigation is needed and then recognize and apply effective mitigation strategies that address cognitive state deficiencies or bottlenecks to improve performance. This process is referred to as *mitigation management*.

Mitigation management. To a great extent, especially in the early phases of augmented cognition research, mitigation strategies have been derived from the notion of adaptive user interfaces (see Chapter 5). As Fuchs et al. (2006, p. 193) observed, mitigation strategies "are generally context-insensitive (i.e., the 'mitigation' is advantageous no matter if a problem is present or not) and often resemble general design improvements rather than dynamic mitigation of cognitive state problems." In a related vein, initial augmented cognition research on neural signatures was largely tied to more general arousal measures, as opposed to more specific correlates. Fuchs et al. (see Chapter 5) also pointed out that such task and context insensitivity may contribute to confusing or disrupting the user when the mitigation trigger is based solely on cognitive state, without accounting for task state. Stanney and Reeves (2005) referred to this as *inconsiderate* application of mitigation strategies. Thus, a major challenge is to perform the neuroscience and behavioral research to derive quantitative measures of operator workload that take cognitive state and context-sensitive task status into account to enable management of when, what, and how to apply mitigation support.

Specific strategies for selecting mitigations have been articulated by Fuchs et al. (see Chapter 5) in a multistage mitigation selection framework that specifies interventions that are gauged to the amount of overload. Figure 5.5 represents this concept. When there is evidence that cognitive resources are available, mitigation strategies should focus on optimizing processing capacity without unduly distracting the operator. More intrusive mitigations are prescribed as cognitive resources become overloaded, using staged levels of mitigation from modality augmentation, to cueing, to interventions that control pacing or sequencing of tasks, to decluttering. At the highest stage, task sharing (collaboration) is suggested.

It is one thing to know what interventions are needed to address problems with cognitive load, and quite another to provide a convincing demonstration of enhanced performance. The reason is that there are many obstacles and pitfalls to be overcome in experimental methodology when dealing with the kinds of relevant, real-world tasks that are the focus of the community of augmented cognition practitioners.

Valuable Information: Practitioner's Skill Set
The augmented cognition practitioner must understand how to apply neuroscience and behavioral measures of operator workload and cognitive state, how to recognize and apply appropriate cognitive context-sensitive task management strategies, and how to experimentally demonstrate their effects on performance.

Research methodology issues. When complex real-world tasks are studied, methodological challenges must be understood and addressed. With real-world or highly realistic tasks that are used to gain external validity, methodological challenges that threaten internal validity are often encountered. For example, in these conditions it is often difficult to design and carry out a controlled experiment with rigorous experimental and control groups. Risks to external and internal validity are highlighted in the questions exhibited in Table 6.1.

A major challenge is the ability to control independent variables—especially task difficulty—when dealing with the realities of small sample sizes (few experimental participants) that lead to within-subjects designs in which the experimental variable is manipulated for each participant. Given that under these conditions it is not possible to use the same problem or task under varying experimental conditions (because the participant will already have seen the problem), it becomes necessary to devise ways to measure the difficulty of tasks in order to control that variable. Greitzer (2005b) described methodological and measurement problems associated with experimentally manipulating cognitively complex intelligence-analysis tasks with relevant real-world problems. Other challenges, shown in Table 6.1, include the need for random sampling of participants to ensure that results will generalize to the population of interest, and the use of appropriate methodological controls to overcome extraneous factors such as learning and order effects.

Table 6.1. Questions About Experimental Validity

- *Are selected tasks representative of operational tasks?* The more similar the experimental tasks are to operations in the field, the greater the external validity of the results. In the design of experiments there is always a trade-off between the desire to maintain high external validity by using real-world tasks and the desire to maximize internal validity by controlling extraneous variables that may confound the results.

- *Are selected experimental and control conditions comparable in difficulty?* It is often the case that tasks used in experimental and control conditions are identical, so that the only difference in task requirements is attributable to the experimental manipulation. However, this is not always possible. In such cases, there is a need to control for important factors such as task difficulty.

- *Are the tasks sufficiently understood to enable evaluation of the quality of the product?* In most practical, laboratory-based research, tasks are sufficiently defined to enable a "ground truth" assessment of the right answer. However, in complex and problem-solving tasks that are typical of real-world decision making and analysis, this is not always the case. A good example is the open-ended sort of problem solving and prediction characteristics of most intelligence analysis tasks.

- *Are there enough participants for the study?* This is a common problem in experiments that use experts or operational personnel as participants or that involve the use of neural and other physiological sensor data collection. A number of studies published in the augmented cognition literature have used caveats about the interpretation of results pending additional data that should be obtained from larger samples of individuals.

- *Are the participants selected randomly?* A nonrandom sample reduces the external validity of a study—the degree to which the results generalize to the target population. As a practical matter, random selection is not always an option. For many laboratory tasks, student populations are used because of the greater availability of these research participants. A study with a nonrandom sample is not fatally flawed; it still can identify facts about the sample (which may or may not apply to the population). The best method for obtaining a representative sample is to use random sampling.

- *Do the participants represent the broader population of operational personnel?* If there is concern (e.g., because of lack of random sampling), the study plan may be improved by including testing in both the laboratory and the field (with operational personnel) to verify that results obtained in the laboratory also reflect the target population.

- *Do the tasks and associated experimental/control manipulations properly represent the operational context (external validity) and address the research hypotheses (internal validity)?* Be sure that the tasks studied represent expected field operations and that the manipulations have operational significance.

- *Does the design of the study account for or control learning effects?* As participants gain experience with the task, their performance will naturally improve. Time (or amount of practice) is a factor that a properly controlled experiment should address. One way to address learning effects is to randomize the order of experimental conditions. Another way is to experimentally manipulate and vary the sequence of conditions. A third method, especially if learning effects are not of interest, is to train participants well beyond initial exposure to the tasks until their performance reaches a stable, or *asymptotic,* level.

- *Does the design of the study account for or control order effects?* That is, we need to make sure that the manipulations or interventions introduced in the study do not all occur in the same order for all participants; this is related to the learning effect issue addressed earlier.

Beyond the methodological issues relating to sampling and control of independent variables, practitioners must be well versed in the use of dependent variables (performance measures) that are critical in demonstrating the impact of their technology. Performance measures are discussed next.

Behavioral/performance metrics. The goal of applying augmented cognition to complex decision-making tasks compels us to develop evaluation methodologies and performance measures that more directly reflect cognitive processes underlying these activities. A body of research in cognition leads to a well-studied collection of performance measures and analytic approaches that may be profitably employed or better exploited by augmented cognition practitioners. These include temporal measures such as latencies or durations of tasks and performance quality, or proficiency measures such as accuracy (e.g., percentage correct). Such performance metrics may be qualitative or quantitative. User satisfaction ratings are examples of qualitative measures that reflect a necessary, but not sufficient, requirement for a tool to be successfully deployed. Measures based on product or output quality are needed to address questions about the effectiveness of a tool.

Quantitative measures of proficiency and/or temporal factors represent current practice in the use of behavioral/performance metrics. To assess the effectiveness of a mitigation strategy, one must examine the impact of mitigation on specific aspects of a task: What aspects or phases of the operational task are most affected by the mitigation? How does the mitigation affect this performance? These effects should be operationally defined in measures such as time taken to perform an activity, accuracy of the result, and completeness of the result.

An excellent example of quantitative measures of effectiveness is provided by the experiments on neurotechnology for image analysis conducted by Mathan and colleagues (Mathan, Whitlow, et al., 2006; Mathan, Ververs, et al., 2006). These experiments examined the use of EEG signals (event-related potentials, or ERPs) as part of a triage system when recorded in conjunction with rapid serial visual presentation (RSVP) of images. In these studies, a clear pattern of spatiotemporal activity was discernible starting at about 150 ms after the onset of the stimulus. Using these differences to drive a triage system that cued human image analysts to examine specific subsets of the overall image, the RSVP study by Mathan, Ververs, et al. (2006) showed a five-fold, statistically significant reduction in the time required to detect targets at high accuracy levels for the neurophysiologically driven image triage method, compared with conventional image analysis techniques.

In its basic embodiment, this study used one independent variable and two dependent variables: The independent variable was type of image analysis/triage method (the experimental condition was the RSVP neurophysiologically driven image triage and the control condition was conventional image analysis); the dependent variables were accuracy (percentage correct detections) and time to find all targets. Stimulus materials obtained from the National Geospatial-Intelligence Agency (NGA) were representative of real-world tasks. This study also represents current best practices in terms of experimental methodology and validity. In particular (comparing with criteria listed in Table 6.1), note the following:

- The task studied (image analysis) is representative of operational tasks for the target population of image analysts.
- Experimental and control conditions (the image analysis problem) were tied to the same imagery data and were therefore comparable in difficulty.
- The tasks studied were very close to real-world tasks for the image analysis problem.
- The number of participants in the study was sufficient (17 participants were studied in the baseline experiment and 6 participants served in the main RSVP study).
- The participants were engineers (baseline study) or graduate students (RSVP study), which potentially decreases the external validity of the study to the extent that the participants' background, experience, and capabilities were not representative of the target population (image analysts). Plans for a follow-up study using trained image analysts were discussed.
- The tasks and experimental conditions faithfully represented the appropriate operational context, in that the RSVP-based triage methodology was the same methodology hypothesized to improve image analysis performance, and the non-RSVP control condition of performing the analysis on a complete image corresponds precisely to the current image analysis method. Varying target images were used in the study, including golf courses (which are not particularly representative targets) and an oil tanker or oil storage depots (which are more representative).
- The design of the study was carefully crafted to control for learning effects. For example, participants first were given a training phase in which to become familiar with the target-detection/image-analysis task and the RSVP procedure. A performance phase followed in which the experimental manipulations occurred and performance data were collected for use in statistical analyses.
- The study was designed to control for order effects using the customary method of randomizing the order in which stimuli were presented (selecting target images at random from a set); that is, it ensured that participants did not receive the various stimulus conditions in the same order.

Finally, the study was exemplary in its choice of dependent variables, which included both accuracy and speed (completion time) measures and had a sufficient number of trials to permit the computation of detection performance (such as hit and false positive rates, as described in Table 6.2, p. 157).

Test Your Knowledge

Wilson and Russell (2006) described an experiment on adaptive aiding based on operator functional state (OFS) as measured by physiological sensor readings. An artificial neural network was used to determine when a participant was cognitively overloaded, and adaptive aiding was managed using two conditions: only when the OFS (a) indicated a high mental workload or (b) was initiated at the first instance of high mental workload and continued through the conclusion of the task.

A simulated uninhabited air vehicle (UAV) task was used as the experimental task, in which participants monitored the progress of four UAVs in pre-planned bombing missions. Participants had to perform visual searches of downloaded target images to locate and mark six targets before the vehicle reached the weapons release point. If the command was not given before the target reached the weapons release point, the weapons could not be released.

Two levels of image complexity were used; the more difficult images contained more distractors and required more complex target priority decisions. Presentation order for easy and difficult images was balanced to avoid sequence effects.

The mitigation applied was to decrease the speed of the vehicle by 50% when overload was indicated. Among the results obtained was that both aiding conditions (applying/removing aiding when the physiologically determined OFS warranted it, versus leave-on aiding) yielded roughly the same performance, improving the objective performance measure of number of weapons released by 42%–50% over the unaided control condition. Both aiding conditions were characterized as having aiding turned on for roughly the same amount of time (48%–52%), even though the physiologically controlled aiding condition had many more, and shorter episodes of aiding compared with the leave-on aiding condition.

1. Given that the aiding condition was speed-related (i.e., control of the UAV's speed), can you think of another control condition that might have been interesting to include in the study?[1]

2. With vehicle slowing as the mechanism for mitigation, there is a trade-off between the employment of aiding and the amount of time available to accomplish the task. The more time that aiding is on, presumably the less time is available for the participant to get the UAV over its target and perform the image recognition task required to release the weapon over the target. In principle, then, slowing of the vehicle could have a negative effect on performance under certain conditions. More complex forms of mitigation may yield different results. What are some other possibilities?[2]

Other Approaches

Signal detection theory. A large body of human performance research can be tapped to extend the analytic methods used in augmented cognition research: In particular, signal detection theory (Green & Swets, 1966) may be profitably applied to tasks involving detection and classification. As will be seen, this applies not only to human performance in detection or classification tasks that are to be enhanced but also to the performance of augmented cognition system components (such as cognitive state assessors) that are expected to mediate mitigation functions.

A mathematical framework for describing and studying decisions that are made in uncertain situations, signal detection theory is well suited for assessing and comparing the performance of human operators under differing conditions. Detection, classification, memory, and even decision-making (e.g., diagnostic) performance can be de-

---

[1] One idea: Add a condition in which the participant is able to control the speed of the UAV; that is, slow the vehicle when perceived workload is high and accelerate it when workload is low.

[2] Consider designing mitigation functions that support the operator in ways other than slowing the vehicles; for example, support for prioritizing or image recognition for the targets (this is speculative, given that the article does not provide sufficient detail about the targeting task to determine the feasibility of this type of mitigation).

scribed in terms of four performance scores or probabilities, as shown in a table that is sometimes referred to as a *confusion matrix*. The rows correspond to the participant's response (e.g., "signal present" versus "signal not present"). The columns reflect the true state (SIGNAL versus NO-SIGNAL). Probabilities or proportions of the participant's responses in each cell of the table are referred to using labels such as "true positive" (response = signal present when true state = SIGNAL); "false positive" (response = signal present when true state = NO-SIGNAL); and so on. (see Table 6.2 for definitions of terms).

From data collected based on participant yes/no responses and/or confidence ratings about the presence or absence of the signal, analyses derived from signal detection theory may be employed to discern levels of proficiency or expertise. Most relevant are the probabilities of true-positive responses and false-positive responses, often called *hits* and *false alarms*, respectively.

Table 6.2. Terms in a Confusion Matrix

		True State	
		SIGNAL	NO-SIGNAL
Participant's Response	"Signal present"	True Positive "Hit"	False Positive "False Alarm"
	"Signal not present"	False Negative "Miss"	True Negative "Correct Rejection"

A receiver operating characteristic (ROC) curve plots the hit rates against the false alarm rates (see Figure 6.1). For yes/no experiments in which the participant's response is that the signal is either present or not, hit and false alarm probabilities generate a single point on the ROC curve. This point reflects the decision or cutoff point of the decision maker. For experiments in which participants are asked to indicate their confidence that a signal is present, multiple points on the ROC curve may be generated (Pollack & Decker, 1958; see also textbooks on mathematical psychology, e.g. Wickens, 2002).

Because 1 minus the hit rate is the false negative rate, the ROC curve can also be viewed as the trade-off between the false-negative and false-positive rates for every possible decision cutoff. Good performance is characterized by low false-positive rates (false alarms) and low false-negative rates (high hit rates) across a reasonable range of cutoff values. Therefore, desirable performance is reflected in ROC curves that are farthest from the diagonal.

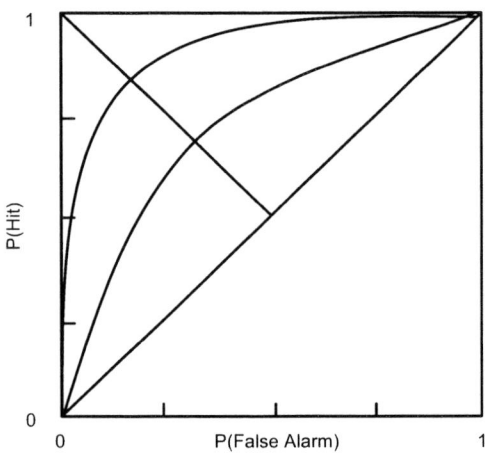

The ROC curve is a graphical representation of decision-making performance under uncertainty. Good performance is reflected by a ROC curve that climbs rapidly toward the upper left hand corner of the graph.

Figure 6.1. ROC curve.

The diagonal line is the expected ROC curve that would be obtained for random performance. For example, if an observer said yes randomly 80% of the time no matter what, the hit rate when the signal is actually present would be 80% and the false alarm rate would also be 80%.

In addition, the theory is used effectively to separate two important characteristics of the receiver: its sensitivity or ability and its bias. The sensitivity measure (d') is reflected in the distance from the diagonal or area under the curve. (More specifically, according to the theory, d' is the difference between the z-transformed probabilities of hits and false alarms when the signal and noise distributions are assumed to be Gaussian.) The bias measure is reflected in the position of a point along the ROC curve, and performance may be influenced by manipulating the bias of the observer by such means as varying the reward and/or punishment for correct responses and errors.

Signal detection theory may therefore be applied to differentiate between levels of operator performance (proficiency/expertise) obtained under varying conditions (such as with and without augmented mitigation aids). An important additional use of the theory is to examine the performance of augmented cognition systems themselves in mitigation management, as described in the next exercise.

Applied Exercise Using Signal Detection Theory Concepts
To gain some insight into the use of concepts from signal detection theory, consider an experiment reported by Polythress et al. (2006) that tested EEG-based measures of cognitive load for a task involving a prototype Tactical Tomahawk Weapons Control System. The analytic method employed in this study utilized the computation of correlations between the EEG measures and other derived measures, including MIDS and an expert's ratings of workload. One such correlational analysis showed that one of the EEG workload measures (% Time Excessive EEG Workload) correlated 0.85 with the MIDS total load measure across five phases of the study by

EVALUATION METHODS AND METRICS

> Polythress et al. (Phases were defined in terms of specific subsets of 26 tasks that made up the experimental scenario.)
>
> Correlations are useful in describing the general agreement between the different indices, but consider what an application of signal detection theory—or even simply a derivation of a confusion matrix and computation of hit and false alarm rates—might provide. (With sufficient data, the application of signal detection theory could also yield calculations of the d' sensitivity measure of the mitigation system or its constituent cognitive-state assessors and distinguish between sensitivity and response bias, as reflected in the placement of decision thresholds for mitigation action.)

Let us see this in action. The Polythress et al. (2006) paper did not include data for individual participants and tasks, but some average data were provided that enable a limited analysis for illustrative purposes. Table 6.3 shows data extracted from this report (Table 6.4 in Polythress et al., 2006, p. 39), consisting of mean EEG % Time in Excessive Workload and the average MIDS total load index scores. To use signal detection concepts, we must establish a criterion for deciding whether a signal is present, which in this case refers to high workload, as determined by the expert ratings (e.g., MIDS).

The MIDS total load scores ranged from 2.9 (lowest workload rating, assigned to Task 12) to 131.4 (highest workload rating, assigned to Task 7). We will operationally define *high workload* for this experimental scenario to correspond to an MIDS score that is greater than 10.0. This is what we will take as defining the "signal" to be detected by the EEG cognitive state assessor.

To classify any task based on the EEG measure, we must set a decision criterion for the EEG output. As described in signal detection theory, the level at which one places this criterion determines the hit and false alarm rates of the detector: If the criterion is placed too high, there are few or no false alarms but the hit rate is depressed; if the criterion is placed too low, both hits and false alarms increase in number.

Consider the rule of classifying the task as high workload if the EEG measure exceeds a value of 29%. For this case, we obtain the confusion matrix and hit/false-alarm rates shown in Table 6.4.

Table 6.3. Data Used in the Applied Exercise

Task	EEG % Time Excess Workload	MIDS Total Load	Task*	EEG % Time Excess Workload	MIDS Total Load
	29	26.2	13	16	5.4
2	36	22.1	14	33	8.9
3	36	46.6	15	28	10.0
4	30	19.1	16	29	6.3
5	32	37.6	17	45	18.2
6	34	10.6	18	37	15.2
7	17	131.4	20	50	28.4
8	26	64.8	21	33	10.2

Table continues on p. 160

Table 6.3 continued

Task	EEG % Time Excess Workload	MIDS Total Load
9	16	36.2
10	34	30.0
11	33	5.0
12	28	2.9

Task*	EEG % Time Excess Workload	MIDS Total Load
22	34	34.3
23	44	13.0
24	31	50.7
25	22	3.2

*Note: Polythress et al. (2006) excluded Tasks 19 and 26 from the analysis because of excessive missing data.

Table 6.4. Results of the Applied Exercise

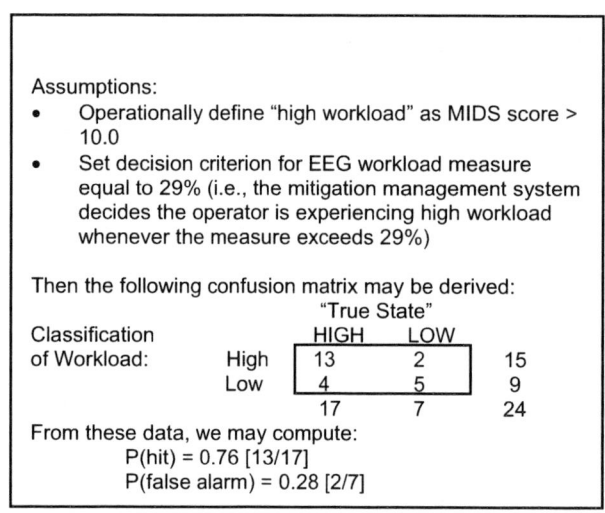

Assumptions:
- Operationally define "high workload" as MIDS score > 10.0
- Set decision criterion for EEG workload measure equal to 29% (i.e., the mitigation management system decides the operator is experiencing high workload whenever the measure exceeds 29%)

Then the following confusion matrix may be derived:

		"True State"		
Classification		HIGH	LOW	
of Workload:	High	13	2	15
	Low	4	5	9
		17	7	24

From these data, we may compute:
P(hit) = 0.76 [13/17]
P(false alarm) = 0.28 [2/7]

Now consider what happens if we adjust the decision criterion. If you repeat the calculations with the decision criterion set to 25%, you will find that the hit and false alarm rate estimates change accordingly, P(hit) = 15/17 = 0.88 and P(false alarm) = 5/7 = 0.71. Try other decision criteria (higher or lower) and see their effects on the estimated hit and false-alarm rates. For example, setting the criterion to 35% yields P(hit) = 0.35 and P(false alarm) = 0; a criterion of 31% produces P(hit) = 0.65 and P(false alarm) = 0.28. Adjustment of the criterion to improve the hit rate has an accompanying effect of increasing the false alarms; adjustments to decrease the false-alarm rate will also decrease the hit rate. There is always such a trade-off in the hit rate and the false-alarm rate. If you plot the empirically derived hit and false alarm rates, the resulting points roughly trace out the ROC curve.

For another thought exercise, consider how adjusting the decision criterion may be used to improve mitigation management. Given the potential for mitigations to disrupt the decision maker, under some circumstances you may wish to select a threshold that keeps false alarms low (to minimize unnecessary mitigation interventions), whereas other circumstances may allow you to relax the threshold and permit slightly more mitigation interventions at the expense of a higher false alarm rate. These manipulations may be done on an ad hoc basis, but it is possible, with appropriate assumptions about the distribution of signal and noise (e.g., Gaussian distributions), to use the signal detection theory methods to compute d' and to automatically derive threshold values that correspond to a specific false alarm rate, and therefore to dynamically control workload or apply decision support/aiding.

Recently, more attention has been directed to the use of signal detection theory in applications of augmented cognition. Bosse et al. (2007) discussed the use of the theory and estimates of a human observer's sensitivity (d') to determine whether or not to engage user support in a visual attention task. (However, they erroneously claim that d' is obtained by subtracting the false alarm rate from the hit rate, which is true if z-transformed probabilities are derived, but not for the basic empirically observed probabilities.)

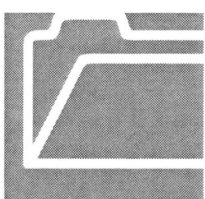

Valuable Information: Analysis Implications
Signal detection theory offers an extremely powerful way to test the performance of diagnostic systems—human or machine. We have described its use in assessing the performance of human decision makers, but it is also suitable for evaluating the performance of cognitive state gauges, as illustrated in the previous exercise.

Mental chronometry methods. Quantitative mental chronometry methods offer a new way to use a very old analytic paradigm of inferring stages of cognitive processing from differences in reaction times. Originally, this paradigm was described in studies by 19th-century Dutch medical scientist Franciscus Donders (1818–1889) and improved and generalized to the *additive factors method* by Sternberg (1966) to assess mental processes in letter recognition. Later employed by Posner (1978), such mental chronometry methods have been used to infer neurophysiological correlates of more complex cognitive processes.

In these neuroscience applications, researchers study temporal characteristics and spatial relationships of brain activation during mental activity by examining data from functional imaging techniques, such as functional magnetic resonance imaging (fMRI). Techniques have since been developed to measure ERPs within fMRI (and EEG) when participants are asked to determine whether a presented digit is above or below the numeral 5. Sternberg's additive theory predicts that this problem is composed of separate stages of processing (encoding, comparing against the stored representation, response selection, and error checking). The fMRI image identifies specific brain locations for these processes.

Recently, researchers studying augmented cognition have used the method in EEG studies of image analysis (Mathan, Ververs, et al., 2006) and spatial orientation (Viirre et al., 2006) with direct implications for augmented cognition mitigation management. Viirre et al. showed significant differences in EEG activity by subtracting EEG power in two types of trials to conclude that significantly more alpha errors and fewer beta errors were observed when participants exhibited spatial disorientation. This can not only be used in real time, enabling neural markers to trigger mitigation in such tasks, but can also be applied to enhance training by enabling a training management system to inform feedback and improve or speed the training process.

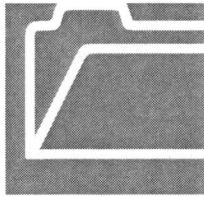

Valuable Information: Application to Training and Skill Development

Since the earliest days of psychology, it has been observed that practice can serve to automate a skill. William James stated in 1890 that "actions originally prompted by conscious intelligence may grow so automatic by dint of habit as to be apparently unconsciously performed" (1890, p. 5). Posner, DiGirolamo, and Fernandez-Duque (1997) reviewed neuroscience research in the development of expertise and offered descriptive models that posit that different brain mechanisms underlie skill development. As operators or decision makers gain proficiency, their cognitive load diminishes—thus, cognitive load measures, and corresponding interpretation of task difficulty, are heavily dependent on learning.

Providing complementary information to behavioral data, neurophysiological measures have been shown to distinguish between novices and experts. For example, Kennedy et al. (2005) observed that in 73% of the neuroimaging studies they reviewed, brain regions that were highly active during early stages of practice exhibited decreased activation levels later in practice. Moreover, particular parts of the brain may be active during the early stages of practice, whereas different parts of the brain may be involved later in practice. This finding has been demonstrated in various studies, including basic laboratory tasks (Ciesielski & French, 1989), marksmanship (Deeny et. al., 2003; Kerick et. al., 2004), and verbal and spatial working memory tasks, as well as a more complex video game (Smith et al., 1999). These important and promising results are significant from the augmented cognition perspective in assessing the impact of decision aids; they are also of critical importance in the learning/training field, in which such correlates are needed to help manage the application of training mitigation strategies.

An interpretation of these neuroimaging results is that we should expect to see reduced activity at about the time when the learner achieves some degree of understanding or automaticity in a particular stage of learning. As the learner progresses to the next higher stage, presumably we should see increased neural activity followed by another period of reduced activity. This "signature" could be one of the triggers that tell us to introduce new concepts or increase the complexity or tempo of the task.

A cognitive neuroscience research challenge is to correlate neurophysiological measures with stages of learning or expertise so that training mitigations may be applied "just in time"—that is, to improve the efficiency or throughput of training. Training mitigation strategies are needed that are informed by triggers that identify not only *when* to advance or apply mitigations but also *how* to facilitate this process. A traditional (associationist or behaviorist) view of learning is that knowledge and skills are gained continuously and incrementally through practice—exemplified by the "learning curve" described by Hermann Ebbinghaus (1885/1913). For this relatively mechanistic characterization, which originally evolved from learning of nonsense syllables, a focus of training is on repetition and practice.

A challenge is to develop convergent methods from neuroscience and cognition/learning to identify both the optimal times for and types of training mitigations to accelerate learning.

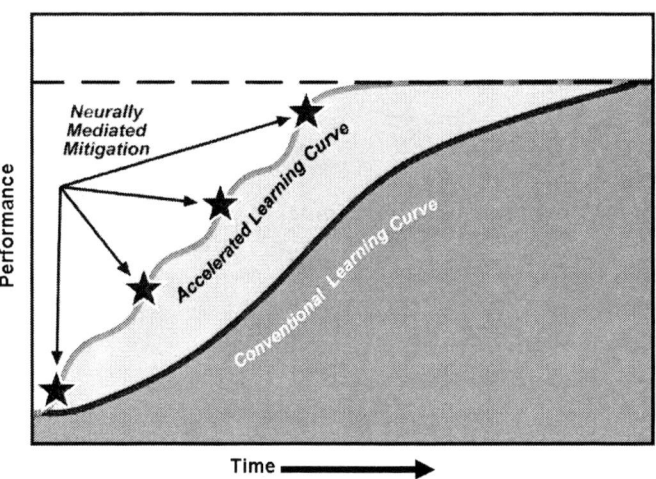

Figure 6.2. Conceptual illustration comparing a conventional learning curve with an accelerated learning curve.

In contrast, a cognitive view emphasizes principles of organization (relating new information to previous experience, e.g., in episodic memory) and the development of understanding through active problem solving. The active learning approach (including simulation and game-based learning) can speed the acquisition of expertise (Lesgold et al., 1992; to be discussed later). There is a need to monitor neurophysiological measures during active learning scenarios to better understand the learning and neurophysiological impacts of different types of mitigations. The ultimate aim is to speed learning by (a) applying mitigations at the right time and (b) applying the most effective mitigations for the given cognitive/learning state to enable the delivery of tailored mitigations at points in the learning process when the learner is most receptive. (See Figure 6.2 for an illustrative comparison of conventional and accelerated learning curves.)

The augmented cognition community is turning its attention to the need to specify neurophysiological measures and their efficacy in the management of training. Palmer and Kobus (2007) considered possible applications of augmented cognition systems in educational and training applications, including the use of neurophysiological measures in conjunction with behavioral data to distinguish *slips,* or accidental or unintended responses, from other types of mistakes; the use of physiological measures and closed-loop augmented cognition systems (such as eye tracking) to adapt presentation rate in reading text; and the use of neurophysiological measures of cognitive capacity to modulate the emphasis on verbal versus visuospatial modes of presentation. Stevens, Galloway, and Berka (2007) showed that fluctuations in EEG-derived measures of cognitive workload, engagement, and distraction could be linked to observable events during problem solving to infer times when students are experiencing difficulty or bottlenecks in the learning process.

Another challenge is to establish or demonstrate that augmented cognition–based training mitigations lead to faster acquisition of expertise compared with traditional methods (such as on-the-job training). One methodology to demonstrate such results is a longitudinal, cross-sectional paradigm that compares performance of augmented cognition condition learners at different times in the training process with baseline participants selected to represent varying levels of expertise, operationally defined in terms of years of (for example) service and classified according to relevant expertise levels (novice, apprentice, journeyman, expert).

An excellent example of this methodology is provided in the celebrated study by Lesgold et al. (1992), who demonstrated the acceleration of expertise in an experiment comparing the performance of expert technicians with that of newly assigned Air Force technicians who did or did not receive training through an intelligent multimedia training system (called Sherlock). Thirty-two newly assigned technicians and 16 experts were included in the evaluation. The new technicians were randomly divided into two groups of 16, one of which received 25 h of training with the Sherlock system (the experimental group) and the other 16 serving as a control group that did not receive the training. Pretests and posttests were given to both groups, and the 16 experts also took the posttest. The results are shown in Figure 6.3.

Several observations are evident from the figure (and validated with statistically significant results): First, the average skill level for the experimental group that received 25 h of training was equivalent to that of the advanced technicians, who had an average of four years of on-the-job experience. Second, performance on the pretest by both experimental and control groups (as well as posttest performance by the untrained control group) was equivalent and significantly below that of the trained participants and experts. Although this study did not investigate neuroscience-based mitigations (its focus was on innovative computer-based training using a guided-discovery, problem-based learning approach), it represents an excellent methodological model that may be followed in augmented cognition research applied to training.

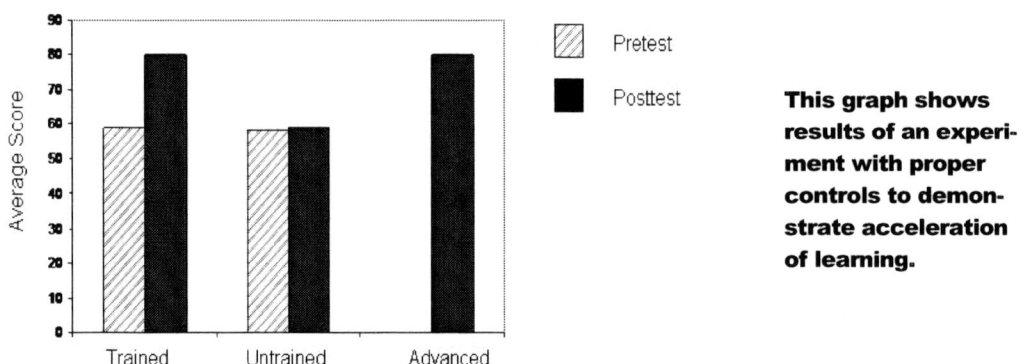

This graph shows results of an experiment with proper controls to demonstrate acceleration of learning.

Figure 6.3. Acceleration of expertise shown in Sherlock Experiment.
(After Lesgold et al., 1992). Comparison of the experimental group receiving training with an untrained control group and advanced/experts provides a convincing demonstration of accelerated expertise acquisition.

EVALUATION METHODS AND METRICS

Review: Testing and Validation of Mitigation Strategies

Although early work in augmented cognition tended to define mitigation in general terms, recently, researchers have focused on specific mitigation strategies that take into account task variables and context in addition to cognitive state (see Chapter 5). Similarly, evaluation methodologies and performance measures to assess the impact of mitigations must be developed to more directly reflect underlying cognitive processes. Methodological challenges are to design experiments using variables, experimental manipulations, task contexts, and subject populations that produce results that are both internally and externally valid. As a reminder, eight questions about experimental validity were discussed.

Choices of performance measures and analysis techniques are also important factors determining the success and interpretability of experiments. Commonly used performance measures are temporal data, such as latencies or durations of tasks, as well as performance quality. An excellent example of a temporally based performance measure is provided in the image analysis/triage experiment that used the RSVP method to utilize the human brain as a preprocessor and demonstrated a fivefold reduction in time required to conduct image analysis.

Other established but less frequently used analytic methods were discussed to promote their use in augmented cognition research: Signal detection theory was described and examples were provided to motivate the use of this powerful approach to test and evaluate the performance of detection and/or diagnostic systems. It was observed that techniques of signal detection theory may be profitably applied to assess the performance of cognitive state gauges (computer-based systems) as well as human operators.

Another analytic method that lately has been used prominently is subtractive or additive factors methods, which derive from early work in mental chronometry; these are now being applied in neuroscience research to study temporal characteristics and spatial relationships of brain activation during mental activity.

Finally, a brief discussion was provided about methodological considerations specific to evaluating the effects of training mitigations.

Lessons Learned

Here, and summarized in Table 6.5, are collected thoughts on lessons learned from the research, theory, and methodologies discussed above and from contemporary practice in augmented cognition. As pointed out earlier in this chapter, the field of augmented cognition has roots in the origins of psychology. Greitzer and Griffith (2006) provided a historical perspective recalling related early work in human factors, its evolution to human-systems interaction (HSI), and implications for framing research and system development in augmented cognition. Arguing for even more advanced augmented cognition technology focusing on more complex decision-making tasks, such as those required in intelligence analysis, they suggested how the distinction offered by Kahneman between two human cognitive systems (System 1, intuition, vs. system 2, reasoning) may provide a framework to guide the development of augmented cognition mitigation strategies—for example, focusing on cognitive-state assessors of System 1 states and providing mitigation support for System 2 processing.

Table 6.5. Lessons Learned About Foundations and Methodologies

- Utilize/extend cognitive theories to drive the definition of mitigation strategies.

- Keep the human in charge (Pilot's Associate [e.g., Miller & Dorneich, 2006]; others [e.g., see Greitzer & Griffith, 2006]).

- Anticipate or seek applications in other HSI areas, including personnel and environment, safety, and occupational health (Arnold et. al., 2006).

- Employ integrated approaches to mitigation management, taking into account task performance and context in addition to neurophysiological cognitive state assessors (Fuchs et al., 2006; Greitzer, 2005a; Miller & Dorneich, 2006; also see Chapter 5).

- EEG and fNIR are the most operationally feasible sensor technologies for brain monitoring (Kruse & Schulman, 2006; also see Chapters 1 and 2).

Arnold et al. (2006) summarized research from the field of human-systems integration to show how augmented cognition could be extended beyond the traditional scope of human factors in systems design to applied HSI disciplines. For example, they pointed out correspondences between augmented cognition studies and HSI disciplines in examples of research on deception detection and the classification of individual differences (personnel), identification of skill level (training), and measures of alertness, spatial disorientation, and gravity-induced loss of consciousness (environment, safety and occupational health).

Another important perspective is offered by Miller and Dorneich (2006), who provided lessons learned and similarities in goals and methods for developing human-adaptive information and automation management technologies within the Pilot's Associate research programs in the 1980s and 1990s. Among the lessons learned in the Pilot's Associate research program are the importance of keeping the human "in charge" and customizing the user interface and user-computer interaction designs to make adjustments based on capabilities and limitations of the mitigation systems (e.g., user override, notification/control of pending mitigations); the need for an explicit task-based framework to coordinate and guide development, particularly aimed at incorporating other nonphysiological sensors (task-based measures and context information) to inform mitigation management and avoid interventions that have deleterious effects of increasing cognitive load; the importance of learning and tuning of sensors; and the need to support adaptation to individual differences in perception, cognition, and behavior. By recognizing this historical perspective and learning from the large body of research that has preceded the augmented cognition program, we will profit from its broad set of lessons learned.[3]

Lessons learned in the young field of augmented cognition are just emerging. The Technical Integration Experiments (St. John et al., 2004; see Chapter 9) conducted in the first phase of DARPA's Augmented Cognition Program provided a major source of information about sensors that worked or had promise versus those that did not work as well or that had practical drawbacks in terms of size, portability, intrusiveness, or ability to produce outputs in real time. Reviewing these findings, Kruse and Schulman (2006) observed, in their presentation at the 2006 Augmented Cognition International Conference, that the most operationally feasible sensor technologies for brain monitoring are EEG (see Chapter 1) and fNIR (see Chapter 2). From a methodological perspective, the insights of Fuchs et al. (2006) in the paper titled "Physiological Sensors Cannot Effectively Drive System Mitigation Alone" echo the lessons of Pilot's Associate, described earlier, which suggests the need for integrated approaches to cognitive state assessment that incorporate analyses of both neurophysiological and behavioral data, and that also take into account contextual information about the task (see also Chapter 5).

Best Practices

The intent of this chapter has been to describe the most useful methodologies, metrics, and analytic techniques that represent best practices for meeting the augmented cognition practitioner's objectives for evaluating the effectiveness and impact of mitigations. The following discussion (with a summary in Table 6.6) represents my impressions and observations of best practices to date, allowing that any list of best practices will have to be updated frequently in light of the exciting contributions that are constantly emerging from this young field.

Among the most significant is the mitigation selection framework proposed by Fuchs et al. (2006), which offers a structured, systematic, theoretically grounded approach for defining progressively more substantial and intrusive mitigations, as warranted by cognitive state assessment and where cognitive state is determined by convergent analyses from behavioral and task-specific data as well as physiological measures (see Chapter 5). Among the graded mitigations described are subtle cueing, intrusive cueing, task sequencing, decluttering displays, and task sharing.

The MIDS cognitive workload measurement method developed by Design Interactive, Inc. (Hale et al., 2005) represents current best practice in the elicitation of expert judgments on cognitive load, perhaps advancing the earlier classic work of NASA's Task Load Index. A distinct advantage of the use of MIDS in recent augmented cognition research is that it enables one to predict cognitive workload for experimental tasks on a second-by-second basis (Polythress et al., 2006).

---

[3] As philosopher George Santayana wrote in his five-volume publication, *The Life of Reason*, "Those who cannot remember the past are condemned to repeat it." (Santayana, G. [1905-1906]) *The Life of Reason*. (Chapter XII, "Flux and Constancy in Human Nature"). New York: Dover Publications, Inc. Available from Project Gutenberg, http://www.gutenberg.org/etext/15000.

Table 6.6. Best Practices

1. **Use a rigorous, theoretically grounded approach to selecting mitigation strategies** (such as the mitigation selection framework described in Chapter 5).

   Why it is important: It provides a structured, systematic method for applying progressively more substantial and intrusive mitigations as warranted by cognitive state assessment.

2. **Use the MIDS cognitive-workload index to assess cognitive load and validate neurophysiological cognitive state assessors.**

   Why it is important: It uses expert judgment rather than subjective judgments of operators, and it may be applied in real time to predict and manage workload (Hale et al., 2005; Polythress et al., 2006).

3. **Use mental chronometry approaches (additive factors, subtractive methods) to infer temporal and spatial correlates of cognitive processes with brain activity.**

   Why it is important: Techniques derived from mental chronometry/additive-factors methods have proven effective in inferring neurophysiological correlates of more complex cognitive processes (cf. Posner, 1978; Viirre et al., 2006)]

4. **Exploit the strengths of human information processing as well as overcoming its limitations.**

   Why it is important: A specific example of best practice in human-system integration for augmented cognition is the use of the RSVP method for preprocessing/triage of imagery data (Mathan, Ververs, et al., 2006).

5. **Use real-time computational methods to derive cognitive workload measures "on the fly."**

   Why it is important: This work is essential in providing computational foundations for deploying augmented cognition systems in decision support and training environments.

6. **Consider applying signal detection theory to evaluate not only human performance but also the performance of cognitive state assessors in augmented cognition applications.**

   Why it is important: Signal detection theory is an immensely successful quantitative/evaluation methodology in human factors/systems research that may provide significant benefit if more fully exploited in augmented cognition applications.

7. **Understand and apply proper scientific methodologies and controls to ensure internal and external validity in studies; this demonstrates the efficacy and effectiveness of augmented cognition technologies.**

   Why it is important: Proper experimental methods and procedures are essential to the acceptance of results from augmented cognition experiments, evaluation studies, and demonstrations of system effectiveness.

Employing rigorous experimental designs is often challenging; techniques that enable analysts to analyze and interpret data sets for a single participant at a time are recommended because the determination of cognitive overload ultimately must be done on a single-individual basis. In this area of computational methods, researchers who employ modern equivalents of classic mental chronometry methods in neuroscience studies have been successful in identifying temporal and spatial correlates of cognitive processes with brain activity (cf. Posner's [1978] early work, and later research in augmented cognition by Viirre et al. [2006]).

Another best-practices concept is an approach that considers the human and the computer as a system for which augmented cognition technology is to be applied to optimize performance of the integrated whole. Greitzer and Griffith (2006) have provided a historical perspective on the well-studied field of human-systems integration/resource allocation and provided some guidelines about where to focus human and computer support. A specific and creative example of HSI and decision support comes from the application of augmented cognition to imagery analysis in experiments by Mathan and colleagues (Mathan Ververs, et al., 2006; Mathan, Whitlow, et al., 2006). This work represents a novel and powerful utilization of preconscious brain responses as a basis for the preprocessing or triage of imagery data. It exploits the superb pattern recognition capabilities of the human brain and combines this method with cueing of human analysts to achieve a fivefold speed-up of imagery analysis. Significant methodological contributions of this work include the use of the RSVP technique, previously described, which generates readily interpretable, quantifiable task-time performance measures that yield statistically significant results.

Real-time computational methods to support the derivation of cognitive workload measures, validate cognitive-state assessors, and determine neurophysiological correlates of cognitive processes are also emerging best practices. Notable work in that area includes data-intensive computational approaches that evaluate and compare methods based on linear (e.g., regression methods and linear discriminant analysis) or nonlinear (e.g., artificial neural networks) methods (Craven et al., 2006; Russell et al., 2005; Wilson & Russell, 2006). This work is essential in providing computational foundations for deploying augmented cognition systems in decision support and training environments.

It may be premature to consider as a best practice the application of signal detection theory to augmented cognition system evaluation. The use of signal detection theory and associated analytic tools represents perhaps the most successful methodology developed in the field of mathematical psychology. Although signal detection theory has not been widely used in augmented cognition practice, in this chapter I argue that, from a measurement and evaluation perspective, it would be constructive to more fully exploit this methodology in augmented cognition. Such a systematic approach to decision making separates the sensitivity of the system from decision thresholds that modulate hit and false alarm rates. Its application to the evaluation of cognitive state assessors and mitigation management strategies would likely provide more insights into the selection and tuning of augmented cognition systems.

Finally, as pointed out earlier in this chapter, augmented cognition practitioners need to understand and apply proper scientific methodologies and controls to ensure internal and external validity in studies and thereby demonstrate the efficacy and effectiveness of augmented cognition technologies. Use of random sampling, operationally valid experimental conditions, and appropriate experimental controls are all essential to the acceptance of results from augmented cognition experiments, evaluation studies, and demonstrations of system effectiveness.

Guiding Principles

A major goal that has been expressed in the augmented cognition field is to develop automated systems that can evaluate an operator's cognitive state and then control the nature and extent of sensory input to the operator in order to optimize performance (Raley et al., 2004; Schmorrow & Kruse, 2004). These and even broader objectives are being addressed to augment more complex cognitive processes, including memory, decision making, and learning. From my perspective (e.g., Greitzer, 2005a; Greitzer & Griffith, 2006), one guiding principle is to promote the study of more complex cognitive tasks involving memory, reasoning, and decision making for augmented cognition applications. This method seeks to move the augmented cognition domain past its contemporary roots in human-computer interaction to meet its historical roots in human-machine symbiosis.

To achieve this ultimate objective, augmented cognition systems must themselves possess a certain level of cognitive proficiency in order to be effective under changing conditions (Smith & Henning, 2006); that is, they are or should be *cognitive systems* that utilize plausible computational models of human cognitive processes as a basis for human-machine interactions (Forsythe & Xavier, 2006). The notion of augmented cognition applications as cognitive systems is a key principle for guiding research both in assessing the operator's cognitive state and in managing mitigation strategies and associated communications with the human operator. This perspective in turn has implications for methodologies that are chosen to develop cognitive models and to test and evaluate performance. In particular, it highlights the need for "a systematic development process rooted in science concerning human cognition" (Forsythe & Xavier, 2006, p. 9), informed by neurophysiological evidence, and validated by experimental research that addresses relevant cognitive measures of effectiveness.

It is hoped that this chapter has been useful in discussing issues and providing practical guidance to practitioners on the employment of methodologies for modeling cognitive state and adaptive mitigation strategies and for employing effective performance measurement/analytic techniques.

Table 6.7. Guiding Principles

- Augmented cognition systems should be **cognitive systems** that utilize plausible computational models of human cognitive processes as a basis for human-machine interactions, and that possess a certain level of cognitive proficiency in order to be effective under changing conditions.

- Cognitive science should guide practitioners in validating cognitive state assessment, defining mitigation management strategies, and evaluating their impacts.

- Our community should continue to identify methodologies and metrics that have proven effective in earlier research and adapt them for augmented cognition research.

Parting Message

In all scientific discovery, great insights and breakthroughs come about after painstaking study, which includes taking care to be informed about methods and findings from preceding research while seeking out new facts, relationships, and insights. Although some have interpreted unexpected scientific discoveries as serendipity, students of the history of science assign true importance to the investigator's contribution to the occurrence of serendipity. This was most elegantly stated by Louis Pasteur: "Chance favors the prepared mind." In this chapter, I have discussed methods, techniques, and findings from the field of augmented cognition research and its historical antecedents to facilitate the discovery and testing of new and more effective mitigation and mitigation management strategies. When this background is synthesized and integrated with material presented in other chapters in this volume, I hope it will better prepare a growing community of augmented cognition practitioners for future discoveries.

References

Arnold, R. D., Cohn, J. V., Stripling, R., Nicholson, D., & Stanney, K. (2006). Augmented cognition for human-systems integration. In D. D. Schmorrow, K. M. Stanney, & L. M. Reeves (Eds.), *Foundations of augmented cognition* (2nd ed., pp. 303–312). Arlington, VA: Strategic Analysis, Inc.

Berka, C., Levendowski, D. J., Cvetinovic, M. M., Petrovic, M. M., Davis, G., Lumicao, M. N., Zivkovic, V. T., Popovic, M. V., & Olmstead, R. (2006). Real-time analysis of EEG indexes of alertness, cognition, and memory acquired with a wireless EEG headset. *International Journal of Human-Computer Interaction, 17*(2), 151–170.

Bosse, T., van Doesburg, W., van Maanen, P. P., & Treur, J. (2007). Augmented metacognition addressing dynamic allocation of tasks requiring visual attention. In D. D. Schmorrow & L. M. Reeves (Eds.), *Foundations of augmented cognition* (3rd ed., pp. 166–175). Heidelberg, Germany: Springer.

Bush, V. (1945, July). As we may think. *The Atlantic Monthly, 176*(1), 101-108. Accessed on August 3, 2008, at http://www.theatlantic.com/doc/194507/bush

Ciesielski, K. T., & French, C. N. (1989). Event-related potentials before and after training: Chronometry and lateralisation of visual N1 and N2. *Biological Psychology, 28,* 227–238.

Craven, P. L., Belov, N., Thomas, M., Berka, C., Levendowski, D. J., & Davis, G. (2006). Cognitive workload gauge development: Comparison of real-time classification methods. In D. D. Schmorrow, K. M. Stanney, & L. M. Reeves (Eds.), *Foundations of augmented cognition* (2nd ed., pp. 66–74). Arlington, VA: Strategic Analysis, Inc.

Deeny, S. P., Hillman, C. H., Janelle, C. M., & Hatfield, B. D. (2003). Cortico-cortical communication and superior performance in skilled marksmen: An EEG coherence analysis. *Journal of Sport & Exercise Psychology, 25*(2), 188–204.

Donders, F. C. (1969). Over de snelheid van psychische processen [On the speed of mental processes; W. Koster, trans.). In W. G. Koster, *Attention and performance II* (pp. 412–431). Amsterdam: North Holland.

Ebbinghaus, H. (1913). *Memory: A contribution to experimental psychology.* New York: Teachers College, Columbia University. (Original work published 1885, available online at http://psychclassics.yorku.ca/Ebbinghaus/index.htm)

Forsythe, C., & Xavier, P. G. (2006). Cognitive models to cognitive systems. In C. Forsythe, M. M. Bernard, & T. E. Goldsmith (Eds.), *Cognitive systems: Human cognitive models in system design* (pp. 1–33). Mahwah, NJ: Erlbaum.

Fuchs, S., Hale, K. S., Berka, C., Levendowski, D., & Juhnke, J. (2006). Physiological sensors cannot effectively drive system mitigation alone. In D. D. Schmorrow, K. M. Stanney, &

L. M. Reeves (Eds.), *Foundations of augmented cognition* (2nd ed., pp. 193–200). Arlington, VA: Strategic Analysis, Inc.

Green, D. M., & Swets, J. A. (1966). *Signal detection theory and psychophysics.* New York: Wiley.

Greitzer, F. L. (2005a, August). *Extending the reach of augmented cognition to real-world decision making tasks.* Paper presented at the Augmented Cognition International Conference/HCI International Conference, Las Vegas, NV.

Greitzer, F. L. (2005b). Toward the development of cognitive task difficulty metrics to support intelligence analysis research. In *Fourth IEEE International Conference on Cognitive Informatics, ICCI 2005* (pp. 315–320). Los Alamitos, CA: IEEE Computer Society.

Greitzer, F. L., & Griffith, D. (2006, October). A human-information interaction perspective on augmented cognition. In D. D. Schmorrow, K. M. Stanney, & L. M. Reeves (Eds.), *Foundations of augmented cognition* (2nd ed., pp. 261–267). Arlington, VA: Strategic Analysis, Inc.

Hale, K., Axelsson, P., Samman, S., & Stanney, K. M. (2005). Validation of predictive workload component of the Multimodal Information Design Support (MIDS) system. In *Human Factors and Ergonomics Society 49th Annual Meeting* (pp. 1162–1166). Santa Monica, CA: Human Factors and Ergonomics Society.

Hart, S. G., & Staveland, L. E. (1988). Development of a multi-dimensional workload rating scale: Results of empirical and theoretical research. In P. A. Hancock & N. Meshkati (Eds.), *Human mental workload.* Amsterdam: Elsevier.

James, W. (1890). *The principles of psychology.* New York: Holt. (Reprinted edition from Dover Publications, Inc., New York, 1950.)

Kahneman, D. (1973). *Attention and effort.* Englewood Cliffs, NJ: Prentice Hall.

Kennedy, R. S., Drexler, J. M., Jones, M. B., Compton, D. E., & Ordy, J. M. (2005). Quantifying human information processing (QHIP): Can practice effects alleviate bottlenecks? In D. K. McBride & D. Schmorrow (Eds.), *Quantifying human information processing* (pp. 63–122). Lanham, MD: Lexington Books.

Kerick, S. E., Douglass, L. W., & Hatfield, B. D. (2004). Cerebral cortical adaptations associated with visuomotor practice. *Medicine & Science in Sports & Exercise, 36*(1), 118–129.

Kruse, A., & Schulman, J. (2006), Neurotechnology for intelligence analysis. In D. D. Schmorrow, K. M. Stanney, & L. M. Reeves (Eds.), *Foundations of augmented cognition* (2nd ed., pp. 27–31). Arlington, VA: Strategic Analysis, Inc.

Lesgold, A., Eggan, G., Katz, S., & Rao, G. (1992). Possibilities for assessment using computer-based apprenticeship environments. In J. Regian & V. Shute (Eds.), Cognitive approaches to automated instruction (pp. 49–80). Hillsdale, NJ: Erlbaum.

Licklider, J. C. R. (1960). Man-computer symbiosis. *IRE Transactions on Human Factors in Electronics, HFE,* 4–11.

Mathan, S., Ververs, P., Dorneich, M., Whitlow, S., Carciofini, J., Erdogmus, D., Pavel, M., Huang, C., Lan, T., & Adami, A. (2006). Neurotechnology for image analysis: Searching for needles in haystacks efficiently. In D. D. Schmorrow, K. M. Stanney, & L. M. Reeves (Eds.), *Foundations of augmented cognition* (2nd ed., pp. 3–11). Arlington, VA: Strategic Analysis, Inc.

Mathan, S., Whitlow, S., Erdogmus, D., Pavel, M., Ververs, P., & Dorneich, M. (2006). Neurophysiologically driven image triage. *CHI '06 Extended Abstracts on Human Factors in Computing Systems. Proceedings of the 2006 Conference on Human Factors in Computing Systems* (pp. 1085–1090). New York: ACM Press.

Miller, C., & Dorneich, M. C. (2006). From associate systems to augmented cognition: 25 years of user adaptation in high criticality systems. In D. D. Schmorrow, K. M. Stanney, & L. M. Reeves (Eds.), *Foundations of augmented cognition* (2nd ed., pp. 344–353). Arlington, VA: Strategic Analysis, Inc.

North, R. A., & Riley, V. (1988). W/Index: A predictive model of operator workload. In G. McMillan et al. (Eds.), *Applications of human performance models to systems design* (pp. 81–89). New York: Plenum Press.

Plato. *The Dialogues of Plato (Meno dialogue).* B. Jowett, trans. Oxford: Clarendon Press, 1875. Accessed online at The Internet Classics Archive by Daniel C. Stevenson, Web Atomics, http://classics.mit.edu/Plato/meno.html

Palmer, E. D., & Kobus, D. A. (2007). The future of augmented cognition systems in education and training. In D. D. Schmorrow & L. M. Reeves (Eds.), *Foundations of augmented cognition* (3rd ed., pp. 373–379). Heidelberg, Germany: Springer.

Pollack, I., & Decker, L. R. (1958). Confidence ratings, message reception, and the receiver operating characteristic. *Journal of the Acoustical Society of America, 31,* 1500–1508.

Polythress, M., Russell, C., Siegel, S., Tremoulet, P. D., Craven, P., Berka, C., Levendowski, D. J., Chang, D., Baskin, A., Champney, R., Hale, K., & Milham, L. (2006). Correlation between expected workload and EEG indices of cognitive workload and task engagement. In D. D. Schmorrow, K. M. Stanney, & L. M. Reeves (Eds.), *Foundations of augmented cognition* (2nd ed., pp. 32–44). Arlington, VA: Strategic Analysis, Inc.

Posner, M. I. (1978). *Chronometric explorations of mind.* Hillsdale, NJ: Erlbaum.

Posner, M. I., DiGirolamo, G. J., & Fernandez-Duque, D. (1997). Brain mechanisms of cognitive skills. *Consciousness and Cognition, 6,* 267–290.

Raley, C., Stripling, R., Kruse, A., Schmorrow, D., & Patrey, J. (2004). Augmented cognition overview: Improving information intake under stress. In *Proceedings of the Human Factors and Ergonomics Society 48th Annual Meeting* (pp. 1150–1154). Santa Monica, CA: Human Factors and Ergonomics Society.

Russell, C. A., Wilson, G. F., Rizki, M. M., Webb, T. S. & Gustafson, S. C. (2005). Comparing classifiers for real time estimation of cognitive workload. In D. D. Schmorrow (Ed.), Foundations of augmented cognition (Vol. 11, pp. 396–404). Mahwah, NJ: Erlbaum.

Schmorrow, D. D., & Kruse, A. A. (2004). Augmented cognition. In W. S. Bainbridge (Ed.), *Berkshire Encyclopedia of Human Computer Interaction* (pp. 54–59). Great Barrington, MA: Berkshire Publishing Group.

Singer, A. (2005). *The future of augmented cognition.* Short film available at http://www.augmentedcognition.org/video.htm.

Smith M. E., McEvoy L. K., & Gevins A. (1999). Neurophysiological indices of strategy development and skill acquisition. *Cognitive Brain Research, 7*(3), 389–404.

Smith, T. J. & Henning, R. A. (2006). Social cybernetics of augmented cognition – A control systems analysis. In D. D. Schmorrow, K. M. Stanney, & L. M. Reeves (Eds.), *Foundations of augmented cognition* (2nd ed., pp. 45–54). Arlington, VA: Strategic Analysis, Inc.

Stanney, K., & Reeves, L. (2005, March). *Mitigation strategies and performance effects.* White paper outbrief from a working session at Improving Warfighter Information Intake Under Stress, AugCog PI Meeting, Chantilly, VA.

St. John, M., Kobus, D. A., Morrison, J. G., & Schmorrow, D. (2004). Overview of the DARPA Augmented Cognition Technical Integration Experiment. *International Journal of Human-Computer Interaction, 17*(2), 131–150.

Sternberg, S. (1966). High speed scanning in human memory. *Science, 153,* 652–654.

Sternberg, S. (1969). The discovery of processing stages: Extensions of Donders' method. *Acta Psychologica 1969, 30,* 276–315. Also published In W. G. Koster (*Ed.*), *Attention and performance II.* Amsterdam: North Holland, 1969 (pp. 276–315).

Stevens, R. H., Galloway, T., & Berka, C. (2007). Exploring neural trajectories of scientific problem solving skill acquisition. In D. D. Schmorrow & L. M. Reeves (Eds.), *Foundations of augmented cognition* (3rd ed., pp. 400–408). Heidelberg, Germany: Springer.

Viirre, E., Wing, S., & Huang, R. (2006). EEG markers of spatial disorientation. In D. D. Schmorrow, K. M. Stanney, & L. M. Reeves (Eds.), *Foundations of augmented cognition* (2nd ed., pp. 75–84). Arlington, VA: Strategic Analysis, Inc.

Wickens, C. (1984). Attention, time-sharing, and workload. In C. Wickens (Ed.), *Engineering psychology and human performance* (pp. 291–334). Columbus, OH: Charles E. Merrill.

Wickens, C. D. (1992). *Engineering psychology and human performance.* (2nd ed.). New York: HarperCollins.

Wickens, T. D. (2002). *Elementary signal detection theory.* New York: Oxford University Press.

Wilson, G. F., & Russell, C. A. (2006). Psychophysiologically versus task determined adaptive aiding accomplishment. In D. D. Schmorrow, K. M. Stanney, & L. M. Reeves (Eds.), *Foundations of augmented cognition* (2nd ed., pp. 201–207). Arlington, VA: Strategic Analysis, Inc.

Wundt, W. (1880). *Grundzuge der physiologischen Psychologie* [Foundations of physiological psychology] (2nd ed.). Leipzig: Engelmann.

Engineering Control System Theory in the Behavioral Sciences

Peter M. Young and Patricia A. Aloise-Young
Colorado State University

Mathematical systems and control theory have been used in many engineering disciplines with great success. These systematic analysis and design tools are now starting to be applied in the behavioral sciences. We give an introduction to this exciting new area.

Introduction

Mathematical system theory takes a rigorous quantitative view of dynamic systems. The system under consideration is *modeled;* that is, a mathematical description of its behavior is developed. This can be based on first principles, experimental data, or a combination of techniques. The foundations for these mathematical descriptions are very general (e.g., ordinary differential equations) so that the resulting approaches are very widely applicable. Thus, many types of physical systems can be studied. These tools have been widely employed in many engineering disciplines, including modeling and analysis of electrical, mechanical, chemical, and aerospace systems, among others.

In this context, control system theory considers the analysis, operation, and design of such systems under *feedback*. **The main idea here is that we wish to modify the behavior of, or *control,* a certain system**. We wish our control to be resistant to external influences, such as modeling errors and disturbance signals. Hence we require that our "controller" be fed back measurements that allow it to assess the impact of its control on the system's behavior, and then adapt its control appropriately to achieve the desired performance. These are also referred to as *closed-loop* systems, because feeding back the measurements essentially closes a loop around the system under consideration (i.e., control action – system – measurement – controller – control action). These topics will be explored more fully in subsequent sections, but for now we merely note that a great deal of ongoing theoretical and applied research has focused on closed-loop systems. **As a result, there is a wide range of powerful and systematic tools for the advanced analysis, simulation, and design of such systems, which can be brought to bear on any problems considered in this framework**. For example, in the area of augmented cognition, modeling and simulation techniques have already demonstrated successful control schemes that facilitate the rapid scanning of data and accurate decision making, beyond unaided human performance (see Tremoulet et al., 2006, for an example involving weapons system operation).

In recent years, there has been rising interest in employing engineering control systems theory in some nontraditional application areas. These have included areas such as economic and biological systems. Note that for systems such as these, the types of models that one can derive may be different from those for more traditional settings. Moreover, the models often necessarily are low in fidelity, with limited information and/or potential for experimentation and significant individual variations. In addition, these approaches are now being employed in application areas from the behavioral sciences, in which there are "humans in the loop." **This is an especially challenging situation because of the inherent complexity of the system and the limited availability of data for modeling**. The situation is further compounded by the fact that, from an engineering viewpoint, these systems are nonlinear and time-varying, with significant uncertainties in the underlying dynamics. For example, researchers have faced tremendous challenges in attempting to develop systems using nerve signals to control artificial limb function, although encouraging preliminary results have been achieved using nonlinear signal-processing approaches (cf. Fagg et al., 2007).

Despite these difficulties, there is still tremendous potential for engineering control systems theory to play a major role in the development of new tools in the behavioral sciences, such as augmented cognition systems. The opportunities afforded by these systematic approaches—with a priori predictions of stability, robustness, and performance, as well as tools to optimize designs with these in mind—justify the research efforts that will be required. In this chapter, we give an overview of some of the promising initial research in this area and illustrate the utility of these tools for practitioners whose goal it is to use feedback to improve cognitive performance.

Scenario

Imagine looking at your display and seeing icons that appear and disappear, seemingly at random. You have no idea what is going on or how to deal with it, much less execute the task you are supposed to be engaged in—the reason you are sitting at the computer in the first place.

OR

ENGINEERING CONTROL SYSTEM THEORY

You are trying to work at your computer, but you are overwhelmed. There are so many tasks to deal with that you hardly know where to start. You are doing your best, but you are slowly losing the battle as more and more tasks go unfinished, and you are starting to make errors with the ones you do complete.

OR

You are working hard. In fact, you are working at your limit, but the workload is barely manageable. You can just about cope with the tasks on screen. It seems you are just about to complete something when something new comes up. It is tough, but you are doing OK.

These three scenarios demonstrate the potential of augmented cognition. **They correspond, respectively, to poor feedback control, no control, and well-designed feedback control.** Drastically different results can be obtained from the same hardware (sensors [see Chapters 1 and 2], displays, etc.) and software (interface design, mitigation strategies [see Chapters 5 and 6], etc.)—the difference being the feedback algorithm that controls them. For augmented cognition to realize its true potential, this feedback algorithm design must be done right, and this will require a rigorous, systematic approach based in engineering control system theory.

Applied Exercise
Taking your own domain as an example, think of three situations that might correspond to no control, poor control, and well-designed feedback control. As we discuss the tools necessary for implementing well-designed control systems, you will have an opportunity to revisit these examples and further apply your knowledge.

General Approach and Associated Toolkit

In order to introduce the framework and some basic concepts of control theory, let us consider a simple illustrative example: a person driving a vehicle. During this task, the person has access to numerous measurements, such as (a) binocular cues from the road of current speed and direction, as well as upcoming terrain and potential obstacles; (b) haptic feedback of forces (e.g., feeling of acceleration); (c) gauges indicating speed, fuel, RPM, directions (GPS), etc.; and (d) auditory feedback of engine/road noise. These measurements are processed by the person, and the person responds to these measurements by executing control actions based on them—namely, accelerating via gas pedal, braking as required, adjusting steering continuously, and changing gears as necessary.

As stated earlier, this represents a complex multi-input–multi-output (MIMO) control task. It is common practice in engineering to break down such tasks into smaller, more manageable pieces, so let us employ that practice here. We will reduce our consideration, initially, to a single piece of this control problem: the speed control loop. In other words, we consider only a single measurement—the speedometer—and a single control action—the gas pedal. This conceptually reduces the control problem under consideration to a single-input–single-output (SISO) problem.

Test Your Knowledge
Can you think of other MIMO control systems, either related to driving or in an area of interest to you? Identify the measurements available to the controller, and also the potential control actions. Note how these might change depending on the desired task (e.g., consider maintaining a safe following distance in the aforementioned driving example). Are there individual SISO loops you can isolate in these problems?

SISO Control System Models

In the remainder of this section, we take you through a series of control system models at various levels of abstraction. You will see how the nature of the model changes with the application task at hand. These models each utilize a common set of terminology that will be unfamiliar to many readers. The terms, their definitions, and their meaning within the driving example are given in Table 7.1.

Table 7.1. Terminology for Control System Models

System P: *Plant–System to be controlled.* In this case, a model of the vehicle dynamic response from gas pedal to resulting speed.

System K: *Controller–Implements desired control scheme/algorithm.* In this case, the driver provides this functionality.

Signal r(t): *Reference input–Reference setting trajectory that we wish the output to hold/track.* In this case, the desired speed.

Signal e(t): *Error signal–Difference between desired output and actual (measured) output.* In this case, the difference between the speedometer reading and the desired speed.

Signal u(t): *Control signal–Output of controller, which is input to plant.* In this case, depressing the gas pedal.

Signal y(t): *Output–Output of plant that we wish to control.* In this case, the speed of the vehicle.

Closed-loop systems. Our SISO abstraction of the vehicle speed control problem, as a classic feedback block diagram, is illustrated in Figure 7.1. Each rectangular block in the diagram represents a system processing an input signal and generating an output signal, with the signals represented as lines with arrows to indicate the direction of information flow. The summing junction, represented as a circle, is a particular system that performs only one simple task: to sum/difference (as indicated) an incoming group of signals and output the result.

ENGINEERING CONTROL SYSTEM THEORY

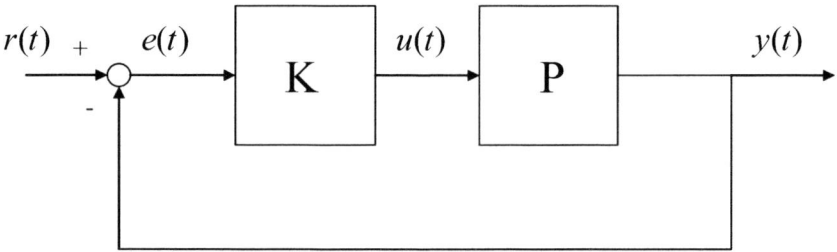

Figure 7.1. Closed-loop control system.

The closed loop that is generated by feedback is now apparent in this diagram. The measured speed y(t) is fed back, differenced from the desired speed r(t) to generate the error signal e(t), which is processed by the driver, K, to generate the gas pedal setting u(t), which causes the vehicle, P, to move at a certain speed y(t).

 **Valuable Information: Measurement Issues**
In this example, we do not draw any real distinction between the actual speed and the measured speed, though in many control applications, measurement may be difficult and/or sensor dynamics significant, so this distinction can be important. In particular, for augmented cognition problems, acquiring certain measurements (e.g., cognitive state) can be a very challenging task, and measurement speed/accuracy may be a significant component of both the plant model and overall achievable system performance.

Control theory now provides a set of mathematical and computational tools for analyzing such closed-loop systems. The types of models and tools used depend greatly on the application area and physical systems under consideration. In this example, for instance, one might employ a continuous-time Linear Time-Invariant (LTI) model of the vehicle speed response (i.e., system P), which could take the form of an ordinary differential equation (ODE):

$$a_2 \frac{d^2 y(t)}{dt^2} + a_1 \frac{dy(t)}{dt} + a_0 y(t) = b_2 \frac{d^2 u(t)}{dt^2} + b_1 \frac{du(t)}{dt} + b_0 u(t)$$

This second-order ODE represents the dynamics from gas pedal input u(t) to output speed y(t), with appropriate choices for the coefficients $a_2, a_1, a_0, b_2, b_1,$ and b_0. In fact, this second-order model can provide a simple approximation to a wide variety of physical systems, if the coefficients are chosen appropriately.

The choice of model structure and the selection of appropriate parameters within that structure have received a great deal of attention over the years. Many approaches exist, including ad hoc tools, systematic techniques, and optimization-based methods. They may employ a combination of first-principles modeling and black-box techniques, which essentially try to fit measured data. This body of work constitutes the field of *system identification,* and the interested reader is referred to Ljung (1999) and the references therein for an overview of the many tools and techniques that make up this area.

Given a setup as described earlier, we could now attempt to model the decision-making and motor response of the human in the loop (represented by the controller, K) and use that to analyze closed-loop properties. At this level of abstraction, it might be a very simple model: For instance, simply model the person's response as a time delay. One could then simulate how different individuals, with different reaction times, would fare under a variety of scenarios and conditions.

Taking our example of the driver a bit further, the next building block is to extend our analysis tools to develop design approaches. In traditional applications of automatic control theory, the goal is to replace the human with an automated agent, so that, for example, a cruise control system would replace the driver as the controller K in this closed-loop system. In this case, our design goal is to arrive at the controller K, which implements the control algorithm. This algorithm will determine how to process the error signal e(t) to generate the control signal u(t), so as to have the desired effect on the output y(t).

Applied Exercise
Taking the example from your own domain that you generated at the beginning of this chapter, determine what r(t) and y(t) would be. Note that there may be more than one possibility for a given example physical system depending on the desired task chosen.

Open-loop systems. Consider again the closed-loop system in Figure 7.1. One could ask the fundamental question, "Why bother with feedback?" If the goal is simply to have y(t) track r(t)—that is, as closely as possible follow

$$y(t) \approx r(t)$$

—then one could achieve this objective simply by choosing

$$K \approx P^{-1}$$

and not feeding anything back (i.e., open-loop control), as illustrated in Figure 7.2.

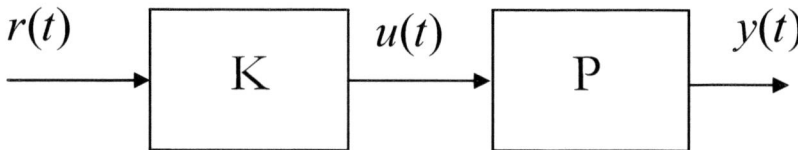

Figure 7.2. Open-loop control system.

There are some technical difficulties with doing this—namely, invertibility and causality/stability—but these could be tackled with an approximate causal stable inverse. In any case, they are not the real issue. The fundamental issue is that of uncertainty. For example, suppose we base our cruise control design on Figure 7.2 and dial in our speed set point. The controller calculates u(t) based on r(t), and, assuming it is accurate (i.e., assuming our model is accurate), we move at the desired speed. If the model is inaccurate, we will not travel at the desired speed, and, moreover, once the car starts to go uphill, it will lose speed in any case. **The open-loop controller cannot react to this disturbance because no measurements have been fed back.**

Valuable Information: Understanding Uncertainty
This open-loop architecture fails because of uncertainty in the problem. This takes the form of both uncertain systems (e.g., modeling errors) and uncertain signals (e.g., environmental disturbances), which cannot be accurately predicted a priori. In an augmented cognition problem system, uncertainty might take the form of (necessarily) inaccurate models of the human operator, and signal uncertainty could arise from measurement noise. In order to address these issues, we need to extend Figure 7.1 to the more realistic closed-loop system illustrated in Figure 7.3.

Control systems incorporating uncertainty. Figure 7.3 shows a more complex control system in which the modeling of Plant P includes uncertainty.

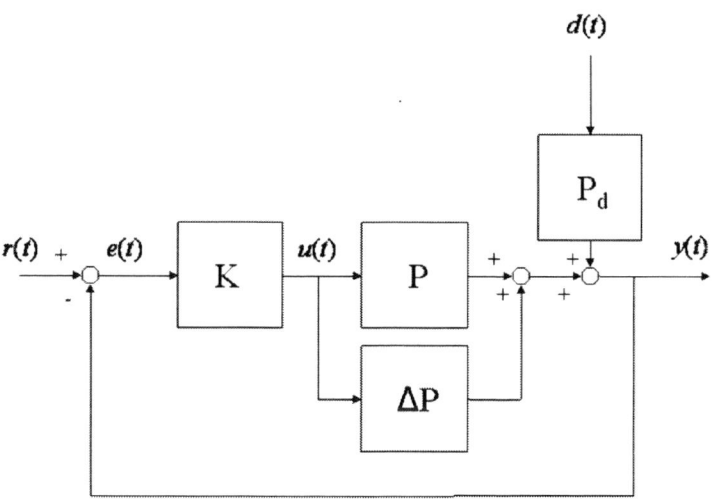

Figure 7.3. Perturbed closed-loop system.

This system now shows uncertain dynamics in the form of model uncertainty ΔP (e.g., errors in our engine model caused by component aging), as well as an unknown disturbance signal d(t)— for example, varying terrain slope that we are traversing. The disturbance signal d(t) may be filtered by some dynamics P_d that we can model. For instance, we can model the effect of any given known slope on the vehicle speed via a dynamic model P_d because the physics of increased load attributable to slope are well understood; however, we do not know what the upcoming slopes d(t) will actually be in practice.

The perturbed closed-loop system in Figure 7.3 can form the basis for a robust controller design; that is, a controller algorithm defining the mapping from e(t) to u(t):

$$u(t) = K(e(t))$$

such that the system output y(t) is largely insensitive to both d(t) and ΔP and instead accurately tracks r(t) despite the modeling errors and disturbance signals. **In other words, the cruise controller, based on a robust feedback design, will hold the desired speed even though the model used to design it was inaccurate and despite whatever terrain the vehicle encounters (e.g., an uphill slope).** Of course, the design of such controllers can be a mathematically challenging task, involving tools from optimization, perturbation theory, operator theory, linear algebra, complex analysis, and other fields. Those details are beyond the scope of the current chapter, but the interested reader is referred to Ogata (2002) and the references therein as a starting point. Rather, here we are concerned with the core concepts of feedback control and how we can put behavioral science problems in this framework, so that the aforementioned tools can be brought to bear on these problems to yield new systematic approaches to tackling them.

In that regard, let us revisit the driving problem, and its system theoretic abstraction in Figure 7.3, from a different viewpoint. Suppose now that we wish to consider an augmented cognition system, providing an advanced human-computer interface to assist a person driving a vehicle. The plant P now becomes a model of the human operator and vehicle response. The input to this plant could be a piece of information from the display, and the output could be the operator decision/action and the resulting vehicle performance. The goal of the controller K is now to display the information in such a way as to optimally load human operators so that they perform at their peak. The results of the operators' actions, plus additional measurements regarding their cognitive state (eye movement, EEG, etc.), are fed back to the controller to allow it to dynamically adjust the display as desired.

Test Your Knowledge
Consider the augmented cognition vehicle-operator assistance example described in the paragraph above. In order to relate it to robust control theory tools, identify the corresponding signals and systems in Figure 7.1, and then in Figure 7.3, for this example.

Dynamic modeling. Note that the type of model used here may be quite different from our earlier (cruise control) example, and, hence, the mathematical tools will need to be tailored to these model types. The underlying framework is the same, however, and the basic building blocks of feedback control theory still apply. For instance, the approach taken to modeling such a system in Young, Clegg, and Smith (2004) models the human information-processing system as three processing stages, as first described in Newell (1990). Information from the system displays first passes through a perceptual processor and is then available to the cognitive processor. The cognitive processor then makes an appropriate decision, and that decision is implemented by the motor processor, with a resulting action on the vehicle controls.

Note that the time taken by each block adds time delay to the model. However, it does much more than that: It also implies a certain bandwidth for the system, because signals that vary more rapidly than the time constant of the system (i.e., high-frequency signals) do not pass through it. Hence, the processing blocks act as low-pass filters, allowing only signals that are below the system bandwidth to pass through. For example, you do not perceive the flicker on a computer monitor because it occurs at a high frequency (e.g., 100 Hz), whereas the perceptual processor has a time constant of about 100 ms, which corresponds to a bandwidth of only about 10 Hz. Taking this approach to modeling each of the human perceptual, cognitive, and motor processing blocks results in a dynamic model of the form shown in Figure 7.4.

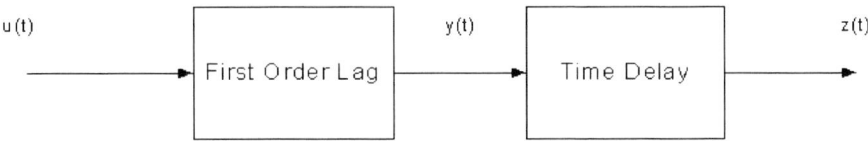

Figure 7.4. Block model for each component.

The dynamic models associated with each component ("First-Order Lag" and "Time Delay") of the foregoing block model are given, respectively, as

$$\tau \frac{dy(t)}{dt} + y(t) = u(t)$$

$$z(t) = y(t - \tau)$$

for each processing block, with overall input u(t) and output z(t), and the time constant τ taken from the relevant processing time in the human information-processor model.

This dynamic system model also captures the frequency domain representation of our system. In order to see this, we must first transform the system model from time to frequency domain, via a Laplace transform. This is defined for a time domain function f(t), which has the real-valued independent variable t representing time, as follows:

$$F(s) = L\{f(t)\} = \int_0^\infty f(t)e^{-st}dt$$

The resulting frequency domain function F(s) has the Laplace frequency variable s (which takes values in the complex plane) as its independent variable. This transform enjoys the following well-known property when applied to differential operators:

$$L\left\{\frac{df(t)}{dt}\right\} = sF(s) - f(0)$$

The Laplace transform is linear, so it can be applied term by term to a summation. Hence, taking Laplace transforms (to map to frequency domain) on both sides of our ODE model for the First-Order Lag, ignoring initial conditions, and rearranging yields

$$Y(s) = G(s)U(s)$$

where the function G(s) is given as

$$G(s) = \frac{1}{1 + s\tau}$$

This is known as the *transfer function* of the system (see Phillips, Parr, & Riskin, 2008, for an overview of transform methods for signals and systems, and see Ogata, 2002, for an overview of the application of these techniques to dynamic systems and feedback control). A key point is that the differential equation model in the time domain has been transformed into a simple multiplication operator in the frequency domain. That multiplication operator, G(s), is both complex-valued and frequency-varying. The function G(s) captures the frequency response of our system, in both magnitude and phase.

Because transfer functions operate by multiplication, we can cascade the models for the individual blocks. Furthermore, given that these are linear models, they "commute," so we can change the order of cascade and, hence, accumulate our time delays into a single block. This now provides us with a quantitative dynamic model for the human, as illustrated in Figure 7.5, which is suitable for use in a feedback-control loop.

Applied Exercise
What is the overall transfer function and time delay for the cascade of systems shown in Figure 7.5? Is this an open- or closed-loop model?

ENGINEERING CONTROL SYSTEM THEORY

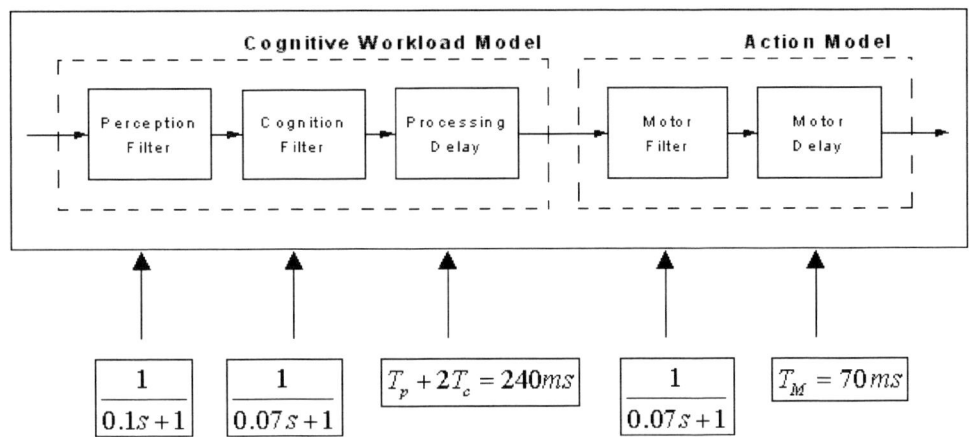

Figure 7.5. Dynamic control system model of the operator.

Behavioral science applications. The model in Figure 7.5 was used by Young et al. (2004) to demonstrate simulation results for a closed-loop augmented cognition system, as well as to carry out robustness analysis. We review some of those results in the next section, but for now we move on with our control-modeling concepts.

Note that we started with a person driving a car, then considered an automatic cruise control system, and then an augmented cognition system assisting the driver. We can move still further up this abstraction hierarchy by taking a bigger-picture view of the driving problem. What if we wanted to change the behavior of the driver in a fundamental, long-term sense? For example, suppose we had an overly aggressive driver, prone to road rage, and we wished to modify that behavior. One could consider an intervention strategy whereby the driver receives a series of counseling sessions. This is an expensive procedure. Furthermore, in many intervention scenarios, there can be negative consequences from overdoing the number of sessions (e.g., resentment over what is perceived as unnecessary repetition, perception of the intervention as browbeating) in addition to the unnecessary extra expense involved. These are some of the reasons people have started to consider adaptive intervention approaches (Collins, Murphy, & Bierman, 2004).

In contrast to the more typical standardized interventions, in which each program recipient receives the same services, in an adaptive intervention, the services are tailored to each individual. Not only are decisions about which services to administer based on the individual's status at the beginning of the intervention, but there is also periodic monitoring of the impact of the intervention on both outcomes and mediating variables. This could take the form of surveys, or, in the case of road rage, one might even consider evaluation via a virtual reality driving simulator, which could put the driver through various (potentially stressful) scenarios. Based on these data, the clinician could then make decisions about the amount, type, and/or timing of the next phase of the intervention. Note that once again a feedback loop has been closed.

This scenario again fits into the framework of Figure 7.3, with the plant P now representing a model of the driver's behavior and the controller K representing the decision process undertaken by the clinician. Furthermore, one can now consider an automatic control system by replacing the clinician with a controller K, implementing an automated decision policy. Indeed, some researchers have started to investigate this approach. For example, one might consider a model, P, for the driver's reaction to the counseling as follows:

$$A(k+1) = A(k) - E(k)I(k) + D(k)$$

where $A(k)$ represents the aggressive tendencies of the driver at sample (intervention) number k, $E(k)$ a measure of the intervention's effectiveness, $I(k)$ the intervention, and $D(k)$ the driver's tendency to slip back into old driving habits. This is a difference equation representing a discrete-time system, because we are able to sample/measure the subject only at a number of discrete time points. It still falls into the framework of Figure 7.3, as described earlier, but now with a different flavor of mathematical model, and, hence, it requires appropriate mathematical tools. In this case, there is a well-developed feedback control theory of discrete time or digital systems that would help us to design an appropriate digital controller.

Indeed, Rivera, Pew, Collins, and Murphy (2005) looked at a number of feedback control strategies as candidates for automating the decision process in an adaptive intervention scheme and demonstrated the efficacy of automated control versus a traditional clinical approach. The control approaches they considered include rule-based control, proportional-integral-derivative (PID) control, and model predictive control (MPC). **The MPC approach is an advanced, optimization-based tool from robust control theory that has seen wide application in the process control industry**. Indeed, we believe that robust control approaches will be a mainstay for control problems arising from the behavioral sciences because of the presence of humans in the loop. We will always have models for the human that are subject to a great deal of variation and uncertainty; hence the need for a robust approach.

Review: SISO Models
In this section, we have described a series of single-input–single-output (SISO) models for the control of a variety of plants P. The models included an open-loop control system that does not utilize feedback at all (Figure 7.2) and a simple closed-loop model that takes feedback into account but does not directly account for errors in the modeling process and external disturbances (Figure 7.1). Finally, a closed-loop system that allows for both uncertainty in the model and unknown external disturbances was discussed (Figure 7.3). This final model motivates the use of powerful, robust control methodologies to guarantee the (robust) stability of the system. However, given the complexities of the human information-processing system, augmented cognition systems are likely to require this type of methodology to be successful.

Lessons Learned

The augmented cognition model described in the preceding section has indeed been used for closed-loop controller analysis and design as well as for simulation studies. In particular, the system considers utilizing a cognitive workload assessor (CWA) to take a measurement of cognitive workload (i.e., via brain sensors, see Chapters 1 and 2). A systems interface director (SID) then takes this measurement and uses it to change the inputs to the human through some type of adaptive automation (AA; i.e., via mitigations; see Chapters 5 and 6).

The CWA, SID, and AA schemes have been active areas of research in the augmented cognition community for some time. For example, Schmorrow, Stanney, Wilson, and Young (2006) modeled the CWA as a simple first-order lag with a time constant of $\tau = 1s$. The SID and AA were modeled via a scheme whereby the workload was reduced (e.g., by passing lower-priority tasks to automated agents) upon detecting cognitive work overload, with the goal of maintaining the human working at maximum capacity. The details of that scheme constitute the *feedback control algorithm,* which will be discussed shortly. The entire augmented cognition, closed-loop, human-machine system model has been implemented in the Matlab/Simulink simulation environment and is illustrated in Figure 7.6.

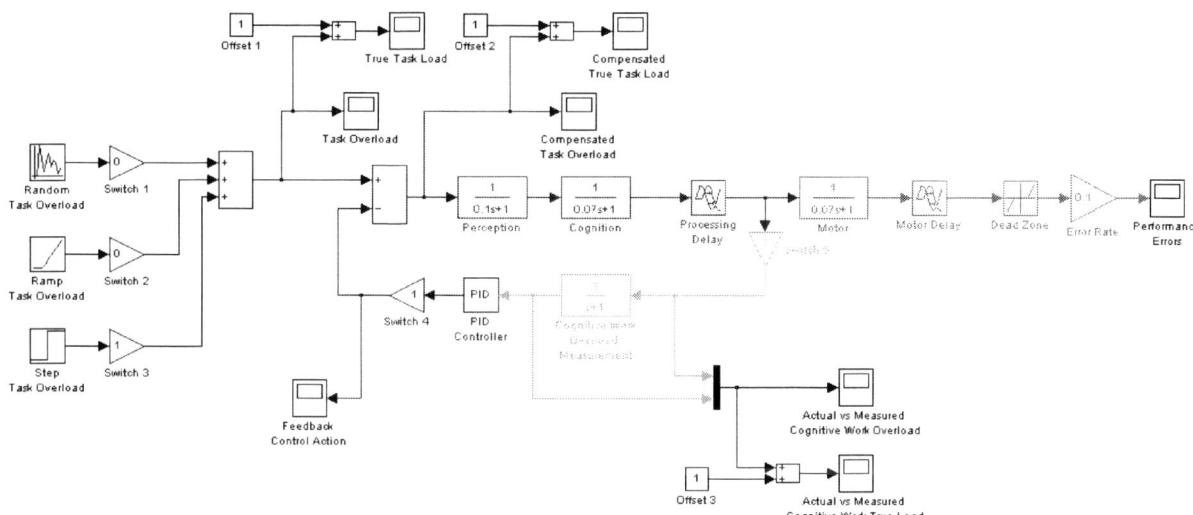

Figure 7.6. Matlab/Simulink model of a closed-loop augmented cognition system.

The various pieces of an augmented cognition system can be seen in this model: the human, including the cognitive processes and the motor processes; the CWA system model that will be detecting the state of the human; and the SID/AA systems that will actually alter the input to the human. The rest of the model shows task inputs to the system, displayed outputs at various points (e.g., actual versus measured cognitive workload), and a simple model of performance errors resulting from cognitive overload. The feedback loop being closed is now apparent in this simulation model, which drives the need for a systematic control theory approach.

Initial simulation studies with this model show how dynamic instability can result from introducing feedback within the system. That is, rapid detection of a cognitive state under high workload might cause input to be removed, which would reduce the workload, and hence information is added, which would once again result in high workload, and so the cycle repeats. As a simple illustration, an operator might find the display rapidly cycling through cluttered and decluttered states as a result of these changes in workload being created by the automation. This is illustrated in Figure 7.7, which shows simulation results from a task overload situation.

The input to each of these simulations is the same—namely, the operator is initially fully loaded (and making no errors), and then a step increase in workload is introduced 1 s into the simulation. This results in a task overload from that point on, with subsequent performance errors. Note that each of these simulations also uses the same system model, so that the only difference is how (or whether) the feedback control is applied/designed.

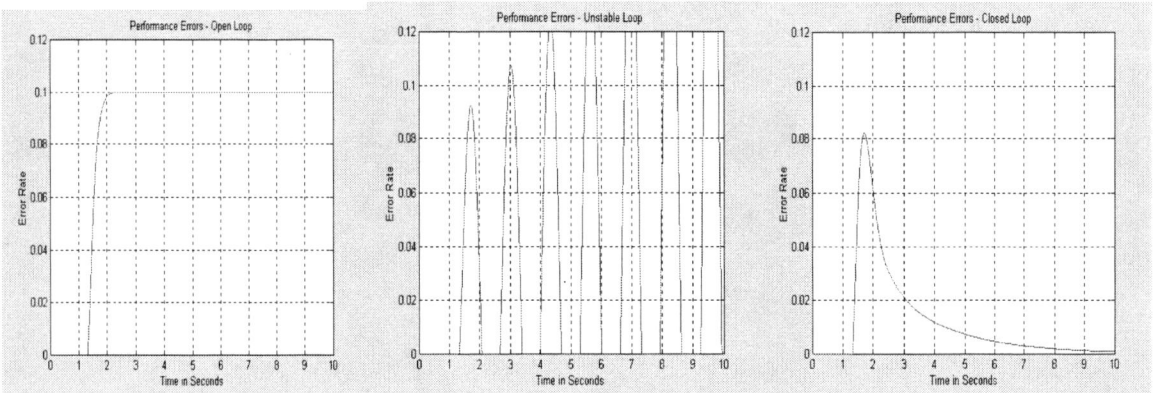

Figure 7.7. Simulation results for a closed-loop augmented cognition system.

Starting from the left, **the first plot of Figure 7.7 shows the resulting performance errors for an open-loop simulation (i.e., with the augmented cognition system disabled)**. As the workload of the task increases, the plot shows how the number of errors quickly rises to a certain level and stays there. **The next panel shows a poorly designed SID/AA controller**. This utilizes simple proportional control; for example, the control action c(t) that reduces the task workload to the operator is just directly proportional to the measured overload m(t). Thus, the controller is of the form

$$c(t) = Km(t),$$

and the control designer simply chooses the proportionality constant or controller gain K, which determines how much feedback is implemented. High-gain controllers, with large K, use a large-magnitude feedback signal, which aggressively tries to drive the control loop to the desired point. If K is chosen too aggressively, however, the closed-loop system will approach (or even exceed) the stability margins. In this example, the gain K is chosen poorly, resulting in instability of the type described earlier, with the input being rapidly reduced and then increased, which leads to highly fluctuating performance from the operator.

Test Your Knowledge
Now reconsider the (control system) examples you thought of earlier. How would scenarios of no control, poor control (e.g., very high gain), and good control be generated? How would the resulting behaviors manifest themselves?

A simple proportional control algorithm can deliver only limited performance improvements, even when designed correctly. For instance, one could never get the steady state errors down to zero with this type of control. It is limited because it utilizes the same gain for all frequencies (and hence all signals), so that one does not have sufficient degrees of freedom to exploit any trade-offs in the design.

A very common type of controller used in engineering applications is the PID controller. This generates a corrective action from the measurement of the form

$$c(t) = K_P m(t) + K_I \int_0^t m(\tau)d\tau + K_D \frac{dm(t)}{dt}$$

This type of controller has been considered for several behavioral science problems, including the augmented cognition results reviewed here from Young et al. (2004) and Schmorrow et al. (2006) as well as the adaptive intervention studies in Rivera et al. (2005). There are now three constants to be chosen (designed): K_P, K_I, and K_D. These correspond to the amounts of proportional, integral, and derivative feedback used in the closed loop.

Note that because the integral action effectively includes memory, it allows for better compensation at low frequency and, hence, improved steady-state performance. The derivative action essentially includes anticipation, and this allows for improved high-frequency performance, resulting in better transient response and improved stability properties. The overall controller has frequency varying gain, which allows for the design trade-offs to be more properly exploited. **The final, right panel of Figure 7.7 shows a functional closed-loop augmented cognition system that uses a well-designed PID controller to deliver closed-loop stability and good performance.** It can be seen clearly that errors—even at maximum—never reach the level of the open-loop (automation-free) system and that they quickly drop to minimal levels (asymptotically approaching zero) without any undesirable oscillatory transient response.

The power of systematic control-theoretic approaches for behavioral science problems is further illustrated in Rivera et al. (2005). The authors consider an adaptive intervention scheme inspired by the FAST Track program (Conduct Problems Prevention

Group, 1992) and aimed at preventing the development of conduct disorders in children via an intervention based on family counseling. The measurement used is the level of parental functioning (PF), as assessed via a questionnaire. The analysis considers a heuristic, rule-based approach intended to model a clinician deciding on appropriate dosage levels for the intervention. This is compared with an automated scheme based on feedback control theory utilizing a PID controller.

In one simulation, the conventional approach fails to attain the goal through 36 months of the intervention. That is, in control terms, the steady-state error does not approach zero, and this undesirable phenomenon results from the simple nature of the decision rules applied to this intervention. In contrast, the PID-based controller is able to take the system to the goal, illustrating once again the power of systematic engineering control theory for these types of behavioral problems. We refer the interested reader to Rivera et al. (2005) for details of the approach and for another tutorial introduction to control theory applied to behavioral science problems.

Another behavioral area that holds promise for the application of control systems theory is in interventions to increase medication compliance. For example, compliance with medical regimes for HIV/AIDS is often in the range of 57% to 77% (e.g., Spire et al., 2002). This is particularly problematic in that not only will the individual's disease progress, but intermittent use of the medication contributes to the development of drug-resistant strains of the virus.

Interventions to increase compliance with HIV/AIDS medication face a number of obstacles. The combinations of medications are complicated, they often have unpleasant side effects, and the individual must continue to take them indefinitely. Moreover, unlike patients with many other chronic diseases (e.g., diabetes, asthma, and hypertension), many patients diagnosed with HIV are symptom-free. Therefore, there is not the usual physiological feedback system associated with stopping medication. In contrast, because of decreased side effects, HIV-positive individuals may actually feel better when they stop taking their medication.

Recently, interventions have been developed that attempt to increase the HIV-positive individual's motivation to comply with the medication regime. Adaptive interventions can utilize a wide variety of information sources to customize the information and other resources provided to the patient. Physiological information can include levels of medication in the blood stream, T-cell counts, and the presence of drug-resistant strains. Behavioral surveys can measure risky behaviors, including sexual activity and recreational drug use. Finally, community data can be used, including community norms around HIV/AIDS. All these variables are time-varying, which makes this a very complex system and one ideally suited to the application of control systems methodology.

The dynamic model of these systems is fundamental to the execution of the rigorous control techniques that have been described in this chapter. There are so-called black-box methods, which eschew the dynamic model in favor of some online learning. These approaches can have their merits under certain circumstances, but they also have their limitations. In particular, they need to allow sufficient time to learn, especially when the underlying system dynamics are complex (as they are for augmented cognition systems). For problems in the behavioral sciences, there will be little time to learn (e.g., in an adaptive mitigation intervention [see Chapters 5 and 6], in which each measurement is expensive and time-consuming and one may be able to perform only a limited number of actions or measurements). In any case, these approaches usually do not come with any a priori guarantees of stability, much less performance, which makes them unattractive for application, because they run online in real time on mission-critical physical systems. Some stability theories are starting to appear for learning-based systems (cf. Kretchmar, Young, Anderson, Hittle, Anderson, & Delnero, 2001). However, these are usually based on combining the learning approach with more traditional controls tools, and, hence, once again they are ultimately model-based.

Engineering control theory will play a crucial role in the future development of augmented cognition systems. There is great potential for an exciting interplay between engineering control theory and various aspects of the behavioral sciences. Moreover, the challenging nature of the problems, and the unique viewpoints that each party brings to the table, should drive advances in both areas and fuel further collaborative efforts.

Best Practices

The power of control-theoretic approaches for many disciplines, including the behavioral sciences, has been outlined in this chapter. In order to get the most from these control methodologies, particularly as applied to behavioral science problems such as augmented cognition systems, we recommend following a few guidelines, as outlined in Table 7.2.

Table 7.2. Best Practices in Using Control System Engineering for Augmented Cognition.

- Start with a simple model. It may be tempting to throw in everything you know about the system. However, that may unnecessarily complicate the design process and the resulting controller. It is better to start simply and add model features only as the design dictates that you actually need them.

- Adopt a performance-driven approach. At the end of the day, closed-loop performance is what actually matters. Hence, any strategies/costs should be evaluated with that in mind (see the foregoing text for an example—a high-order model might be slightly more accurate but does not offer enough advantage to justify the cost).

- Examine and exploit trade-offs. Feedback control is all about trade-offs (e.g., "I might have to choose between speed and accuracy for a sensor"). It is very often possible to further improve almost any given quantity, but the designer must weigh that against the cost.

- Better measurements are almost always a good idea. Control system performance, particularly for augmented cognition systems, can be limited by the quality of available measurements. Unless the cost is prohibitive, better measurements are likely to enhance closed-loop performance.

Design Guidelines

In the previous sections, we have focused on the utility of robust control methodology for the design and implementation of interventions to augment human decision making and performance. These methodologies are part of an overarching design strategy that includes (a) determining system specifications, (b) developing a system model, (c) designing the controller, and (d) iteratively evaluating the design. In this section, we provide an overview of the design strategy required to implement robust control methodology on the problem of interest.

Step 1: Determine system specifications. As stated earlier, we favor a systematic feedback control design approach. The starting point for such a design is to determine your system specifications. This means specifying, in as detailed and quantitative a fashion as possible, the **desired operation of your system**. The more specific you can be at this stage, the better. The reason is that although we can all agree on a long list of features for a system that would be good (high speed and accuracy, low cost, high reliability, etc.), it is often true that those features contradict one another in the sense that delivering very high levels on one of them necessarily detracts from others. Engineering, and, in particular, feedback control is all about managing these trade-offs. Hence it is important to accurately specify what you really want—and only what you really want—so that a good/optimal trade-off can be achieved.

In the case of augmented cognition systems, a feedback loop is closed around a human operator. However advanced the sensors, data fusion, interface design, and other items may be, that fundamental fact remains. All feedback loops experience a natural tension between stability and performance. In order to get higher performance, one has to push the control system harder, and this inevitably draws one closer to the stability boundaries. This was seen in its simplest form for the proportional control system described earlier (i.e., the augmented cognition system performance shown in Figure 7.7), in which higher gain delivered faster response but at the expense of more oscillation. The trade-off between stability and performance is not always as transparent as that, but it is always there in some form or another. Hence, it is of paramount importance that we be explicit in specifying which trade-offs are acceptable and which are desirable.

Step 2: Develop a system model. Now it is time to model your system. Note that this step comes second because, in fact, the desired specifications can dictate choices in the modeling approach. They can guide the selection of model granularity or abstraction. Consider the very different types of models in our examples of the operator driving a vehicle. There are many techniques for system identification (i.e., extracting or abstracting models from data). Once again, there are difficult decisions to be made with regard to trade-offs. It is, of course, desirable to have the model match physical reality as closely as possible. However, such models might be expensive to generate, in terms of the types and amount of data needed. Furthermore, increased accuracy often brings with it the expense of increased mathematical complexity, which could make the associated analysis and design problems intractable. Note also that for robust analysis and design, one also needs to consider uncertainty characterization as part of the modeling process. This means that we would like as much information as possible about the unknown or unmodeled dynamics and disturbance signals. Of course, we cannot get complete and precise information, else these would not be uncertainties, but any information we can obtain (e.g., "How does the disturbance size vary with frequency?") might be exploited to provide better analysis and design tools.

Step 3: Design the controller. Next comes the controller design phase. First we need to combine and convert the specifications and the model into a design framework. Note that this may also involve some simplifications for tractability. In a typical robust/optimal controller design setting, this step means deciding on the appropriate topology (i.e., where uncertainties, disturbances, measurements, and control signals enter the system) as well as the desired robustness specification and performance criterion, often specified in terms of weighted penalty signals.

It should be noted that these steps are of tremendous importance. Although we are in some sense only posing the problem, it is here that we enforce decisions on the trade-offs discussed earlier using engineering judgment. From this point on, it is largely a mathematical exercise to "solve" for the controller that optimally exploits these trade-offs as specified by our choices for performance and robustness criteria. The actual design itself is then typically carried out via some computer-aided tool, usually employing numerical optimization algorithms.

Step 4: Iteratively evaluate the design. After Step 3, one would then test the design in simulation (simulation models often can be more complex and higher in fidelity than design models), as well as carry out any desired robustness analysis prior to actual implementation and testing. Indeed, each of these tests may trigger a redesign. For this reason, the whole process should be viewed as an iterative one, rather than a one-shot design. It is a common mistake to view computer-aided "optimal" designs as one-shot because they are "optimal." That optimality is, of course, only with respect to whatever criterion you have specified (which imparts your decisions about the various design trade-offs). Changing those criteria redefines the notion of optimality and, hence, the iterative design process.

Review: Robust Control Design Guidelines
In this section, we have described a series of steps that the practitioner would need to take in order to place an augmented cognition problem within the control system framework. In Table 7.3, these steps are reviewed and the practitioner is provided with examples of the types of questions that should be asked during each step of this process.

Table 7.3. Robust Control Design Guidelines

STEP 1: Determine System Specifications
Specify, in as detailed and quantitative a fashion as possible, the desired operation of your system. What aspect(s) of performance is/are most important to achieve? <ul><li>Is reliability or peak performance more important?</li><li>Is one variable more important to optimize than others?</li><li>What (numerical) level of performance is desired?</li></ul>
STEP 2: Develop a System Model
Specify a mathematical model of the relations between the elements of your system (e.g., perceptual, cognitive, motor system models), drawn from both theoretical and empirical research. <ul><li>What are the expected value ranges for parameters of interest?</li></ul> For robust analysis and design, one also needs to consider uncertainty characterization as part of the modeling process. <ul><li>Where is error in the model most likely?</li><li>What variables have, by necessity, been omitted from the model?</li></ul>
STEP 3: Design the Controller
Combine and convert the specifications and the model into a design framework. <ul><li>Where will uncertainties, disturbances, measurements (e.g., brain sensor readings; see Chapters 1 and 2), and control signals (e.g., mitigations; see Chapters 5 and 6) enter the system?</li><li>What is the range of acceptable outcomes? What is the most desirable outcome?</li><li>How often will input/output be measured?</li><li>How often will the system be adjusted? Is there a maximum amount the system should be adjusted at one time?</li></ul>
STEP 4: Iteratively Evaluate the Design
Test the design both in simulation and in real-world tests.
The model and the controller may both be adapted based on observed performance.

Parting Message

Feedback control systems are in widespread use in many areas of technology. However, the ability of robust control methodology to guarantee stability in the face of uncertainty makes it particularly well suited to problems in the behavioral sciences. Augmented cognition models will have to include the human operator in some form, whether it is a detailed brain model for use with an augmented cognition sensor or a higher-level behavioral model. Whatever advances one might anticipate in our understanding of these complex issues, it is highly likely that augmented cognition models will contain significant uncertainties in the dynamic response. These could come from a combination of gaps in our knowledge, difficulty of obtaining measurements, and/or individual variation among operators. Hence, it becomes of paramount importance to systematically address this model uncertainty so that stability is achieved and performance is guaranteed. We firmly believe, for all the foregoing reasons, that a rigorous, systematic (robust) control-theoretic approach to feedback controller analysis and design is the only methodology that will enable practitioners to deliver the high performance they seek.

References

Conduct Problems Prevention Group. (1992). A developmental and clinical model for the prevention of conduct disorder: The FAST Track Program. *Development and Psychopathology, 4*, 509–527.

Collins, L. M., Murphy, S. A., & Bierman, K. (2004). A conceptual framework for adaptive preventive interventions. *Prevention Science, 3*, 185–196.

Fagg, A., Hatsopoulos, N., de Lafuente, V., Moxon, K., Nemati, S., Rebosco, J., Romo. R., Solla, S., Reimer, J., Tkach, D., Pohlmeyer, E., & Miller, L. (2007). Biomimetic brain machine interfaces for the control of movement. *Journal of Neuroscience, 44*, 11842–11846.

Kretchmar, R., Young, P., Anderson, C., Hittle, D., Anderson M., & Delnero, C. (2001). Robust reinforcement learning control with static and dynamic stability. *International Journal of Robust and Nonlinear Control, 11*, 1469–1500.

Ljung, L. (1999). *System identification* (2nd ed.). New York: Prentice-Hall.

Newell, A. (1990). *Unified theories of cognition*. Cambridge, MA: Harvard University Press.

Ogata, K. (2002). *Modern control engineering* (4th ed.). New York: Prentice Hall.

Phillips, C. L., Parr, J., & Riskin, E. (2008). *Signals, systems and transforms* (4th ed.). New York: Prentice Hall.

Rivera, D., Pew, M., Collins, L., & Murphy, S. (2005). Engineering control approaches for the design and analysis of adaptive time-varying interventions (Tech. Report 05-73). Philadelphia: The Methodology Center, Pennsylvania State University.

Schmorrow, D., Stanney, K. M., Wilson, G., & Young, P. (2006). Augmented cognition in human-system interaction. In G. Salvendy (Ed.), *Handbook of human factors and ergonomics* (3rd ed., pp. 1354–1383). New York: Wiley.

Spire, B., Duran, S., Souville, M., Leport, C., Raffi, F., Moatti, J. P., & APROCO Cohort Study Group. (2002). Adherence to highly active antiretroviral therapies (HAART) in HIV-infected patients: From a predictive to dynamic approach. *Social Science and Medicine, 54*, 1481–1496.

Tremoulet, P., Barton, J., Craven, P., Gifford, A., Morizio, N., Belov, N., Stibler, K., Regli, S., & Thomas, M. (2006, October). *Augmented cognition for tactical Tomahawk weapons control system operators*. Paper presented at the 2nd International Conference on Augmented Cognition, San Francisco, CA.

Young, P., Clegg, B., & Smith, C. (2004). Dynamic models of augmented cognition. *International Journal of Human-Computer Interaction, 17*, 259–273.

Chapter 8

Design Platform Methodology for Augmented Cognition

Mark Austin[1] and Colby Raley[2]
[1]University of Maryland-College Park
[2]Strategic Analysis, Inc.

As augmented cognition technologies mature, systems transition from the military to the commercial domain. Platform-based design will become indispensable in system-level development, evaluation, and tuning to end-user capability.

Introduction

In this chapter, we explain how the principles of platform-based design will benefit those who design and develop augmented cognition systems, particularly systems with a commercial end use. We begin with overviews of augmented cognition and engineering system development, including system development models and the pathway from industrial-age to information-age systems, which also includes augmented cognition. We then describe a variety of techniques for handling system complexity in design, the enhancement of reuse via system modularity, and a general approach to platform-based design. Finally, we propose a methodology and platform architecture for the definition, architecture-level design, and personalized customization of automobile dashboards that are enhanced with augmented cognition.

Augmented Cognition
Augmented cognition researchers are developing new technologies to noninvasively measure the cognitive state of humans (see Chapters 1 and 2) and to use that state information to adapt closed-loop computational systems to accommodate human needs. Mitigation strategies (i.e., scheduling; see Chapters 5 and 6) are used to increase overall performance while maintaining situation awareness and enabling a highly engaged user to take on increasingly difficult tasks (Schmorrow and Kruse, 2004).

As indicated in Figure 8.1, augmented cognition systems gather information associated with the cognitive state of the user, the status of a task, and the state of the surrounding environment. The general flow of information is as follows:

1. Commands influence a task.
2. Sensors detect activity (see Chapters 1 and 2).
3. Sensed and modeled information is combined (see Chapter 3) to create real-time models (see Chapter 7).
4. State and interface information is used by an augmentation manager to determine an appropriate strategy to mitigate information bottlenecks (see Chapters 5 and 6).
5. The augmentation manager affects the interface.
6. The interface communicates with the user.
7. Autonomous agents can complete tasks for which the user is too overloaded (Raley and Marshall, 2004).

Augmented cognition was historically developed by government, military, and academic researchers, but as the technologies have matured, industry applications are becoming feasible. Future commercial applications of augmented cognition may be best developed and supported through frameworks for platform-based design (PBD).

The PBD approach (see Figure 8.2) assumes that good design solutions can be developed through the selection and assembly of components or subsystems that are developed for a particular class of problems. **A platform is an organizational abstraction that places a priority on bridging the gap between functional design and implementation with modular subsystems**. Platforms come in a variety of forms. As illustrated in Figure 8.2, **the goal of platform-based design is to achieve efficiency and customization in the creation of design iterations**.

The creation of design iterations includes synthesis, design, evaluation, and customization. Each iteration is part of an ongoing development process in which, temporally, customer requirements change and implementation technologies advance. From a process perspective, multi-project integration is achieved via a stable basis of processes and tools that support networks of cooperation with project partners and suppliers (Austin, 2004, 2005; Sangiovanni-Vincentelli, 2003; Simpson, Jonathon, & Mistree, 2001).

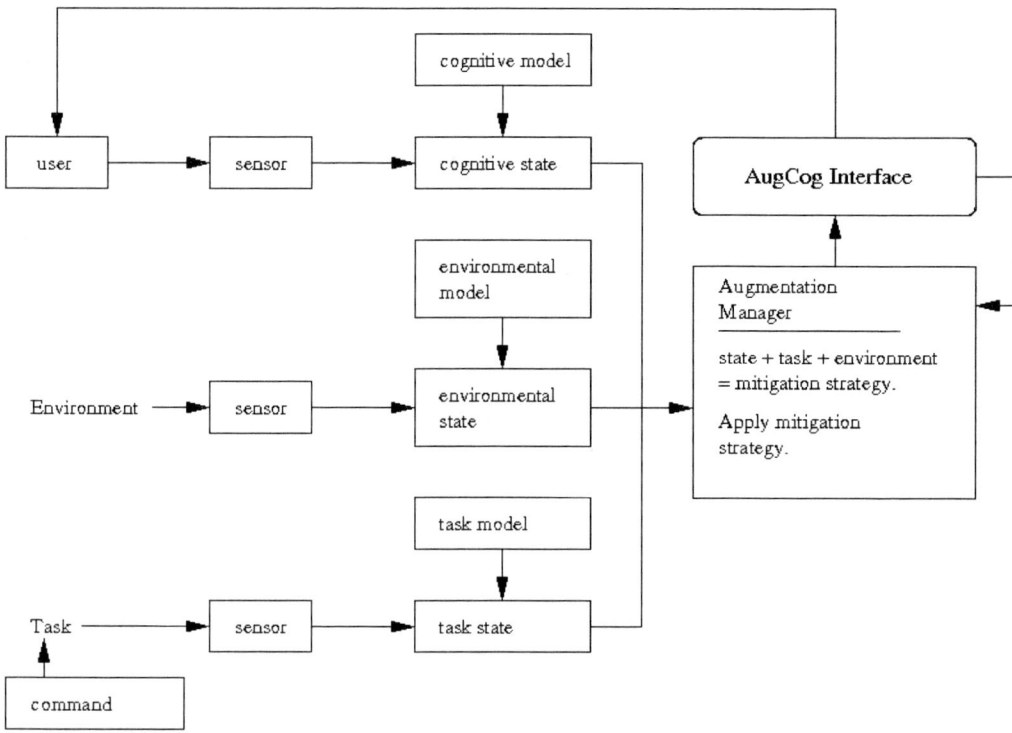

Figure 8.1. System-level architecture for augmented cognition.

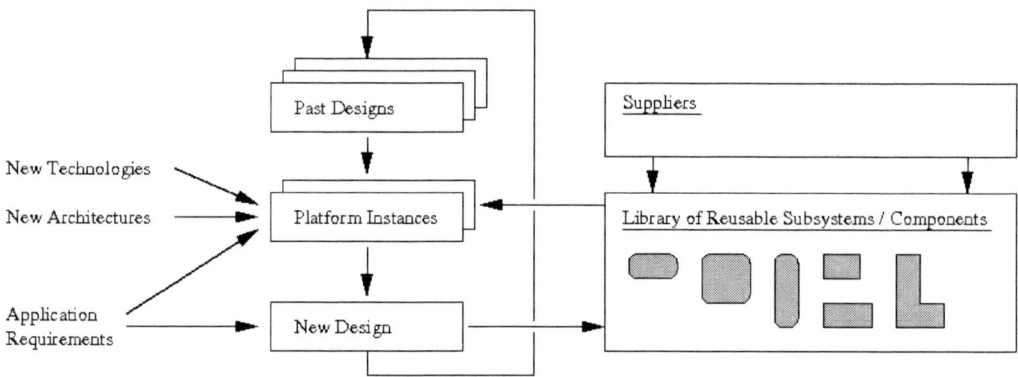

Figure 8.2. Flowchart of activities for platform-based design.

Design Platforms

Platform-based design procedures are important for many areas of engineering and commerce, but for two key reasons they are particularly relevant to augmented cognition systems (Raley, 2005):

1. The technologies required to implement augmented cognition systems are still being developed and are expected to improve in capability in the coming years. Therefore, system architectures will have to provide constant or improving performance while utilizing ever-changing components.

2. Augmented cognition systems will continually be used in new scenarios and situations; therefore, the systems should be easily adaptable to new environments.

Present-Day Platforms

Design platforms are characterized by layers of well-defined design abstraction, problem-domain specific design constraints and cost functions, and a meet-me-in-the-middle design process. To date, design platforms for engineering application have been developed for automotive electronics (Keutzer, Malik, Newton, & Sangiovanni-Vincentelli, 2000; Pinto, Bonivento, & Sangiovanni-Vincentelli, 2006; Sangiovanni-Vincentelli, 2000), autonomous vehicles (Horowitz et al., 2003), spacecraft (Gonzales-Zugasti, Otto, & Baker, 2000), and other applications. A number of computer manufactures (e.g., Dell, Apple) use platform-based design to support the assembly of customized computers that are driven by specifications (e.g., operating system, RAM, storage space) derived from user needs. In the 1970s and '80s it was common practice to write computer programs in low-level assembly languages. These have now been replaced by high-level languages such as Java and integrated development platforms such as Eclipse (Holzner, 2004) and NetBeans (http://www.netbeans.org/). Augmented cognition development efforts are likely to be highly analogous to this complex development of computational systems.

Pathway from Military to Commercial Applications

J. C. R. Licklider (1960) laid the groundwork for human-computer interaction[1] (HCI) in the military by calling for unprecedented cooperation between humans and machines. Two years later, Licklider became the director of the Information Processing Techniques Office at the Department of Defense's (DoD) Advanced Research Projects Agency, where his views on the ways in which humans and computers should work together had a large impact on the development of advanced computing technology in DoD systems. For example, years later, in ARPA's biocybernetics program, researchers explored the use of biologically measurable signals, helped by real-time computer processing, to assist in the control of vehicles, weaponry, or other systems (Beatty, 1978). Performers on the program at the time were investigating the use of electroencephalogram (EEG) measures (specifically, the P-300 response) to measure workload (Isreal et al., 1980).

Recent research in augmented cognition has assessed automation techniques for maximizing warfighter performance and efficiency under stress with an emphasis on prototype experimentation in conjunction with battle experiments (see Chapter 9).

---

[1] Human factors/ergonomics aims to understand the interactions between humans and systems and uses various methods to optimize human well-being and system performance (IEA, 2000). Human-computer interaction focuses on computers as the primary system with which humans must interact, given the continued exponential growth in the use of computational systems in all types of work.

Because of the DoD's Biocybernetics and Pilot's Associate programs, and DARPA's Improving Warfighter Information Intake Under Stress program, most of the platforms explored thus far for augmented systems lie in the military domain (Barker & Edwards, 2005; Dickson, 2005; Horowitz et al., 2003; Ververs, Whitlow, Dorneich, Mathan, & Sampson, 2005; Gonzales-Zugasti, Otto & Baker, 2000). As augmented cognition technologies mature, it can be expected that systems will find their way into a wide array of commercial applications, particularly those in which satisfaction with safety and time constraints are prerequisites to proper functionality. Early indicators of this trend can be found in areas such as displays to support fire emergency response from building environments (Steingart, 2005), enhanced educational training (Nicholson, Lackey, Arnold, & Scott, 2005), bioinformatics (Kuchar, 2005), and new types of electronic systems and dashboards in automobiles (Backs, Shelley, & Lenneman, 2005; Sangiovanni-Vincentelli, 2000). Platform-based design techniques can accelerate such transition into commercial applications by enabling system designs that are simultaneously modular and highly integrated.

Test Your Knowledge
Draw a flowchart of activities for platform-based design. Make a list of design platforms currently in use in a commercial sector of your choice. Why are platform-based approaches to design particularly suitable for augmented cognition applications? As augmented cognition applications become increasingly common in the commercial arena, list additional constraints that are likely to be placed on their development.

Engineering System Development: Established Strategies of Development

Most modern engineering systems are designed, implemented, and tested within the framework of a well-defined development process. Development usually begins at the system level (top level) and works toward a complete description of lower-level design detail. Classical models of system development include the *waterfall model*, the *spiral model*, and the *V-model*. Although the details of implementation in each model are distinct, the models share the common goal of defining the sequence of development steps, which helps members of a development team understand what is expected and when (Holt 2004; Wasson 2006).

A good development model will also simplify the task of working from requirements, to design, to implementation, and testing. Figure 8.3 illustrates the flow of activities in the V-model of development. Requirements state what is desired of the system (i.e., "This is the problem space."). Measures of effectiveness and appropriate test requirements are associated with each level of development, and verification requirements precisely define the method for determining whether or not a specific requirement has been met. Architectures state how the system will accomplish its purpose within a solution space. The result is a "what-how-what-how" development sequence.

DESIGN PLATFORM METHODOLOGY

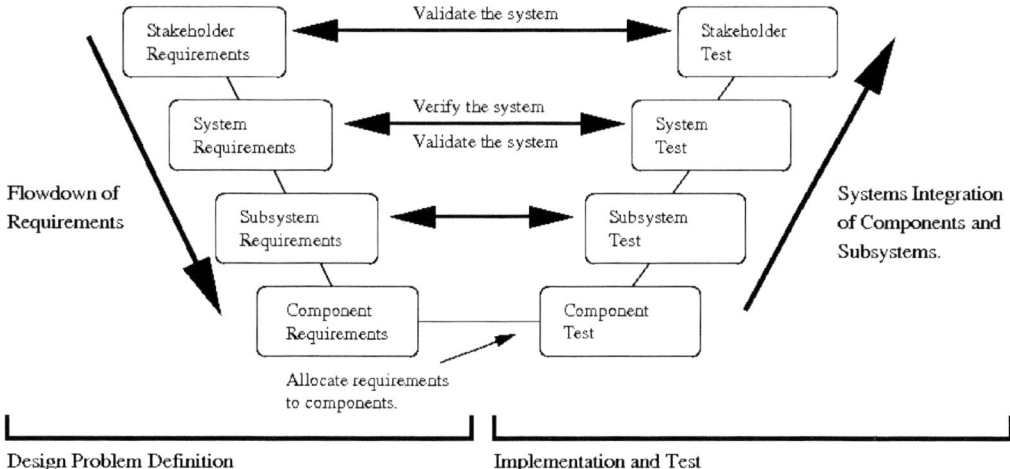

Figure 8.3. Typical V-model of system development.

The rationale for decomposing large, complex problems into networks of smaller, related problems is based on two phenomena: (1) Very large problems commonly appeal to multiple disciplines and technologies, and (2) decomposition strategies result in smaller problems, each of which requires expertise from a smaller number of technologies. Therefore, a small number of specialists is needed to implement the subproblem (O'Grady, 1998). Breaking large problems into smaller ones also encourages an organization size and granularity that has simpler management and control requirements.

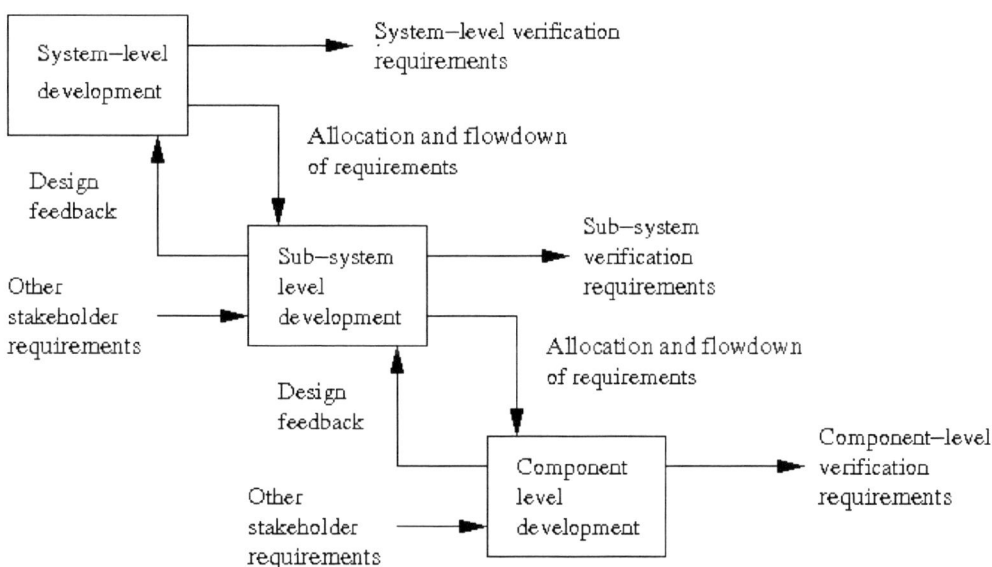

Figure 8.4. Detailed model of requirements flowdown (Martin, 2003).

Figure 8.4 shows that at each stage of the decomposition process, new requirements place additional constraints on the design solution space. Designers working on subsystem elements provide feedback to system-level designers, and designers working on the components provide feedback to the subsystem-level designers. For each requirement, there is an obligation to define precisely how and when that requirement will be verified.

Figure 8.5 is a detailed model of system testing, integration, and delivery. The satisfaction of component-level requirements is a prerequisite for the satisfaction of subsystem-level requirements, and so forth. Timely verification of requirements ensures that corrective action can be taken as early as possible. Testing procedures begin at the component level (where functionality of the component is checked) and work their way up to the system and stakeholder levels (where functionality of the entire system is checked).

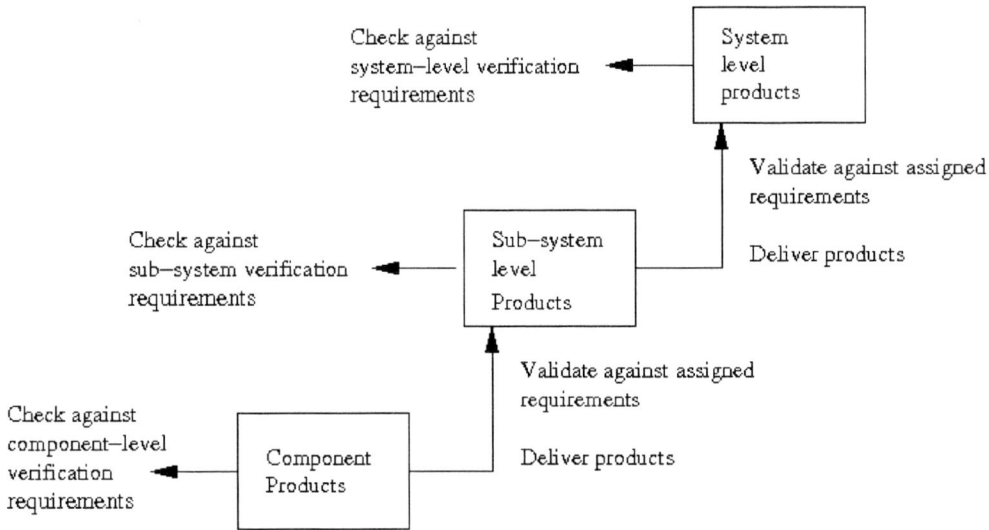

Figure 8.5. Detailed model of system testing, integration, and delivery (Martin, 2003).

Use of Visual Modeling Languages

A key challenge in the front-end development of complex engineering systems is the capture of system functionality: What exactly will the system do? The use of visual modeling languages, such as the Unified Modeling Language (UML; http://www.omg.org/uml), can simplify this stage of development because they provide a suite of high-level diagram types for the representation of required functionality, as well as simplified models of system behavior and structure.

Let us assume that a detailed system description does not exist. Each layer of development begins with an operations concept consisting of detailed scenarios of system functionality, together with simplified models (or fragments) of system behavior and structure. A functional description dictates what the system must do. To create a system description, we employ use-case diagramming notation to indicate elements of functionality and association with external entities (actors, which can be indicated by small human stick figures). Each facet of system functionality can be expanded to textual scenarios that cover the expected functionality of the system. Textual descriptions of functionality are then transformed into activity and sequence diagrams. Activity diagrams show flows (sequencing) of functionality. Sequence diagrams show interaction among objects that is expected to occur during the execution of scenarios. A complete system description will also include statements on minimum levels of acceptable performance and maximum cost. Given that a system does not actually exist at this point, these aspects of the problem description will be written as design requirements/constraints. Further design requirements/constraints will be obtained from higher layers in the development process (see Figure 8.6), and from the structure and communication of objects in the models for system functionality (e.g., required system interfaces).

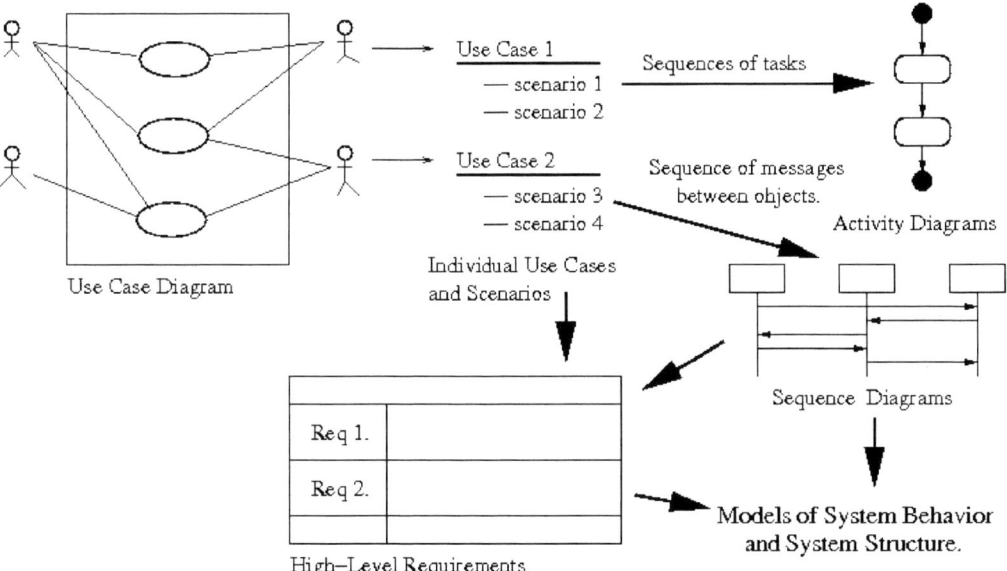

Figure 8.6. Pathway from operations concept to models of behavior/structure to requirements (Austin, 2004).

Pathway from Industrial- to Information-Age Systems

The drive to expand system functionality and improve performance, without sacrificing safety/reliability and economics, is resulting in the replacement of many industrial-age systems (dominated by hardware) with information-age systems (combinations of hardware, sensors, communications, and software) that depend on automation and control for system functionality and performance. Examples of information-age systems now include modern workspaces, automobiles, and aircraft. Augmented cognition technology is poised to improve the performance of these types of systems.

Limitations. From an engineering design perspective, the central design challenge with information-age systems is how to expand system functionality and improve performance through a judicious combination of hardware, sensors, communications, and software. When engineers have limited experience in working with a new domain, such as augmented cognition, simplified empirical procedures for design may not be available. As such, they will need to return to first principles and create and evaluate designs through the use of formal analysis.

Market pressures may require that the system be delivered to market in the shortest possible time. In this regard, the V-model of system development is limited in two key respects:

1. *Slow time to market.* In the V-model of system development, testing occurs toward the end of the development process. The well-known shortcoming of this approach is the excessive cost of fixing errors.

2. *No built-in support for reuse.* The V-model does not contain any support for reuse of components, subsystems, and so forth. This in turn affects the quality of components and subsystems, which must be extensively tested before each implementation.

Methodologies for platform-based design aim to mitigate these limitations.

Handling System Complexity in Design

Design for automation in large-scale system operations is challenging because, in addition to making sure system performance is adequate, questions of appropriate functionality become important. How, for example, do we know that a system will respond to a specific scenario in an appropriate way? Lessons learned from industry (Jackson, 2006; Magee & Kramer, 2006; Sangiovanni-Vincentelli, McGeer, & Saldanha 1996) indicate that there are now many automated engineering systems with complexity approaching the point at which the validation (see Chapter 6) of design correctness will be impossible unless mechanisms for verification are built into the design process itself. These mechanisms are based on three cornerstones (Sangiovanni-Vincentelli, 2000):

1. *Formal models:* ways to capture the design representation and its specification in an unambiguous formal language that has precise semantics.
2. *Abstraction:* mechanisms to eliminate details that are of no importance when evaluating system performance and/or checking that a design satisfies a particular property.
3. *Decomposition:* breaking hierarchical design into subsystems and components that can be designed and verified almost independently.

In the design of augmented cognition systems, these principles can be handled through the use of two mechanisms: formal model-based approaches to design, and separation of concerns. Because augmented cognition systems exist within the context of a surrounding environment, models of the surrounding environment should also be explicitly considered.

Formal Model-Based Approaches to Design

Formal models of system behavior are needed for the rigorous analysis for how a set of components (i.e., the architecture) behaves when working together. As outlined by

Sangiovanni-Vincentelli et al. (1996), formal approaches to design are based on the following components:

1. A set of explicit or implicit equations that involve input, output, and possibly internal (state) variables (see Chapter 7).
2. A set of properties that the design must satisfy. Normally these will begiven as a set of equations over design variables (i.e., inputs, outputs and states).
3. A set of performance indices that evaluate the quality of the design in terms of cost (see Chapter 3), reliability (see Chapter 7), and so forth, given as a set of equations involving design variables.
4. A set of constraints on design variables and on performance indices, specified as a set of inequalities.

The process of formalizing system specifications forces the project participants to state all assumptions, thereby minimizing the likelihood of omissions. Mathematical techniques can be used to analyze properties of a formal specification and to prove the consistency of different specifications.

Emerging approaches to system design (Magee & Kramer, 2006; Sidorova, 2007; Uchitel, 2004) are based on formal methods and the selective use of design abstraction and decomposition, both of which need to have precise and easy-to-understand interpretations. These approaches benefit system design in two ways. First, concepts and notations from mathematics can provide methodological assistance, facilitating the communication of ideas and the thinking process (see Chapter 7). Second, formal methods enable us to calculate properties of a design using formal models for the synthesis of architecture-level representations.

Moreover, the combination of equalities and inequalities provides a distinct separation of optimization and verification problems. Functional and performance requirements and their associated inequality constraints lead to optimization problems, such as optimizing system functionality and performance subject to constraints. Equality constraints lead to formal approaches to system validation (see Chapter 6). As illustrated along the right-hand side of Figure 8.7, the goal is to move design processes forward to the point at which early detection of errors is possible and system operations are correct by construction (Sidorova, 2007).

Separation of Concerns
Complex systems, such as augmented cognition systems, are characterized by many components, concurrent subsystem-level behaviors, and complicated communications and interactions among subsystems, components, and points of contact with the surrounding environment. To facilitate understanding of these design issues, system designers aim to pull a design apart and examine it from orthogonal perspectives. This may not be easy, particularly when design concerns are intertwined. Still, achieving orthogonality is important because it enables us to explore one design consideration without other design concerns being affected. The result is easier exploration of a design space.

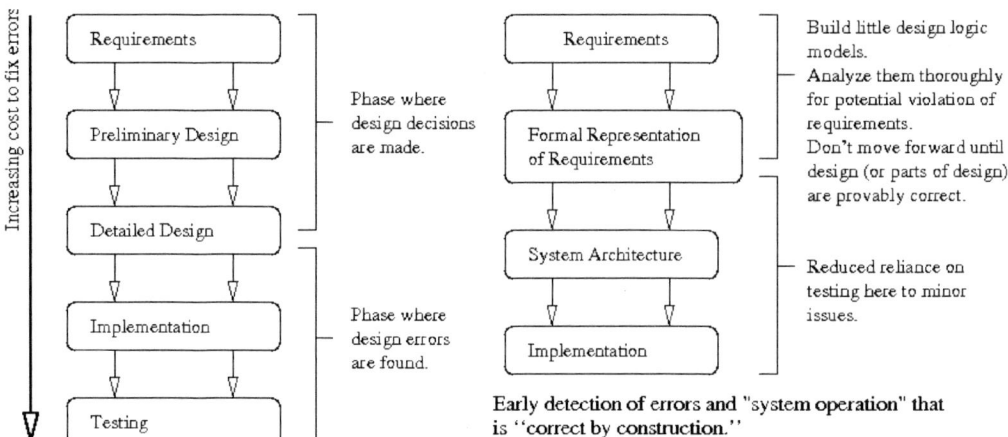

Figure 8.7. Pathways of traditional and model-based system development (adapted from Sidorova, 2007).

Figure 8.8 shows the essential details of the simplification of design through the separation of concerns. The task of understanding and evaluating a design is simplified through the separate consideration of structure, behavior, and communication. System structure covers decomposition of the system into hierarchies of principal subsystems, identification of required connectivity among components, and assignment of geometric properties. System behavior covers the identification of key functions and the ordering and decomposition of these functions. It is the ordering of these functions, coupled with system logic, that produces system behavior. Communication includes the identification of interfaces to encapsulate component/subsystem behavior and the selection of appropriate protocols for communication.

Model of the Surrounding Environment

Unfortunately, a formal design representation by itself is not enough to verify a design for correct functionality and behavior. This is because every system operates within the context of a larger surrounding environment, as shown in Figure 8.9. Engineering system inputs and outputs are connected to parameters in the surrounding environment. The surrounding environment will also place physical or geometric constraints on the system model. In order to fully understand the engineering system, one will also need behavior and structure models for the surrounding environment.

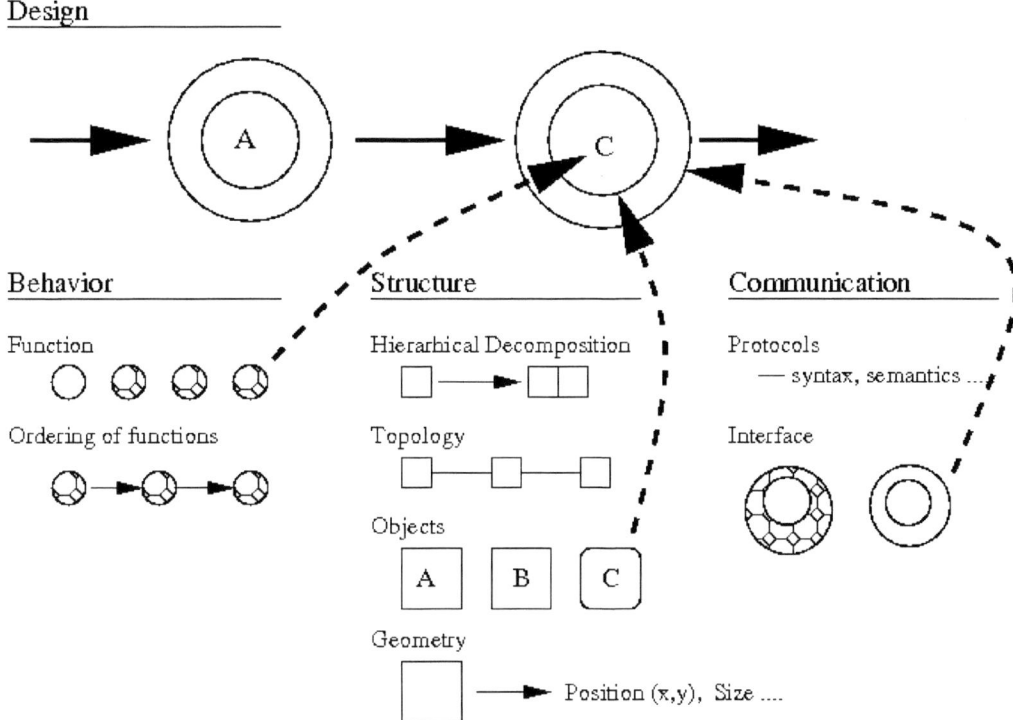

Figure 8.8. Separation of design concerns: structure, behavior, and communication.

Test Your Knowledge
Draw and label a diagram showing the essential elements of the V-model of system development. Make a list of the benefits of formal model-based approaches to design. How do the principles of abstraction, decomposition, and separation of concerns simplify the design process?

Applied Exercise
Suppose that you are driving an automobile through an urban area. Develop a simplified model for the behavior and structure of the automobile being driven. Develop a simplified model for the behavior and structure of the surrounding environment. Where will you put the driver in this modeling setup? How will the engineering system and surrounding environment models interact via inputs and outputs? What kinds of constraints will the surrounding environment place on the engineering system?

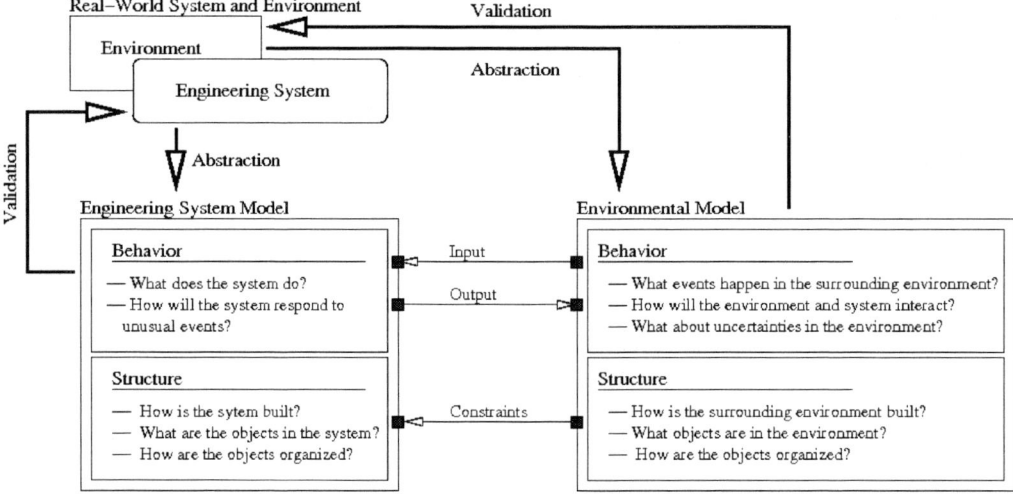

Figure 8.9. Step-by-step procedure for synthesis and validation of models for the engineering system and surrounding environment.

Reuse in System Design: Reconfiguration for System Modularity

A modular architecture has well-defined, standardized, and decoupled interfaces that collectively allow for design changes to be made to one module without generally requiring a change to other modules.

As illustrated in Figure 8.10, there are four types of product architectures:

1. *Modular design:* one function is allocated to one module.
2. *Function distribution:* one function is mapped to multiple modules.
3. *Function sharing:* several functions are allocated to one module.
4. *Integrated design:* several functions are allocated to several modules.

State-of-the-art procedures for the assessment and reconfiguration for modularity are guided by analytical procedures such as design structure matrices (Browning, 1998).

Coupling, cohesion, and module complexity. Methodologies for the design and evaluation of modules are guided by the concepts of system coupling, cohesion, and module complexity. *Coupling* is a measure of the interface complexity (or degree of interdependence) between modules. *Cohesion* is a measure of how well the components of a module are related to one another, or a measure of the functional association of an element within another element. Modules should be kept simple and should hide the details of implementation from the outside environment. Factors that contribute to module complexity are size, the quantity of internal functions, connections within the module, and the quantity of interfaces to the modules.

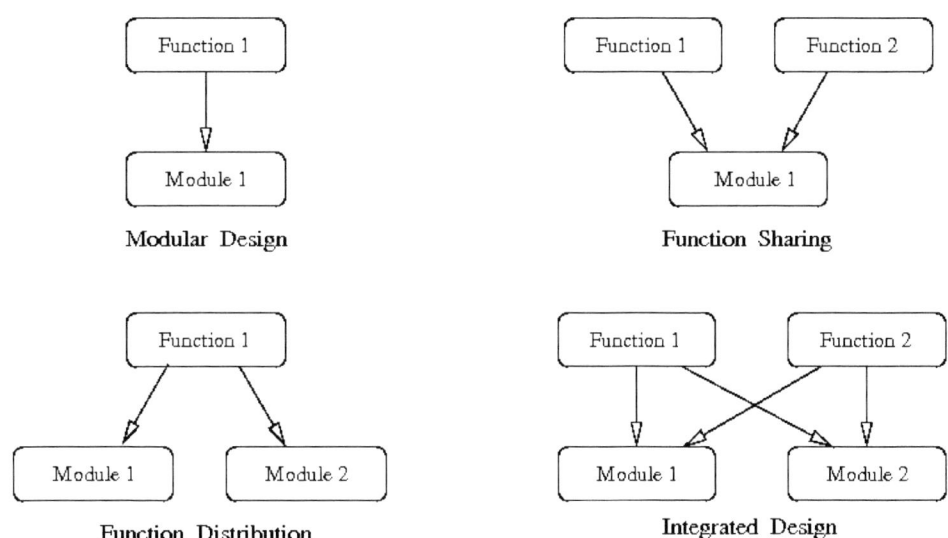

Figure 8.10. Different types of product architecture (Erens & Verhulus, 1997).

The criteria of coupling and cohesion work together. Generally speaking, modules with components that are well related will have the capability of plugging into loosely coupled systems. Functions in the behavior model will be reorganized so that highly interrelated functions are allocated into common modules. By clustering functions that are strongly affected by noise into separate modules, the impact of noise on the total overall system can be minimized. The result is a system architecture that is characterized by minimized coupling among modules and maximized cohesion within modules.

Tension between modularity and integration. Modular systems with sophisticated behavior are created through the composition of simple modules. The key benefit of modularity is ease of understanding in both the functionality of individual modules and the composed system. Although modular system architectures are also amenable to change, achievable levels of performance are often less than optimal.

The underlying premise and drive for integration is very simple: A system will function better when the subsystems work together as a team rather than independently. Hence, as shown in Figure 8.10, highly integrated systems aim for a variety of functionality and superior performance through the mapping of multiple functionalities onto a single architecture. As a result, tightly connected components may be more difficult to understand, modify, and validate.

Because augmented cognition systems have stringent performance requirements, the benefits of system modularity must be balanced against potential gains in performance attributable to the integration of system functionality. The natural way of accomplishing this balance is through the use of modules that are programmable.

Test Your Knowledge
Construct a table showing the strengths and weaknesses of modular design and integrated design. From a performance perspective, how does system integration help augmented cognition system design? How does system integration complicate such design? What role do programmable modules play in such design?

General Approach to Platform-Based Design

To keep the complexities of augmented cognition system design in check, designers need a methodology that promotes the expressiveness of top-down approaches to design, simplicity through the separation of concerns, and the efficiencies of bottom-up methods of implementation. The goals of platform-based design are to satisfy these needs with a meet-me-in-the-middle process whereby semiformal/formal specifications are refined into lower-level abstractions that can be realized through the selection and connectivity of components/subsystems selected from a library (otherwise termed the *platform*). Ideally, a platform should also assist designers in the evaluation of augmented cognition systems (often through the use of simulation) and in the systematic exploration of trade-offs among design choices (often through the use of computational environments for optimization-based design).

Basic Development Process

As indicated in Figure 8.2, platform-based design is an ongoing process whereby new design iterations are created from previous versions. Within a single cycle, there are three basic steps of development: (1) problem definition, (2) synthesis of design alternatives, and (3) design evaluation and optimization.

Problem Definition: Functionality, Performance, and Economics

As illustrated in Figures 8.7, 8.11, and 8.12, the system development process begins with a definition of required functionality: What functions will the augmented cognition system be expected to perform? Who will be involved in the execution of these functions? How will the system handle unexpected events? At this stage of development, visual modeling languages such as UML can be employed for the high-level description and organization of fragments of required system functionality.

Simplified models of system behavior (see Chapter 7) are created by combining and organizing fragments of functionality. Activity and sequence diagrams are useful for expressing required functionality. Simplified models of system structure can also be obtained by identifying and organizing subsystems and objects into networks and hierarchies of system elements. Class and object diagrams are a useful way of expressing system structure requirements.

A real-world implementation will also need to satisfy performance and economic constraints. These aspects can be obtained by looking at elements of system functionality and asking the question, "In addition to working, what levels of performance are required?" Message interactions between objects in the sequence diagram imply interfaces that may need to exist between components. Later in the design process, elements of system functionality and performance will be evaluated through the execution of validation and verification procedures (see Chapter 6).

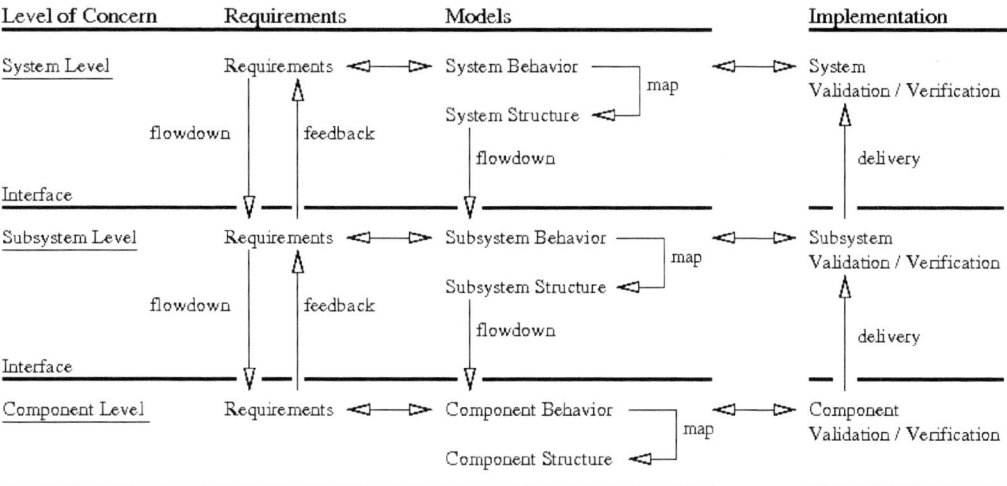

Figure 8.11. Layered model of system specifications, implementation, and testing.

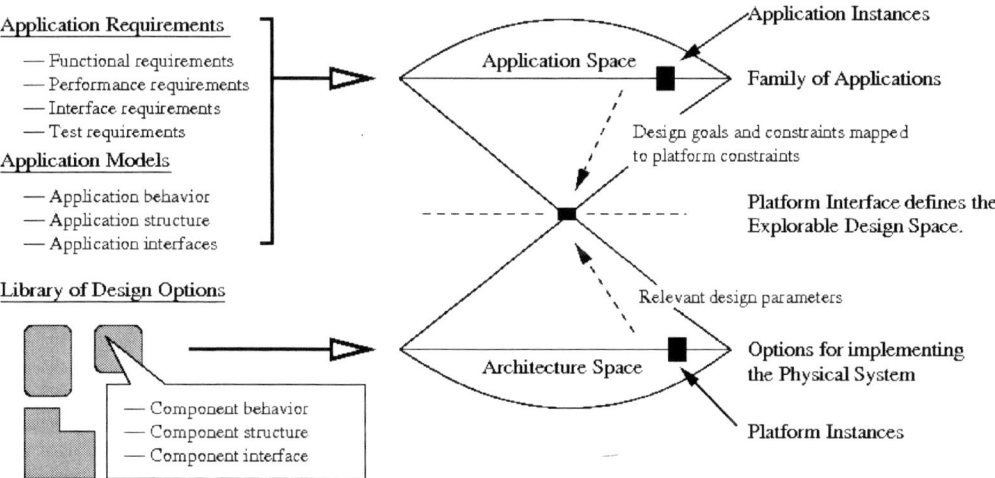

Figure 8.12. Elements of the platform stack.

Because a system does not actually exist at this point, these aspects of the problem description must be written as design requirements and constraints. The end result of the problem definition is layers of requirements and models of behavior and structure, as shown in Figure 8.11.

Synthesis of Design Alternatives

It is at this stage in an augmented cognition system's development that platform-based design procedures can be implemented through the notion of a *platform stack*. Figure 8.12 shows key elements of the platform stack. From the top, the application space is defined by the design problem requirements and accompanying models of system behavior and structure. The architecture space is defined by a library of design options. In traditional approaches to system development, system design alternatives are created through the assignment of system behavior fragments to elements of the system structure, coupled with the imposition of constraints on performance and operation.

In the platform-based approach to design, the same behavior-to-structure mapping occurs, but requirements/constraints accompanying the design specification (i.e., the application space) are matched against design alternatives assembled from components and subsystems in a design library (i.e., the architecture space). The supporting design flows employ clean interfaces (i.e., plug and play) and specifications that support formal analysis.

The platform interface defines a point of contact between the application and architecture spaces. An initial application design helps to define a provisional platform interface, which in turn suggests what the architecture implementation needs to provide. The architecture space is then explored to find an implementation that comes closest to satisfying both the terms of the interface and preset physical requirements/constraints. In practice, some iteration may be required to find a suitable platform interface.

If the gap between required functionality and the identification and implementation of design options is too large, then systematic exploration of the design space will simply be too difficult. The solution to this problem is suggested in Figures 8.2 and 8.11. Instead of modeling the design problem with one platform, one can organize the design process into a stack of platforms. Figure 8.13 shows, for example, applications being implemented from subsystem-level architecture options, subsystem architectures being implemented from components, and components being assembled from hardware and software implementations.

Along the application-to-implementation pathway, design benchmarks, functional blocks, and interconnect structures are essentially requirements that flow down to the lower levels of design. Conversely, along the implementation-to-application pathway, implementation models, parameters and attributes, and model estimates are essentially expressions of design option capability.

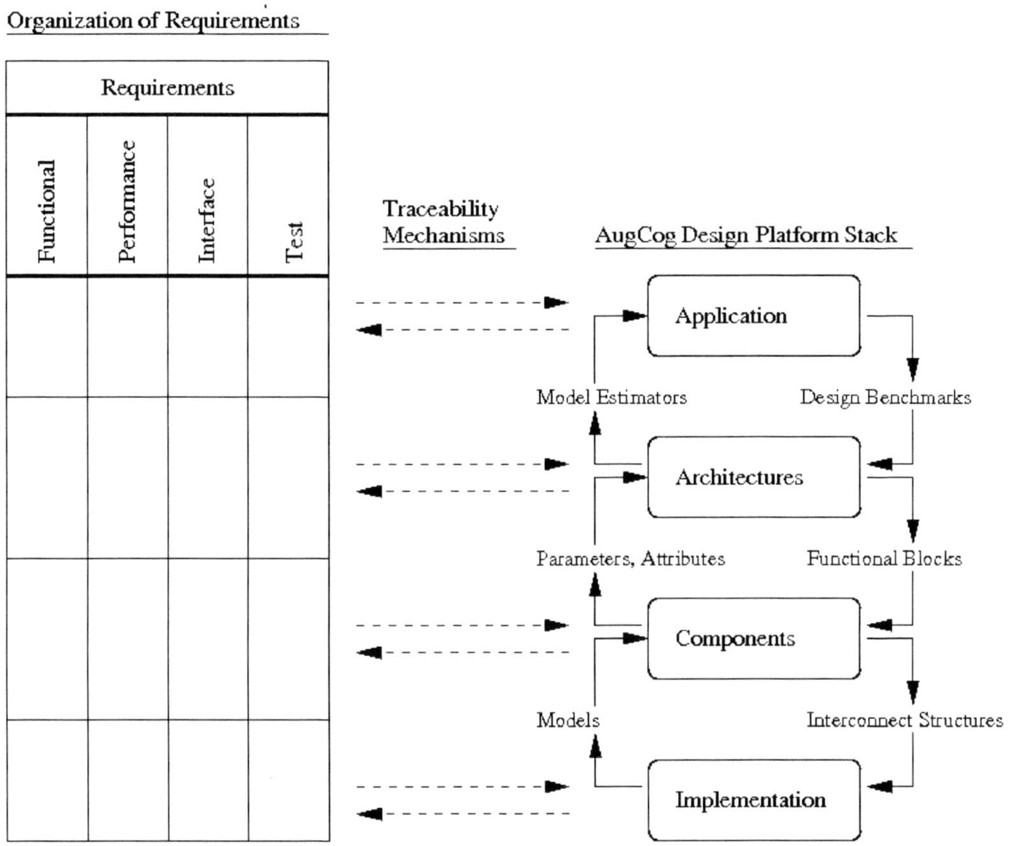

DESIGN PLATFORM METHODOLOGY

Figure 8.13. Definition and organization of requirements traced to layers in the stack of design abstractions.

Design Evaluation and Optimization

Final acceptance will be based on the evaluation of requirements to ensure that functionality, performance, and economic concerns are satisfied. Formal analysis procedures can be used to validate system functionality, as demonstrated by the work of Uchitel et al. (2004). Similarly, simulation procedures are often used to validate system performance.

Related Approaches to Design

A natural question to ask is, "How do platform-based approaches to design compare with other commonly used development models?" To address this concern, Table 8.1 provides a succinct comparison of platform-based design to the spiral, waterfall, and DOD architecture models of development (Green & DiCaterino, 1998; DoDAF, 2007).

Table 8.1. Comparison of PBD to Other System Engineering Methods

Development Method	Problem Definition	System Design Alternatives	Evaluation and Customization
Spiral development	Tailored toward the analysis of risks in development	Generated through mapping of behavior fragments to system architectures	Likely to occur later in the design process as the initial prototypes focus on conveying the system concept and designing the riskiest components
Waterfall development	Completed in the requirements before design begins	Not supported; system design only follows development of initial requirements	Occurs during requirements definition
DoDAF	Developed in the initial steps of creating DoDAF diagrams; also represented in TV-1 diagrams	Required before architecture design; new alternatives require redesign of architecture	Represented as SV-8 and TV-2 diagrams
PBD	Can use visual languages to specify high-level functionality; performance and economic constraints derived from functionality	Generated through selection and connectivity of subsystem- and component-level alternatives from a library	Determined iteratively throughout the design process

Key Points for Platform-Based Design

1. Augmented cognition system development efforts can achieve efficiency and customization in design through platform-based design.
2. Methodologies for platform-based design satisfy this need with a meet-me-in-the-middle process whereby semiformal/formal specifications are refined into lower-level abstractions that can be realized through the selection and connectivity of components/subsystems selected from a library.
3. From the top, the augmented cognition *applications space* is defined by the design problem requirements and accompanying models of system behavior and structure (see Chapter 7). The *architecture space* is defined by a library of design options. The platform interface defines a point of contact between the application and architecture spaces.
4. An initial augmented cognition application design helps to define a provisional platform interface, which in turn suggests what the architecture implementation needs to provide. The architecture space is then explored to find an implementation that comes closest to satisfying both the terms of the interface and preset physical requirements/constraints.
5. Ideally, a platform will also assist augmented cognition system designers in the evaluation of systems (often through the use of simulation) and in the systematic exploration of trade-offs among design choices (often through the use of computational environments for optimization-based design).

DESIGN PLATFORM METHODOLOGY

6. If the gap between required functionality and the identification and implementation of design options is too large, then systematic exploration of the design space will simply be too difficult. Instead of modeling the augmented cognition system design problem with one platform, one can organize the design process into a stack of platforms.

Test Your Knowledge
Draw and label a diagram showing the key elements of platform-based design. How is the application space defined? How is the architecture space defined? Why are multi-layer design platforms sometimes needed?

Applied Exercise
Let us suppose that you want to follow a platform-based approach to renovating an augmented cognition system for a kitchen for seniors (over age 70) with new interactive appliances, adjustable countertops, and so forth. Create a visual model of the functionality that will be supported in the augmented kitchen. Develop a simplified application space by adding requirements and constraints for required functionality and performance (how will the seniors be "augmented"?). What kinds of modules (e.g., based on disability?) are likely to appear in the architecture space?

Scenario: Platform-Based Design of Automobile Dashboards Enhanced by Augmented Cognition

The safe and efficient operation of an automobile requires that drivers remain alert. Remaining alert includes active engagement in sensing, interpreting, and reasoning, combined with determining meaning in a way that increases anticipation of relevant threatening events. These mental processes are the fundamental elements of situation awareness (see Chapter 5). Experimental studies (Parasuraman & Davies, 1977; See, Howe, Warm, & Dember, 1995; Grier et al, 2003) indicate that repetitive and mundane driving tasks can degrade alertness in as little as 15 minutes. A fatigued driver may fail to identify and react to a dangerous situation, or even create a dangerous situation. To combat these problems, we propose that (a) car dashboards be enhanced with augmented cognition technology, and (b) these systems be developed within the framework of platform-based design.

Platforms for the Design of Electronic Systems in Automobiles

A major trend in the automobile industry is innovation to extend system functionality, improve real-time performance, and satisfy safety-critical issues through the use of advanced electronic systems. At Daimler, for example, more than 90% of the innovation in a car can be attributed to the use of electronics. Similarly, BMW reports that electronic systems now account for 30% of an automobile's manufacturing cost (Sangiovanni-Vincentelli, 2003).

From an automobile design and manufacturing standpoint, augmented cognition–enabled dashboard systems will be an addition to the electronic system. As such, their development and economic evaluation will be subject to the same schedules and competitive pressures facing the automobile industry at large.

Present-day electronic systems fall into three categories: power train management, body electronics, and information displays. To keep the complexities of system development in check, carmakers are using platform infrastructures for the design and validation of electronic systems in automobiles. Platforms support the representation and organization of design specifications, synthesis of design solution architectures, performance analysis via simulation, and formal evaluation of system functionality. A well-designed platform will also allow for the efficient exploration of the design space, identification of feasible design alternatives, and generation design variants.

Car Dashboard Enhanced with Augmented Cognition Technology

In a departure from present-day dashboard systems, augmented cognition–enabled dashboard systems will play an active role in working *with* the driver to accomplish driving tasks. Figure 8.14 is a high-level schematic showing just one layer of the step-by-step development, evaluation, and customization of a car dashboard enhanced with augmented cognition technology. As with all design problems, the key steps are problem definition, development of design solutions, and system validation and customization.

Augmented Cognition Problem Definition

The problem definition begins with an articulation of the required functionality from two viewpoints: the driver and the dashboard. The driver will be expected to sense the surrounding environment, including the dashboard, and take appropriate actions to achieve driving tasks. In its supporting role as an attention management system, the augmented cognition–enabled dashboard will monitor automobile data, sense driver activities, understand geographical positions and events in the external environment, and alert the driver of impending danger in a timely and appropriate manner. For both the driver and the augmented cognition system, functionality will be defined in terms of what must be provided and what action and outcomes must be prevented.

With preliminary definitions for the required functionality of the driver and the augmented cognition system in place, the real challenge lies in defining ways in which the driver and augmented cognition system will work together to enhance system safety and performance. A complete system description will specify minimum levels of acceptable performance and maximum cost. And because a dashboard system will not actually exist at this point, these aspects of the problem description will be written as design requirements/constraints.

DESIGN PLATFORM METHODOLOGY

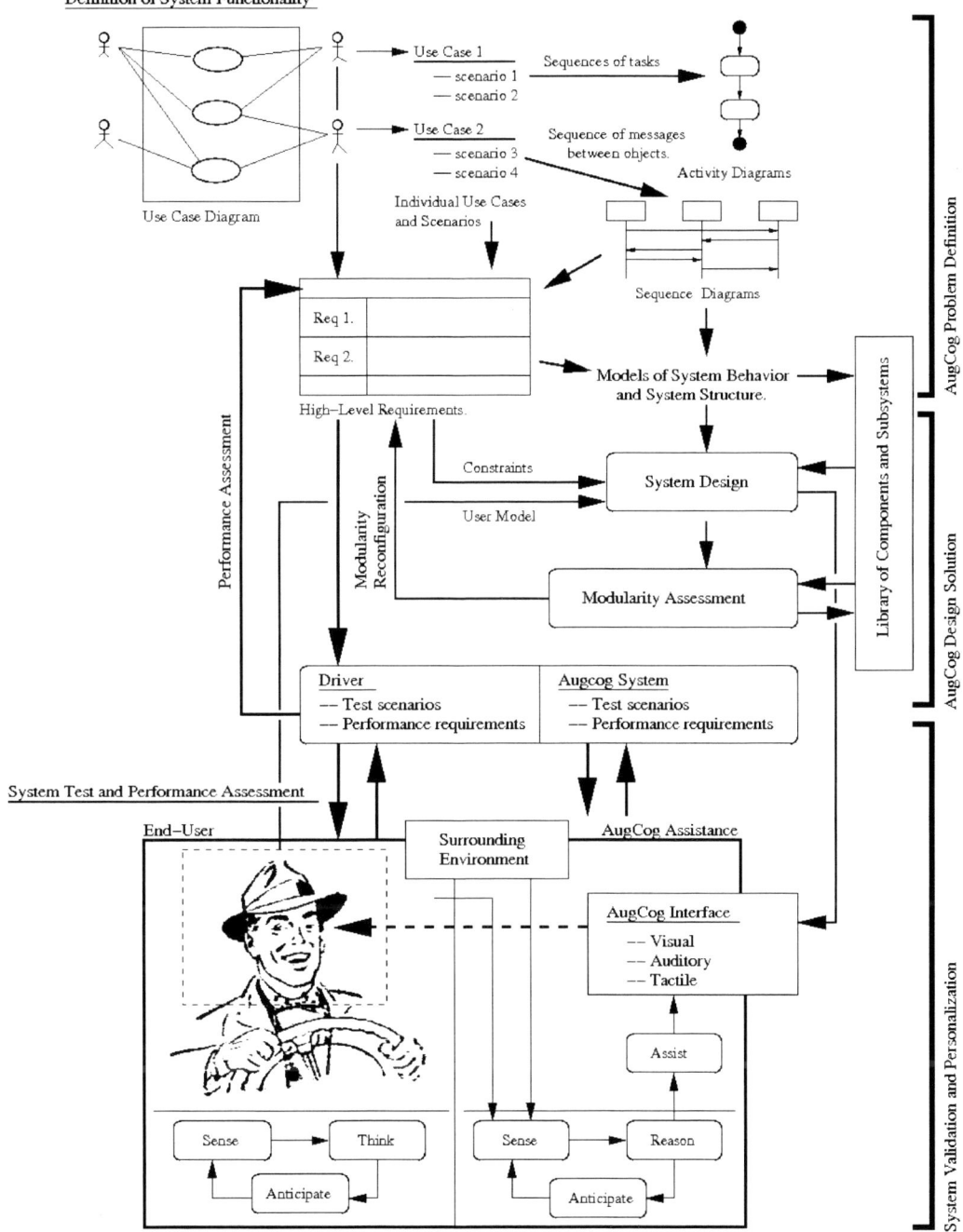

Figure 8.14. Step-by-step development procedure for the synthesis, design, test, and customization of an augmented cognition–enhanced dashboard.

Oncoming traffic monitor. Part 1: Problem definition. To see how these principles can be applied in practice, let us consider the role that augmented cognition assistance might play in the monitoring and appropriate reaction to oncoming traffic. In the ideal setting (see the bottom of Figure 8.14), an alert driver operates in a sense-think-anticipate cycle. He/she detects oncoming traffic, classifies the impending event, and, when required, takes appropriate action to avert an accident. In its capacity as a closed-loop attention management support system, the augmented cognition–enabled dashboard aims to make sure that this happens. Therefore, in addition to detecting oncoming traffic, the augmented cognition–enabled dashboard system also needs to monitor the driver to see that an appropriate response is occurring. When the driver behavior is determined to be inconsistent with an appropriate action, to mitigate danger, the augmented cognition–enabled dashboard system will quickly alert the driver via one or more feedback mechanisms.

Figure 8.15 illustrates, in an abbreviated manner, the definition of the functionality and performance of an augmented cognition–enabled car dashboard through use-case diagrams, textual scenarios, and requirements. Although it is certainly possible to represent all of these concerns in a single (albeit very complicated) use-case diagram, in practice, the functional specification can be simplified through separate use-case diagrams for the driver and augmented cognition system. The driver system has actors—*driver* and *automobile*—and fragments of required functionality—*identify oncoming traffic, react to oncoming traffic*, and so forth. Similarly, actors for the augmented cognition system are the *driver* and *sensors*. Required fragments of behavior are *identify oncoming traffic, monitor driver attention, sense surrounding environment,* and *alert driver*.

As pointed out by Byrne et al. (2004), improved situation awareness of the surrounding environment will occur through the use of geographic information positioning (GPS) systems to pinpoint vehicle location and generate a model of the surrounding terrain.

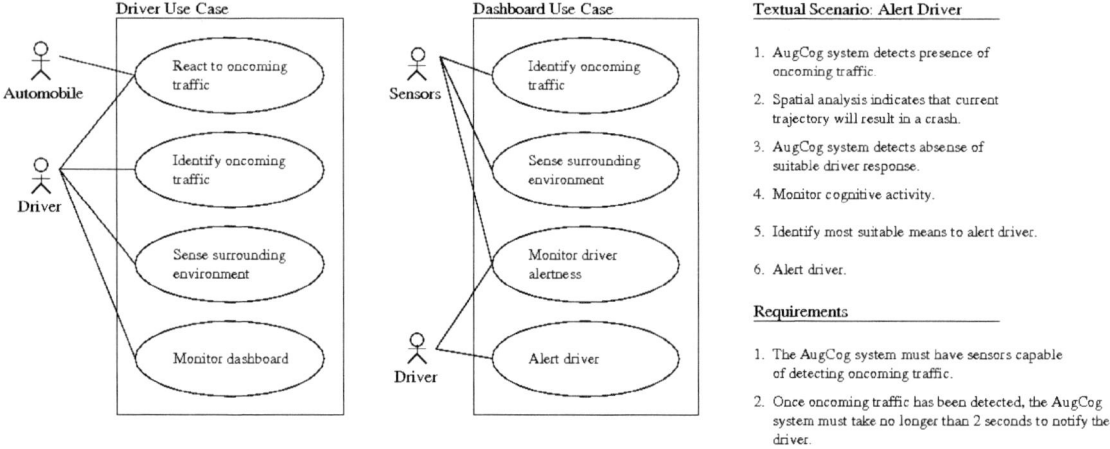

Figure 8.15. Abbreviated functionality and performance of an augmented cognition-enabled car dashboard defined through use cases, textual scenarios, and requirements.

Individual use cases are elaborated into textual scenarios. The right-hand side of Figure 8.15 shows, for example, a step-by-step procedure for identifying and classifying the threat of oncoming traffic and, if required, alerting the driver of an impending event. In a practical implementation, textual scenarios would be written for each use case. Then, the textual scenarios would be parsed to identify (a) objects that will need to be part of the dashboard system and (2) constraints on the ordering of the subsystem/component functionality necessary to generate the required system-level behavior.

To be effective, all these tasks need to be executed quickly. The associated constraints on performance can be represented as requirements, as shown in the lower right-hand corner of Figure 8.15.

Augmented Cognition Design Solutions

The primary benefit of platform-based design in this setting is one of organization. As indicated in Figure 8.13, design specifications (requirements) define an applications space, and libraries of subsystems and components (e.g., sensors and controllers) define a space from which potentially good design solutions can be generated through subsystem/component assembly.

Oncoming traffic monitor. Part 2: Design solutions. Relevant details of the application and architecture spaces are as follows:

1. *Application space.* The application space will be defined in terms of required functionality, performance, and economics for the driver and the augmented dashboard. Detailed specifications for most of these concerns will be elaborated in the problem definition phase and will be represented formally as requirements and constraints. Secondary concerns will be defined in terms of streams of data that will be measured, reasoning processes that need to be in place, and mechanisms for providing feedback to the driver. To be practical, the augmented cognition system will need to track driver behavior through minimal physical interactions, which requires an accurate user model of individual head/body geometry.

2. *Architecture space.* Augmented cognition–enabled dashboard implementations will be a specialization of the system-level architecture for augmented cognition shown in Figure 8.1. Extended functionality and improved performance will be achieved through sensors that measure the state of the driver, sensors that detect impending events in the surrounding environment, gauges that estimate cognitive workload, and systems that alter flows of information and modality of presentation to the driver (Belyavin, 2005; St. John et al., 2006).

Oncoming traffic monitor. Part 3: System validation and custom personalization. Enhanced driver capability is enabled through personalized driver-in-the-loop assessment procedures, which, in turn, are mapped onto the selection and calibration of design options that optimize system-level functionality. For the oncoming traffic monitor scenario, and as shown in the middle and lower half of Figure 8.14, calibration of the augmented cognition system will occur through the execution of baseline tasks by a driving individual.

The driver will respond to events in the surrounding environment through a sense-think-anticipate cycle. The primary purpose of the augmented cognition system is to enhance driver/end user performance, so it, too, will operate through a sense-reason-anticipate cycle. Output from the reasoning module will assist the driver through adjustments to the dashboard interface. From an evaluation standpoint, care must be taken to look for actions that are different, less severe, or incorrect in addition to just looking for the expected or desired result, because the exhaustive testing of all possible scenarios is impractical.

Test Your Knowledge
Briefly describe the ways in which an augmented cognition–enabled dashboard system will assist a driver. What role do user models play in the implementation of augmented cognition–enabled dashboards? What kinds of data and/or information are likely to be part of this model?

Applied Exercise
In this section, we have worked through the details of conceptualizing the design of augmented cognition assistance for the scenario of avoiding oncoming traffic. In what other ways might augmented cognition provide assistance to a driver? What are some of the practical constraints that might need to be addressed in the implementation of a system to support these scenarios?

Design Guidelines for Platform-Based Design

Although commercial applications of augmented cognition are still in the future, lessons learned from present-day practices for platform-based design will be indispensable in their implementation. To this end, Table 8.2 contains a summary of design guidelines derived from present-day practices, starting with definitions of system functionality and ending with the validation/verification and personalization of parameters in design instances.

Parting Message

The purpose of this chapter has been to explain how the principles of platform-based design may benefit those designing and developing augmented cognition systems, particularly systems with a commercial end use. We have described a number of techniques—formal methods, abstraction, decomposition—for handling system complexity in design, the enhancement of reuse via system modularity, and the general approach to platform-based design.

Platform-based approaches to design are appropriate for systems that evolve through iteration. They couple top-down specification of system functionality and performance with procedures for the bottom-up assembly of systems from components. The intersection of these approaches requires an understanding of how augmented cognition–enabled functionality can be supported with appropriate measurement and drive the identification and selection of appropriate components. Stringent levels of system performance in adverse environments will be achievable only if systems are customized (possibly programmable) to the capabilities/limitations of specific end users. An appropriate balance of competing criteria suggests the need for system-level architectures that are both modular and highly integrated (Austin, 2005; Raley, 2005; Raley & Marshall, 2005).

Table 8.2. Lessons Learned from Platform-Based Design That Will Be Useful for Augmented Cognition-Enabled Applications

Aspect of Platform Design	Design Guideline
System functionality	Use-case diagrams provide a high-level representation of system functionality and the connectivity of functionality to external entities.
Performance and cost	Requirements for system performance and cost can be obtained directly from simplified models of system behavior.
Organization of requirements	As indicated in Figures 8.4, 8.11, and 8.13, requirements need to be organized into layers that match abstractions in the augmented cognition design platform. Traceability mechanisms connect requirements to appropriate abstractions in the design platform.
Generation of design alternatives	Design alternatives are generated through the mapping of behavior fragments and constraints onto system structures. System structures are assembled from subsystems and components located in a library.
Systems analysis	Sophisticated techniques for systems analysis and control are justified by life-critical safety risks and the adverse consequences of poor system throughput (Taylor, 2003).
Modularity and integration	Because augmented cognition systems have stringent performance requirements, the benefits of modularity must be balanced against potential gains in performance attributable to the integration of system functionality. The natural way of accomplishing this is through the use of programmable modules.
Validation and verification	Validation and verification procedures (see Chapter 6) are needed for both the end user and the augmented cognition system. These procedures need to be an integral part of the design process rather than a postscript to development.
Personalization	To maximize the performance and reliability of augmented cognition–enabled designs, design instances need to contain parameters that can be calibrated to the capabilities and limitations of each end user.

References

Austin, M. A. (2005). Toward platform architectures for modular cognitive cockpits. In*ceedings of First International Conference on Augmented Cognition*, 2-disc DVD Set, Augmented Cognition International Society, Las Vegas, NV, July 22-27.

Austin, M. A. (codelivered with John Baras) (2004, June 21). *An introduction to information-centric systems engineering,* Tutorial F06, *INCOSE,* Toulouse, France.

Backs, R. W., Shelley, J., & Lenneman, J. K. (2005). Using modes of cardiac autonomic control to assess demands upon processing resources during driving. In *Proceedings of the First International Conference on Augmented Cognition, 2-disc DVD Set, Augmented Cognition International Society, Las Vegas, NV, July 22-27.*

Barker, R. A., & Edwards, R. E. (2005). The Boeing Team fundamentals of augmented cognition. In *Proceedings of First International Conference on Augmented Cognition*, 2-disc DVD Set, Augmented Cognition International Society, Las Vegas, NV, July 22–27.

Beatty, J. (1978). *Concluding report: ARPA Biocybernetics Project.* (Contract No. N00014-70-C-0030). Department of Psychology, Univeristy of California, Los Angeles, CA.

Belyavin, A. (2005). Construction of appropriate gauges for the control of augmented cognition systems. In *Proceedings of the First International Conference on Augmented Cognition, 2-disc DVD Set, Augmented Cognition International Society, Las Vegas, NV, July 22-27.*

Browning, T. R. (1998). *Modeling and analyzing cost, schedule, and performance in complex system product development.* Ph.D. thesis, Massachusetts Institute of Technology, Cambridge, MA.

Byrne, M. D., Kirlik A., Feetwood M. D., Huss D. G., Kosorukoff A., Lin R., & Fick C. S. (2004). A closed-loop ACT-R approach to modeling approach and landing with and without synthetic vision system (SVS) technology. In *Proceedings of Human Factors and Ergonomics 48th Annual Meeting* (pp. 2111–2115. Santa Monica, CA: Human Factors and Ergonomics Society.

Department of Defense Architecture Framework. (DoDAF). (2007). *Volume 1. Definitions and guidelines* (Version 1.5). Available at http://www.defenselink.mil/cio-nii/docs/DoDAF_Volume_I.pdf (Accessed August 23, 2008).

Dickson, B. T. (2005). The cognitive cockpit—A testbed for augmented cognition. In *Proceedings of First International Conference on Augmented Cognition, 2-disc DVD Set, Augmented Cognition International Society, Las Vegas, NV, July 22-27.*

Erens, F., & Verhulst, K. (1997). Architectures for product families. *Computer Industry.* 33, 165–177.

Gonzales-Zugasti, J. P., Otto K. N., & Baker J. D. (2000). A method for architecting product platform. *Research in Engineering Design, 12,* 61–72.

Green, D., & DiCaterino, A. (1998). *A survey of system development process models.* Center for Technology in Government. University of Albany, SUNY, New York.

Grier, R. A., Warm, J. S., Dember, W. N., Mathews, G., Galinsky, T. L., Szalma, J. L., & Parasuraman, R. (2003). The vigilance decrement reflects limitations in effortful attention, not mindlessness. *Human Factors, 44,* 349–359.

Holt, J. (2004). *UML for systems engineering—Watching the wheels* (2nd ed.). London: Institution of Electrical Engineers.

Holzner, S. (2004). *Eclipse* (Coverage of Version 3.0). Sebastopol, CA: O'Reilly and Associates.

Horowitz, B., Leibman, J., Cedric, M. A., Koo, T. J., Sangiovanni-Vincentelli, A., & Sastry, S. S. (2003). Platform-based embedded software design and system integration for autonomous vehicles (Invited Paper). *Proceedings of the IEEE,* 91(1), pp. 198–211.

Isreal, J. D., Wickens, C. D., Chesney, G. L., & Donchin, E. (1980). The event related brain potential as an index of display-monitoring workload. *Human Factors, 22,* 214–217.

Jackson, D. (2006). Dependable software by design. *Scientific American,* 294(6), June, pp. 69-75.

Keutzer, K., Malik, S., Newton, A. R., & Sangiovanni-Vincentelli, A. (2000). System-level design: Orthogonalization of concerns and platform-based design. *IEEE Transactions on Computer-Aided Design of Integrated Circuits and Systems, 19*(12), 1523–1543.

Kuchar, O. A., Reyes-Spindola, J., & Benaroch, M. (2005). Augmented cognition for bioinformatics problem solving. In *Proceedings of First International Conference on Augmented Cognition, 2-disc DVD Set, Augmented Cognition International Society, Las Vegas, NV, July 22-27.*

Licklider, JCR (1960). Man-Computer Symbiosis, IRE Transactions on Human Factors in Electronics, volume HFE-1, pp. 4–11.

Magee, J. L., & Kramer J. (2006). *Concurrency: State models and Java programs* (2nd ed.). New York: Wiley.

Martin, J. N. (2003). *Overview of the EIA 632 standard—Processes for engineering a system* (EIA 632 Working Group Chairman). Raytheon Systems.

Nicholson, D., Lackey, S., Arnold, R., & Scott, K. (2005). Augmented cognition technologies applied to training: A roadmap for the future. In *Proceedings of First International Conference on Augmented Cognition, 2-disc DVD Set, Augmented Cognition International Society, Las Vegas, NV, July 22-27.*

O'Grady, J. O. (1998). *System validation and verification*. Boca Raton, FL: CRC Press.

Parasuraman, R., & Davies, D. R. (1977). *A taxonomic analysis of vigilance performance, in vigilance: Theory, operational performance, and physiological correlates*. New York: Plenum.

Pinto, A., Bonivento, A., & Sangiovanni-Vincentelli, A. L. (2006). System level design paradigms: Platform-based design and communication synthesis. *ACM Transactions on Design Automation for Electronic Systems, 11*(3), 537–563.

Raley, C. (2005). *Development of requirements to incorporate neurophyiological measures in human-computer interface design*. M.S. thesis, University of Maryland, College Park, MD.

Raley, C., & Marshall L. (2004). *Augmented cognition modular design, Class Project in Graduate course ENSE 622: System Requirements Design and Trade-Off Analysis, Master of Science in Systems Engineering Program, Institute for Systems Research, Univeristy of Maryland College Park, MD 20742*.

Raley, C., & Marshall, L. (2005). Modular design for augmented cognition systems. In *Proceedings of First International Conference on Augmented Cognition, 2-disc DVD Set, Augmented Cognition International Society, Las Vegas, NV, July 22-27.*

Sangiovanni-Vincentelli, A. (2003). Electronic-system design in the automobile industry. *IEEE Computer Society*, pp. 8–18.

Sangiovanni-Vincentelli, A., McGeer, P. C., & Saldanha, A. (1996). Verification of electronic systems: A tutorial. In *Proceedings of the 33rd Design Automation Conference*, Association for Computing Machinery, Las Vegas, NV, pp. 106-111.

Schmorrow, D. D., & Kruse, A. (2004). *Augmented cognition* (pp. 54–59). Great Barrington, MA: Berkshire Publishing Group.

See, J. E., Howe, S. R., Warm, J. S., & Dember, W. N. (1995). Meta-analysis of the sensitivity decrement in vigilance. *Psychological Bulletin, 117*, 230–249.

Sidorova, N. (2007). *An introduction to process modeling. Class notes on formal methods*. Dept. of Mathematics and Computing Sciences, Technische Universitei, Eindhoven, Eindhoven, Netherlands.

Simpson, T. W., Jonathon, R. A., & Mistree, F. (2001). Product platform design: Method and application. *Research in Engineering Design, 13*, 2–22.

St. John, M., Risser, M. R., & Kobus, D. A. (2006). Toward a usable closed-loop attention management system: Predicting vigilance from minimal contact head, eye, and EEG measurements. In *Proceedings of Second International Conference on Augmented Cognition*. Augmented Cognition International Society, San Francisco, CA.

Steingart, D., Wilson, J., Redfern, A., Wright, P., Romero, R., & Lim, L. (2005). Augmented cognition for fire emergency response: An iterative user study. *Proceedings of First International Conference on Augmented Cognition, 2-disc DVD Set, Augmented Cognition International Society, Las Vegas, NV, July 22-27.*

Taylor R.M., Bonner M.C., Dickson B., Howells H., Milton N., Pleydell-Pearce K., Shadbolt N., Tennison J,. & Whitecross S., (2001). Cognitive cockpit engineering: Coupling functional state assessment, task knowledge management, and decision support for context-sensitive aiding, Human Systems IAC Gateway, Vol XII, (1).

Uchitel, S. (2004). *Incremental elaboration of scenario-based specifications and behavior models using implied scenarios*. Ph.D. thesis, Imperial College, London.

Uchitel, S., Kramer, J., & Magee, J. (2004). Incremental elaboration of scenario-based specifications and behavior using implied scenarios. *ACM Transactions on Software Engineering and Methodology, 13*(1), 37–85.

Ververs, P. M., Whitlow, S. D., Dorneich, M. C., Mathan, S., & Sampson, J. B. (2005). Aug-Cogifying the Army's Future Warfighter. In *Proceedings of First International Conference on Augmented Cognition, 2-disc DVD Set, Augmented Cognition International Society, Las Vegas, NV, July 22-27.*

Wasson, C. S. (2006). *System analysis, design, and development: Concepts, principles and practices.* Hoboken, NJ: Wiley.

Chapter 9

Practical Considerations for Developing Augmented Cognition Applications

Mark St. John and David A. Kobus
Pacific Science & Engineering Group, Inc.

In this chapter, we provide guidelines for developing augmented cognition applications for operational tasks, from developing operationally relevant augmentation models through an iterative design process of increasingly realistic proxy tasks, along with guidelines for choosing psychophysiological measures that suit the operational context.

Introduction

Developing augmented cognition applications for operational tasks poses many challenges that are not faced in laboratory research. In the laboratory, researchers are free to study artificial tasks with well-rested users and ungainly equipment in order to explore new concepts. In operational environments, however, practitioners must address a variety of constraints arising from the complexities of the operational task, the capabilities and limitations of real users, and the nature of the operational environment and the artifacts used to accomplish the task (Gray & Altmann, 2001; see Chapter 4). All these constraints must be considered and taken into account simultaneously, and the solutions that are developed have real implications for operational tasks and performance. Augmented cognition applications pose additional challenges, including how cognition is measured (see Chapters 1 and 2), how the mitigations operate (see Chapters 5 and 6), and how well the measurement technology suits the operational environment (see Chapter 4).

The goal of this chapter is to guide practitioners through these practical considerations. Rather than attempt to review the burgeoning augmented cognition field, we draw primarily from our own research experiences in the development of decision support tools for air warfare and collaboration tools for distributed team operations and intelligence exploitation. From these experiences, we codify issues and provide some guidelines to practitioners.

We have divided these guidelines into five steps for practitioners to undertake:

1. Develop an augmented cognition model for how the mitigations will operate in the operational task (see Chapters 5 and 6).
2. Develop a laboratory proxy task to iteratively develop and test the application.
3. Select operationally feasible and task-appropriate data collection equipment.
4. Use multiple psychophysiological measures (see Chapters 1 and 2).
5. Validate the augmented cognition model (see Chapters 6 and 7).

General Approach

Step 1—Develop an Augmented Cognition Model

The first step toward developing an augmented cognition application is to develop a conceptual model of how the augmented cognition mitigation will operate in the operational task (see Chapters 5 and 6). The first stage in developing this model is to identify and describe in detail the difficulties users face. Are users sometimes overwhelmed by task demands, and do they fall behind or make errors? Are users multitasked and interrupted during tasks in order to start or complete other tasks? Do users lose vigilance and fail to detect targets or threats? Is the task laborious, or do users spend more time than they need in order to complete the task?

Applied Exercise
Choose an operational task to consider throughout this chapter. List the task/performance requirements; identify the users' procedures; list user capabilities, limitations, and constraints; and identify problems users have completing their tasks efficiently and effectively.

The second stage in developing the model is to describe how augmented cognition could mitigate the difficulties or enhance the operational task. There are two basic paradigms for the way the augmentation could function: adaptively or integrally. Figure 9.1 displays a graphic representation of each paradigm.

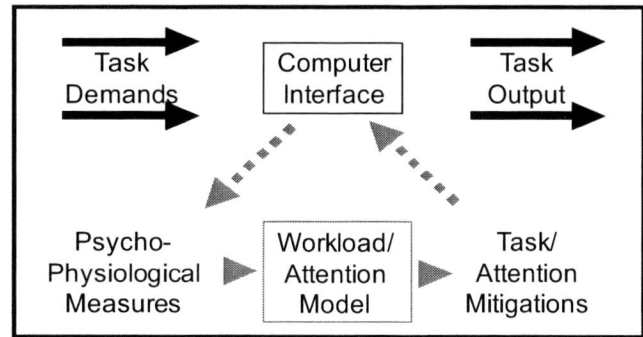

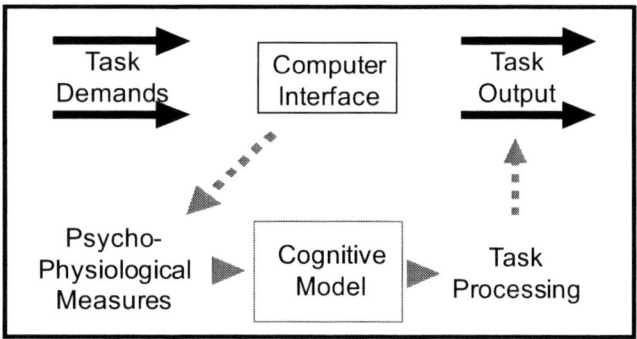

Figure 9.1. Two augmented cognition paradigms.

The Adaptive Paradigm
The adaptive paradigm applies to systems that monitor users for difficulties, such as excessive mental workload or a loss of vigilance. When a difficulty is detected, the application adapts the task or interface to mitigate the difficulty and return users to optimal performance. Once workload or vigilance returns to normal, the mitigation is removed. For the adaptive paradigm, then, the application of the mitigation is contingent on the user's state, rather than continuous (see Chapter 5). If it were useful to apply the mitigation continuously, there would be no need to monitor the user's psychophysiological state, and one would no longer have an augmented cognition application. Instead, it would simply be good interface design. Consequently, when it is time to test the effectiveness of the system empirically, it will be essential to demonstrate that a contingent or adaptive mitigation is better than a continuous mitigation. Engaging the mitigation continuously should produce suboptimal results.

Suppose, for example, a user under low or moderate task load has no difficulty completing all tasks. However, under certain circumstances, the user's workload can become excessive, leading to unacceptable delays and errors. When this situation occurs, an augmented cognition mitigation may be deployed to automate certain secondary activities. The trick is that if the automation is entirely effective, then it should always perform these secondary activities rather than perform them only when the user is overloaded. For the adaptive augmented cognition paradigm to make sense, the automation must be less effective than the user under normal circumstances.

For example, perhaps the automation completes the secondary task, but it reduces the user's situation awareness (see Chapter 5), or the automation uses a less sophisticated analysis than does the user. When the user has the capacity to perform these secondary activities, he or she performs better than the automation, but when the user is in danger of task overload, it makes sense to automate the secondary activities and accept the lower effectiveness in order to allow the user to remain focused on the primary activities. Only under such circumstances would adaptive augmentation make sense.

Furthermore, the adaptive paradigm requires that implementation of the mitigation be controlled by psychophysiological measures of the user's cognitive state (see Chapters 1 and 2). Monitoring changes in task demands alone is not sufficient, though it may be included along with psychophysiological measures (e.g., Fuchs et al., 2006; Van Orden, Limbert, Makeig, & Jung, 2001; see Chapter 5).

In fact, a variety of methods have been examined to control the implementation of automation to mitigate high workload. Automation has been controlled as a function of performance or specific task variables (e.g., Parasuraman, Mouloua, & Molloy, 1996; see Chapter 5), or of users' psychophysiological states (e.g., Mikulka, Scerbo, & Freeman, 2002; Prinzel, Freeman, Scerbo, Mikulka, & Pope, 2003; Wilson, Lambert, & Russell, 2000; see Chapters 1 and 2), or manually by the users themselves (Bailey, Scerbo, Freeman, Mikulka, & Scott, 2006; see Kaber & Endsley, 2004, for a recent review). However, an effective adaptive augmented cognition model should be able to demonstrate that control of the automation based on psychophysiological measures leads to better performance than control by other methods.

Valuable Information
The adaptive paradigm requires that the model satisfy three criteria:

1. The mitigation must be effective.
2. Applying the mitigation adaptively should be more effective than applying it continuously.
3. Using psychophysiological measures to control the implementation of the mitigation should be more effective than controlling the implementation by other dependent variables.

There are many operational tasks in which users may become overloaded and in which adaptive automation controlled by monitoring of the user's state may be valuable. One commonly investigated task is piloting an aircraft. In one commonly used laboratory simulation of piloting tasks, the Multi-Attribute Task Battery (Comstock & Arnegard, 1992), users perform a continuous-tracking task to simulate flying, monitor a number of lights and gauges, and maintain a complex system of pumps and fuel tanks. Several studies have monitored users' mental workload using psychophysiological measures while users perform this simulation (e.g., Smith, Gevins, Brown, Karnik, & Du, 2001; Wilson & Russell, 2003). Some researchers investigated adapting the simulation by temporarily automating the tracking task when mental workload was detected to be excessive (Prinzel et al, 2003; Wilson et al., 2000).

There are many other operational tasks in which sustained attention, rather than overload, is the controlling factor for performance—for example, long-distance driving, baggage screening, perimeter and building security, remote sensing, reconnaissance, and surveillance. For these tasks, an augmented cognition application could monitor the user's cognitive state for a loss of vigilance and then mitigate the loss by alerting the user or modifying the task interface to help sustain attention.

The adaptive paradigm for vigilance tasks requires that the same three criteria be met as for overload tasks. First, the mitigation should be effective in helping the user maintain or sustain vigilance. Second, continuous mitigation, or even implementing the mitigation on a fixed schedule regardless of the user's cognitive state, should be detrimental. For example, triggering an alert while the user is fully vigilant might distract or annoy the user and actually degrade performance. Third, monitoring the user's psychophysiology to provide the trigger for the mitigation should produce better results than monitoring only performance or task variables.

The adaptive paradigm has been evaluated in several empirical studies of vigilance tasks. For example, Berka et al. (2005) measured users' electroencephalograms (EEGs) for losses of vigilance during a simulated driving task, and the system sounded alerts when losses of vigilance were detected. This augmented cognition mitigation decreased the mean time spent drifting, veering, and driving out of lane. In addition, the alerts significantly decreased the number of accidents that occurred in the simulator.

Mikulka, Scerbo, and Freeman (2002) utilized the adaptive model by measuring users' EEGs in a simple target detection task and changing the presentation rate of the stimuli when losses of vigilance were detected. This mitigation improved target detection performance better than randomly timed changes in the presentation rate.

Applied Exercise
Brainstorm adaptive mitigations for the operational task you have chosen to consider (see Chapter 5 for some insights on different types of mitigation). Consider how to test and document all three criteria listed above for the adaptive paradigm.

The Integral Paradigm
Unlike the adaptive paradigm, the integral paradigm applies to systems that monitor and utilize users' psychophysiology as a continuous and integral part of the augmented task. To date, the integral paradigm has been applied primarily to signal detection tasks. A good example is augmented photo analysis (Kruse & Schulman, 2006; Mathan et al., 2006). In this task, users examine complex satellite images for specific objects. In the augmented system, pieces of the image are presented to the user sequen-

tially and very rapidly, and the user's electrophysiology, specifically the P300 evoked potential, is monitored for telltale signs of object recognition. Photos that exhibit a large P300 are then automatically revisited for a more thorough analysis by the users. The argument is that this system allows images to be scanned much faster than normal because the brain's object recognition signal (P300) can be detected in much less time than the user would take to indicate a target detection by pressing a button, and the P300 may be a more sensitive measure of object recognition than a button press.

In this application, the augmentation, which is controlled by specific EEG events, is an integral step in the image analysis process. It is not a mitigation used to return the user to optimal performance or limit the impact of task demands when problematic conditions arise. Rather, it is a key component of the task process that continuously modifies which or when information is presented to users. Consequently, unlike with the adaptive paradigm, the augmentations in the integral paradigm are used continually to enhance the ability of the user in real time while performing a task.

Integral augmentation is still augmented cognition. because the user's psychophysiology is measured and used to control a function in the task. Other cognitive phenomena or other evoked potentials besides the P300 may be incorporated into future applications.

Valuable Information
The integral paradigm requires that the model satisfy three criteria:

1. The augmentation must be effective.
2. The augmentation must be applied continuously.
3. The implementation of the augmentation must be controlled by psychophysiological measures.

Applied Exercise
Brainstorm integral augmentations for the operational task you have chosen to consider. Does the adaptive paradigm or the integral paradigm appear to be more fruitful?

Test Your Knowledge
1. Describe the cognitive difficulties for users of the chosen operational task.
2. Develop an augmented cognition model that specifically addresses the problem.
3. Determine how the model satisfies the adaptive or integral paradigm criteria.
4. Determine how you would ultimately verify the model empirically.

Step 2—Develop a Laboratory Proxy Task to Iteratively Develop and Test the Application

Ideally, practitioners would develop applications within the actual operational task environment immediately. This approach would virtually guarantee that solutions developed and tested would apply to the real task and real users. Unfortunately, the complexity of real tasks, the number of variables, the lack of control, the expense, and the difficulty of obtaining expert users with time to participate are daunting.

Instead, operational applications are more effectively and efficiently developed using an iterative design process that proceeds from simplified laboratory proxy tasks to increasingly realistic simulations and field tests. This process may seem more expensive and time consuming than moving immediately to the operational environment, but experience shows that it is all too easy to make costly mistakes or become overwhelmed by operational details before the application foundations are laid. Iterative design provides many opportunities for user and other stakeholder feedback, and it significantly increases the chances of developing a successful application, and ultimately at a lower cost and with less trouble (e.g., Larman & Basili, 2003; Sy, 2007).

A major question for designing laboratory proxy tasks is how much control—and simplification—should be imposed on the operational task. A spectrum of fidelity is available, from high-fidelity simulations to microworlds to basic laboratory tasks. The right choice depends entirely on what the practitioner is trying to accomplish at each stage of the development process (see Gray, 2002). Practitioners must determine what are the key issues and variables that must be maintained in their operational form, versus what variables should be controlled and simplified without perverting the task.

Regarding the lower end of the fidelity spectrum, St. John, Risser, and Kobus (2006) devised a UAV surveillance task to develop and evaluate an augmented cognition application for vigilance tasks. The task employed standard laboratory vigilance task parameters to present the stimuli, but the stimuli were more complex and realistic than those typically used in laboratory-based vigilance research. These stimuli maintained the UAV cover story and increased the face validity of the task as compared with standard laboratory vigilance tasks, yet they allowed good experimental control over stimulus presentation parameters. The experiment succeeded in obtaining typical vigilance task phenomena, such as the vigilance decrement, but using the more complex stimuli.

The use of more realistic stimuli brought the research a step closer to generalization to the operational task while still maintaining a close connection with the extensive literature on laboratory vigilance phenomena. More important, the low fidelity afforded tight control and measurement so that alternative mitigation strategies could be assessed quantitatively. Future applications can build on these findings and increase the level of fidelity. In this way, iterative development (see Chapter 8) builds each higher level on empirically grounded lower levels.

Somewhat higher on the fidelity spectrum, a proxy task for naval air warfare has been developed by St. John, Kobus, Morrison, and Schmorrow (2004) that could still be performed by participants drawn from the general public. The task was used to integrate, evaluate, and compare a wide range of psychophysiological measures. In the actual operational task, expert users monitor several thousand square miles of airspace for threatening aircraft. Electronic signals and intelligence information are evaluated as the aircraft travel about the airspace. The threatening characteristics of each aircraft may wax and wane over time, and users must continuously determine which, if any, actions are appropriate to take at any given moment. Time pressure and mental workload vary depending on the number of aircraft, especially threatening aircraft that must be monitored.

A cognitive task analysis of the real-world task revealed the cognitive and perceptual requirements the task imposes on real users. A proxy task was then developed to contain many of the same component cognitive and perceptual tasks, but each component was simplified significantly (see St. John, Kobus, & Morrison, 2003, for details). These simplifications were necessary to make the task less complicated so that the public could perform the task without substantial specialized training. These simplifications also made the task very easy to perform. Consequently, to increase task difficulty, the pace was increased substantially by comparison with that of the operational task. The operational task is actually fairly deliberative, taking place on the order of minutes rather than seconds (Morrison, Kelly, & Hutchins, 1996; St. John, Smallman, Manes, Feher, & Morrison, 2005).

These changes were acceptable to meet the overall requirements of the experimental design: to develop a quasi-military task that manipulated a broad range of cognitive and perceptual component processes and that could be performed by public volunteers. Figure 9.2 shows a low- versus high-fidelity comparison of air warfare display consoles. The simplifications, however, made generalization back to the operational task more tenuous.

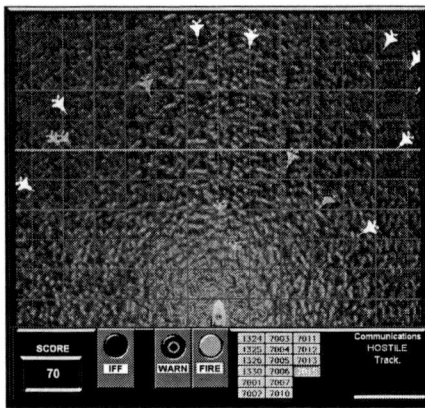 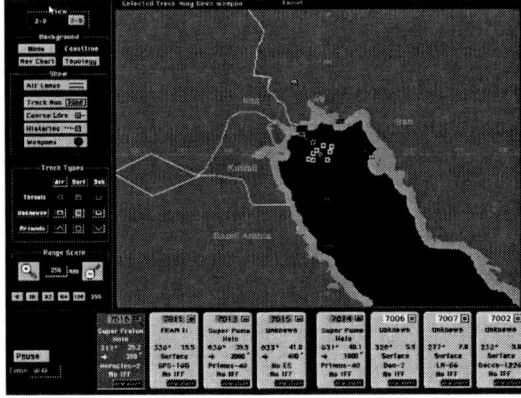

Figure 9.2. Screenshot of a low-fidelity air warfare display (left); screenshot of a high-fidelity air warfare display (right).
(Morrison, Kelly, & Hutchins, 1996)

Further investigations capitalized on the findings and experience obtained from the 2004 experiment and applied them to more realistic tasks. Four industry teams investigated how augmented cognition might be applied in four operational environments: (a) a dismounted infantryman (see Chapter 4), (b) a Tactical Tomahawk Weapons Control System (TTWCS) operator, (c) a light armored vehicle (LAV) operator, and (d) an unmanned aerial vehicle (UAV) operator. These high-fidelity simulations offered good generalization and transfer of lessons learned to the actual operational tasks (see Morrison, Kobus, & Brown, 2006, for a review).

In addition to the choice and design of the laboratory proxy task, the selection of participants may also affect the level of experimental control and the ability to generalize to the operational task. Members of the general public are far more available than expert users, but they have obvious limitations in experience and skill with the operational task. Operational decision making commonly involves recognizing the current situation or problem as similar to some prior experience and then adapting the prior experience to fit the new situation. This process is called *recognition-primed decision making* (Klein, 1993), and it typically requires a large body of experiences on which to draw (Gonzalez, Lerch, & Lebiere, 2003). Naive users clearly do not have this wealth of experience. These individual differences are reflected in the different behaviors in which expert and novice personnel engage to complete similar operational tasks (cf. Kobus, Proctor, & Holste, 2001; Staal, 2004).

In addition to greater knowledge, operational tasks may also require a high degree of skill and automaticity of cognitive and perceptual processes that make naive users unlike expert users (Anderson, 1982; Newell & Rosenbloom, 1993). Further, workload affects novel and highly practiced skills differently (Wickens & Hollands, 2000, chap. 12). If an investigator plans to use naive participants in a proxy task, it may be necessary to train the participants to a reasonable level of competency. Establishing clear performance criteria during training sessions is a useful strategy for this purpose. If this level of competency is lower than that of experts, practitioners need to understand the implications that the lower level implies for evaluating the application. The greater the difference, the more tenuous the generalization to real-world expert performance.

If the mitigation or augmentation is integrally tied to user skill or knowledge, then less than expert-level competence may limit the value of testing the application with naive participants. However, if the mitigation or augmentation is more generic, such as alerting users if they lose vigilance, then relatively naive participants can be tested with confidence that the results will generalize to the operational task and expert users. Again, the iterative design process recommends a ramping up of competence levels from novice to expert as the general level of fidelity rises.

Experts and naive users may also have different motivations and may respond to overload differently. Experts know that the real task must be accomplished despite high workload, and they may transfer that motivation to proxy tasks. They are likely to offer a "can-do" attitude to get the job done. This effect is present especially when real-world context is simulated.

Naive users, on the other hand, may be much more willing to drop a task, or even give up entirely, if the workload is too high. Similarly, users in a low-fidelity simulation may not feel the same high motivation to perform as users in the real task, at least partly because the costs of failure are so different. If a study requires intense motivation and "white-knuckle" performance, then real users and high-fidelity simulations with motivating consequences, or even operational tasks with real consequences, may be needed to evaluate an application. However, many applications may require only the moderate levels of motivation typical of naive users in laboratory proxy tasks.

Finally, experts and naive users may react to stress differently. In a classic article, Baddeley (1972) documented how experts exhibit both lower stress and better performance than novices in dangerous situations. Experts may have lower arousal during dangerous situations, or may have learned to inhibit high arousal, and they may use strategies that are different from those of naive users to mitigate stress factors (see Wickens & Hollands, 2000, chap. 12, for a review).

A related finding is that experts may actually thrive on danger and perform well in its presence (see Klein, 1996, for a review). Highly experienced pilots, for example, report high arousal concurrent with very high performance. This situation is sometimes referred to colloquially as the *pucker factor* (Menza, 2002). These issues are complex, the research literature is voluminous and varied, and there appear to be many factors that influence the impact of stress on operational performance (Wickens & Hollands, 2000, chap. 12).

Valuable Information
If high levels of skill, knowledge, motivation, or arousal, or expert responses to stressors are required to test an application, then high-fidelity simulations, significant incentives, or expert users may be necessary.

Applied Exercise
Lay out a series of proxy tasks that iteratively increase their fidelity to the operational task and environment. Identify the user characteristics essential for evaluating each iteration.

Test Your Knowledge
1. Explicitly lay out a series of development and evaluation goals that iterate toward the operational task and environment.
2. Develop proxy tasks for each iteration.
3. Choose participant populations to match the skills, knowledge, motivation, and stress management appropriate for each proxy task.
4. If you are using naive participants, set objective performance criteria and train the participants accordingly.

Step 3—Select Operationally Feasible and Task-Appropriate Data Collection Equipment

A wide range of psychophysiological measures have been investigated for use in augmented cognition applications (see Chapters 1 and 2). These measures include heart rate and heart rate variability, galvanic skin response (electrodermal response), temperature, blood oxygenation in the body or brain, stress hormones in the bloodstream, head and body posture, mouse click pressure, eyeblink rate, number of eye fixations, pupil size and rate of change, size of eyelid opening (distance between eyelids), EEG and event-related potentials (ERPs), and functional near-infrared brain scans (fNIR).

St. John et al. (2004) reported on a large study in which a wide range of measures were evaluated in a quasi-operational task. Several recent journal special issues on the topic of psychophysiological measurement have been published, including Russo, Schmorrow, and Thomas (2007), Schmorrow and McBride (2004), and Trimmel, Wright, and Backs (2003). Many of these measures focus on different characteristics of the human response to tasks. Therefore, a clear requirement in choosing the best measures, or suite of measures, is that they suit the nature of the task and cognitive augmentation. For example, heart rate has been used primarily to measure general arousal, whereas eye fixations, pupillometry, and EEG theta power have been used to measure mental workload (Hankins & Wilson, 1998). Head posture, eyelid opening, and EEG theta and alpha power have been used to measure vigilance (Mikulka, Scerbo, & Freeman, 2002; St. John, Risser, & Kobus, 2006).

Beyond the choice of measures, there are two competing criteria for choosing equipment for data collection: data precision versus ecological validity and operational practicality (see Chapter 3). Using the most precise and informative equipment available is the first and foremost consideration. Certainly for basic research, data precision is paramount. However, for applied applications, the competing goals of ecological validity and the operational practicality of the equipment take on a larger role. Unfortunately, the most precise and informative equipment tends to be impractical, invasive, constraining, and uncomfortable for extended use in the operational environments that are the ultimate settings for the equipment. The idealized environments of laboratories typically pay little attention to long-term comfort, much less noise, mobility, or invasiveness (see Chapter 4).

The goal of practitioners, on the other hand, is expressly to extend the basic empirical findings of the laboratory to operational uses, considering all the additional variables and constraints imposed by real tasks, experienced users, and task environments. Invasive, uncomfortable, and ungainly equipment is not appropriate, will be rejected by operational users, and will fail in the operational environment. Ecological validity and operational practicality must be taken into account, and these issues typically control the type of measurement equipment that can be used. A primary consideration is that all data collection equipment must be acceptable to operational users during the performance of their tasks.

Of course, one can argue that better-applied equipment is just around the corner, and that there is little reason to limit application development to today's applied equipment when fielding the operational application is still several years away. However, the primary benefit of addressing practicality concerns early in the iterative design process is that important issues of user acceptability and environmental fit will be revealed. Once revealed, they can be addressed and used to move the field forward. A secondary benefit is that the laboratory proxy environment will begin to look like the actual operational environment sooner rather than later. This match may motivate participants by providing context and raising the fidelity of the simulation. It can also provide a vision for the research itself and the generation of new ideas.

Operational constraints that must be considered can be divided into at least four basic categories: comfort, mobility, noise, and safety. Regarding comfort, operational users must wear or otherwise engage with the psychophysiological measurement equipment for extended periods. Typically, experimental sessions in the laboratory may last 1–2 hours, whereas military watch periods last 4–12 hours. In extreme situations, such as sustained operations, users may be actively engaged for days at a time (Kobus, 1989, 1990; Krueger, 1991). Over such long periods, comfort becomes a critical factor. EEG and eye-tracking equipment, because they are typically attached to the head, can cause substantial discomfort from weight, pressure, tightness, and limited vision in some cases. However, remote desktop eye trackers, for which cameras reside on the desktop and nothing touches the user, are improving rapidly and have reached the level of precision at which they can be used effectively for many operational tasks.

EEG equipment still must be attached to the user's head, but recent advances have made the devices much more comfortable (see Chapter 1). For example, wet electrodes are more comfortable, and they do not require abrading the scalp as older electrodes did in order to attain good readings. Dry electrodes appear to be making significant improvements, and they may be practical in the near future. Further, typical laboratory EEG equipment contains tens or hundreds of electrodes, depending on the requirements of a study. Studies that attempt to localize activity to specific brain regions, for example, typically require many electrodes (cf. Ferree & Tucker, 1999). On the other hand, studies that measure global brain activity, such as mental workload or vigilance, can obtain adequate measures from three or even fewer electrodes (Berka et al., 2005; Mikulka, Scerbo, & Freeman, 2002; St. John, Risser & Kobus, 2006). The reduced number of electrodes minimizes the impact on the user, both for comfort and ease of attaching the equipment.

Heart rate sensors are also becoming more comfortable and unobtrusive. Some equipment attaches to fingers, which is minimally invasive, though this can interfere with task performance. Other equipment attaches to toes or the torso or is even built into clothing. The U.S. Army has a long-standing program investigating data collection systems that can be incorporated within clothing (Bonsor, 2005; Wikipedia, 2007).

Mobility is also a serious constraint for many operational tasks (see Chapter 4). Even tasks in which the user sits at a workstation require some movement. Users may need to move their heads and bodies to view multiple monitors and interact with colleagues, and they will almost certainly need to leave their workstations periodically over the course of a shift. A tremendous advantage for EEG data collection in the field is the advent of EEG devices that transmit their signals wirelessly so that users do not need to bear the weight and restriction of wires connecting their heads to a computer. In the

St. John et al. (2006) vigilance study, participants wore a lightweight wireless EEG cap without discomfort for 4 hours over the course of training, lunch, and the experimental task. Further, participants wandered through the building during their lunch breaks and were effortlessly reconnected to the computer system when they returned to the laboratory. See Chapter 4 for additional examples.

Another operational concern is electrical, auditory, and visual noise. Electrical noise from a variety of sources can mask the EEG signals associated with cognitive functions. Electrical artifacts are induced by a variety of sources, including electronics, the movement of electrical cabling, muscle activity, eye activity, and arousal-related changes in the skin conductivity of the user (see Chapter 4). These electrical "artifacts" can be a significant constraint for operational tasks, and they must either be eliminated in the environment or be detected and removed through signal processing so that cognitive signals can be processed correctly. For example, when the idea of collecting EEG data from dismounted soldiers was first proposed in the Defense Advanced Research Projects Agency's (DARPA) Augmented Cognition Program, there were strong objections that EEG could not be collected because of overwhelming electrical interference from muscle activity as users moved. However, if the EEG collection occurs only while the user is stationary, or if specialized filters can be developed to remove the muscle artifact from the electrical signal in real time, this noise problem can be managed.

The processing and analysis of psychophysiological data has, in the past, largely been conducted off line. In 1993, for example, Wilson developed a method for removing several types of electrical artifacts off-line. However, for augmented cognition technologies to work in applied settings, effective and computationally efficient real-time artifact removal is necessary. Fortunately, progress in signal processing is significantly improving artifact removal. For example, Honeywell has developed computational routines to remove artifacts induced by shock, rubbing cables, and gross muscle movement (Dorneich, Whitlow, Ververs, Carciofini, & Creaser, 2004; also see Chapter 4). The techniques and computational power available today make real-time artifact removal feasible.

Auditory and visual noise may also be a problem and must be carefully controlled in the operational setting. Rapid changes in illumination "noise" may interfere with eye tracking, although recent advances in eye-tracking algorithms are making it possible for these systems to overcome this problem. Eye-tracking equipment, for example, is being successfully used in cars, with all of their inherent changes in ambient lighting.

A final note is to begin and end with safety (see Chapter 8). Electrically powered equipment that is physically connected to participants should be powered through an isolation transformer. This requirement, however, is not necessary when wireless technology powered by low-voltage batteries is used. All equipment should be tested frequently to ensure that all parameters affecting participants are within normal safety limits. All equipment that comes in contact with participants should be cleaned and maintained scrupulously. This requirement is especially important for such items as reusable electrodes. Practitioners should follow cleaning procedures established by individual manufacturers.

Applied Exercise
Identify sources of noise artifacts within the operational environment of your chosen task. How might these sources be controlled or filtered?

Test Your Knowledge

1. Identify psychophysiological measures best suited for the task and the proposed augmented cognition model (see Chapters 1 and 2 for insights on measures).
2. Find out which data collection equipment best meets the constraints of the operational environment.
3. Consider what operational constraints will be in play and how to handle them.
4. Identify the artifacts that can be controlled or removed and the methods for use.
5. Safety, first and last.

Step 4—Use Multiple Psychophysiological Measures

Findings from many studies have successfully demonstrated the ability of even a single psychophysiological measure to detect changes in workload or vigilance. Further, some researchers have demonstrated the ability of a single measure to control a mitigation and improve task performance (cf. Berka et al. 2005; Mikulka, Scerbo, & Freeman, 2002; Wilson et al., 2000; also see Chapters 5 and 6). Nonetheless, it appears that significant gains can be achieved by combining multiple psychophysiological measures (see Chapter 3).

Wilson and Russell (2003) found that a combination of EEG, eye, heart, and respiration measures predicted the level of task workload with 88% accuracy across participants. No individual measure achieved greater than 26% accuracy for any individual participant. Marshall (2007) combined a suite of eye-tracking measures, including pupillometry, across a set of tasks; each measure contributed unique information toward predicting task performance. St. John et al. (2006) combined remote eye tracking, head tracking, and wireless EEG measures to monitor users' vigilance. No individual measure accounted for more than 14% of the variance in target detection performance. A combination of measures, however, accounted for 42% of the variance in performance.

These gains in predicting performance may accrue from a variety of sources. It may be that different measures tap different mental processes. It may be that different measures have different time courses, so that some are early predictors of performance decrements whereas others are later predictors of severe decrements. The different measures may also have different inherent temporal sensitivities. For example, head nodding may be used as a measure of sleepiness, but it inherently requires a substantial time window in order to sample a large amount of head movement over time. Frequency of eyeblinks and frequency of saccades similarly require many samples over time, although at shorter time scales than head movements. Other measures, such as EEG and eyelid opening, can be computed in very short time windows.

Further, individual users may be sensitive to different psychophysiological measures (Crosby & Ikehara, 2006; Wilson & Russell, 2003). For example, experts may show smaller changes in arousal than do novices as workload changes. St. John et al. (2006), and many other researchers have found that the predictiveness of each measure varies across individuals. Individually tailored combinations typically predict better than do group-based combinations.

There are a variety of methods for combining multiple measures into a single prediction of, for example, workload overload or loss of vigilance. One of the simplest methods is multiple linear regression. In multiple linear regression, a continuous performance variable, such as lane following while driving or percentage of targets identified per minute in a target detection task, is predicted from a set of psychophysiological measures. Each measure is multiplied by a scalar value and then the measures are added together to predict the performance variable. The multiple linear regression procedure sets the values of the scalars from a sample of measures and performance data.

Rather than predict performance, it is also possible to predict the prevailing level of a task variable, such as workload, from psychophysiological measures. To predict discrete levels of an independent variable, logistic regression or linear discriminant function analysis is more appropriate (see Chapters 4 and 7).

These simple methods, however, do not allow for interactions among measures. Nor do they allow for nonlinear functions between independent and dependent variables (see Chapters 4 and 7). Other modeling tools must be used to provide these options. For example, artificial neural networks (ANNs) have been used extensively for these reasons (Marshall, 2007; Van Orden, Limbert, Makeig, & Jung, 2001; Wilson & Russell, 2003; see Chapters 3, 4, and 7).

All these methods require a fairly substantial data set in order to train the regression or ANN. Care should be taken to sample the space of values for the independent and dependent variables adequately. Establishing appropriate training, validating (see Chapter 6), and testing samples is a complex topic that should be planned for early in the development of an application (see, for example, Smith, 1993).

The choice of measures, suites of measures, and methods for combining them turns on availability of quantitative comparisons between approaches (see Chapter 3). Unfortunately, many earlier research studies reported only qualitative effects because they were focused on exploration and discovery. As the field of augmented cognition matures and advances toward increasingly sophisticated and operational applications, the need for quantitative data and comparisons will grow. Practitioners can help to advance the field by reporting quantitative results and, when possible, quantitative comparisons.

Applied Exercise
1. For the adaptive paradigm, identify performance measures that can be predicted from psychophysiological measures and that can in turn be used to control mitigations.
2. For the integral paradigm, identify cognitive phenomena, such as the P300, that can be detected psychophysiologically and that can be used to control the application.

Review
1. Use multiple measures whenever possible.
2. Be cognizant of task-related and measure-related individual differences.
3. Choose appropriate methods for combining measures.
4. Report quantitative comparisons among measures and methods.

Step 5—Validate the Augmented Cognition Model

Relatively few studies have investigated the effectiveness of complete closed-loop models of adaptive or integral cognitive augmentations. The most likely reason is that the field is relatively new, risks and costs are high for conducting research in the applied environment, and a complete closed-loop application is the culmination of all prior steps. Empirical evaluation, however, is the key for obtaining external validity for augmentations under investigation (see Chapter 6).

Step 1, choosing a real task and developing an augmented cognition model, outlines the requirements for evaluating the model. Few studies have met all the aforementioned requirements, though efforts are progressing rapidly. The proceedings from the annual meeting of Augmented Cognition International (Schmorrow, Nicholson, Drexler, & Reeves, 2007; Schmorrow, Stanney, & Reeves, 2006) provide an overview of the state of the art of augmented cognition and its empirical evaluation.

There is much room for improvement in each stage of the closed loop of augmented cognition models. Many advances in the ways data are collected, artifacts are removed, and data are analyzed continue to be made (see Chapter 4). The sophistication of methods for closing the loop and providing real-time augmentations of operator performance is growing rapidly. Especially in the final augmentation stage, there are tremendous opportunities for increasing the variety and sophistication of adaptive and integral augmentations beyond simple alerts and warnings (see Chapter 5). We hope the advice presented here brings new practitioners to the promising field of applied augmented cognition.

Best Practices

This chapter focuses on practical considerations and best practices for developing a successful augmented cognition application. We described five steps that practitioners must undertake. The first step, developing a model of how the mitigation will operate in the task, is crucial (see Chapters 5 and 6). Carefully reasoning through what difficulties target users have with the task, how their cognition will be measured, and how mitigations will address the difficulties identified is the cornerstone of the application. Take time to get the model right.

The second step, iteratively designing proxy tasks with increasing fidelity to the operational task, represents the core process through which successful development happens. Leaping to the operational environment is no more practical than studying an artificial task in a laboratory, because there are too many constraints to address all at once. Iterative design works to address the practical constraints in an orderly succession and build on previous results (see Chapter 7). Iterative development also provides opportunities for users and other stakeholders to influence the application early, when it is easier to change. The key is to choose proxy tasks carefully so that they preserve important task characteristics while simplifying less important ones.

The third step, selecting operationally feasible psychophysiological equipment (see Chapters 1, 2, and 4), pushes in the opposite direction, toward early practicality and high fidelity to the operational environment. Equipment in this young field presents many constraints, and considering early on how users will interact with it improves the chances for developing a successful application. Again, it is important to consider carefully what aspects of the equipment and environment should be addressed early or simplified and addressed later.

The fourth step, employing and combining multiple measures, is a less critical point but still one that reaps dividends (see Chapter 3).

The final step, evaluating the complete closed-loop application, is imperative (see Chapter 6). Nothing sells like success, and having data to show that an application actually works and enhances user performance is an effective marketing tool as well as a significant contribution to the field. Quantitative results and comparisons among alternative methods are even better, because they increase the sophistication of evaluation and the sophistication of the field.

Parting Message

Strive to achieve a balance between practicality and simplification. Although it is important to consider and address practical issues early in the design cycle, application development is an iterative process of increasing levels of sophistication and fidelity to the operational task and environment. Rome was not built in a day, and neither will be your operational application of augmented cognition.

References

Anderson, J. R. (1982). Acquisition of cognitive skill. *Psychological Review, 89*, 369–406.
Baddeley, A. D. (1972). Selective attention and performance in dangerous environments. *British Journal of Psychology, 63*, 537–546.
Bailey, N. R., Scerbo, M. W., Freeman, F. G., Mikulka, P. J., & Scott, L. A. (2006). Comparison of a brain-based adaptive system and a manual adaptable system for invoking automation. *Human Factors, 48*, 693–709.

Berka, C., Levendowski, D. J., Westbrook, P., Davis, G., Lumicao, M. N., Ramsey, C., Petrovic, M. M., Zivkovic, V. T., & Olmstead, R. E. (2005). Implementation of a closed-loop real-time EEG-based drowsiness detection system: Effects of feedback alarms on performance in a driving simulator. In *Proceedings of the 11th Annual Conference on Human-Computer Interaction* (pp. 651–660). Mahwah, NJ: Erlbaum.

Bonsor, K. (2005). How the future force warrior will work. Available at *HowStuffWorks.com*, http://science.howstuffworks.com/ffw.htm

Comstock, J. R., & Arnegard, R. J. (1992). *The multi-attribute task battery for human operator workload and strategic behavior research* (Tech. Report Memorandum No. 104174). Hampton, VA: NASA Langley Research Center.

Crosby, M. E., & Ikehara, C. S. (2006). Using physiological measures to identify individual differences in response to task attributes. In D. Schmorrow, K. Stanney, & L. Reeves (Eds.), *Foundations of augmented cognition* (2nd ed., pp. 162–168). Arlington, VA: Strategic Analysis.

Dorneich, M., Whitlow, S., Ververs, P. M., Carciofini, J., & Creaser, J. (2004). Closing the loop of an adaptive system with cognitive state. In *Proceedings of the Human Factors and Ergonomics Society 48th Annual Meeting* (pp. 590–594). Santa Monica, CA: Human Factors and Ergonomics Society.

Ferree, T. C., & Tucker, D. M. (1999). Development of high-resolution EEG devices. *International Journal of Bioelectromagnetism, 1,* 4–10.

Fuchs, S., Hale, K. S., Berka, C., Levendowski, D., & Juhnke, J. (2006). Physiological sensors cannot effectively drive system mitigation alone. In D. Schmorrow, K. Stanney, & L. Reeves (Eds.), *Foundations of augmented cognition* (2nd ed., pp. 193–200). Arlington, VA: Strategic Analysis.

Gonzalez, C., Lerch, J. F., & Lebiere, C. (2003). Instance-based learning in dynamic decision making. *Cognitive Science, 27,* 591–635.

Gray, W. D. (2002). Simulated task environments: The role of high-fidelity simulations, scaled worlds, synthetic environments, and microworlds in basic and applied cognitive research. *Cognitive Science Quarterly, 2,* 205–227.

Gray, W. D., & Altmann, E. M. (2001). Cognitive modeling and human-computer interaction. In W. Karwowski (Ed.), *International encyclopedia of ergonomics and human factors* (Vol. 1, pp. 387–391). New York: Taylor & Francis.

Hankins, T. C., & Wilson, G. F. (1998). A comparison of heart rate, eye activity, EEG and subjective measures of pilot mental workload during flight. *Aviation, Space, and Environmental Medicine, 69,* 360–367.

Kaber, D. B., & Endsley, M. R. (2004). The effects of level of automation and adaptive automation on human performance, situation awareness and workload in a dynamic control task. *Theoretical Issues in Ergonomics Science, 5,* 113–153.

Klein, G. A. (1993). A recognition-primed decision (RPD) model of rapid decision making. In G. A. Klein, J. Orasanu, R. Calderwood, & C. E. Zsambok (Eds.), *Decision making in action: Models and methods.* Pp. 138–147. Norwood, NJ: Ablex.

Klein, G. (1996). The effect of acute stressors on decision making. In J. E. Driskell & E. Salas (Eds.), *Stress and human performance.* Pp. 49–88. Mahwah, NJ: Erlbaum.

Kobus, D. A. (1989). Event-related potentials during sustained operations. In G. R. Banta & C. E. Englund (Eds.), *Sustained operations research: A blend of psychology and physiology* (NHRC Report 89-54, pp. 35–46). San Diego, CA: Naval Health Research Center.

Kobus, D. A. (1990). The operational assessment of human performance during sustained/continuous operations. In D. F. Neri & R. E. Gadolin (Eds.), *Sustained/Continuous Operations Subgroup of the Department of Defense Human Factors Engineering Technical Group: Program summary and abstracts from the 9th Semiannual Meeting* (NAMRL Report SR90-1, pp. 41–45). Pensacola, FL: Naval Aerospace Medical Research Laboratory.

Kobus, D. A., Proctor, S., & Holste, S. (2001). Effects of experience and uncertainty during dynamic decision-making. *International Journal of Industrial Ergonomics, 28,* 275–290.

Krueger, G. P. (1991). Sustained military performance in continuous operations: Combatant fatigue, rest and sleep needs. In R. Gal & A. D. Mangelsdorff (Eds.), *Handbook of military psychology* (pp. 256–277). Hoboken, NJ: Wiley.

Kruse, A. A., & Schulman, J. J. (2006). Neurotechnology for intelligence analysts. In D. Schmorrow, K. Stanney, & L. Reeves (Eds.), *Foundations of augmented cognition* (2nd ed., pp. 27–31). Arlington, VA: Strategic Analysis.

Larman, C., & Basili, R. (2003). Iterative and incremental development: A brief history. *Computer, 36,* 47–56.

Marshall, S. P. (2007). Identifying cognitive state from eye metrics. *Aviation, Space, and Environmental Medicine, Special Supplement on Operational Applications of Cognitive Performance Enhancement Technologies, 78,* section II supplement, B165-175.

Mathan, S., Ververs, P., Dorneich, M., Whitlow, S., Carciofini, J., Erdogmus, D., Pavel, M., Huang, C., Lan, T., & Adami, A. (2006). Neurotechnology for image analysis: Searching for needles in haystacks efficiently. In D. Schmorrow, K. Stanney, & L. Reeves (Eds.), *Foundations of augmented cognition* (2nd ed., pp. 3–11). Arlington, VA: Strategic Analysis.

Menza, M. D. (2002, March). The pucker factor. *Approach.* Available at http://www.safetycenter.navy.mil/media/approach/issues/mar02/pucker.htm

Mikulka, P. J., Scerbo, M. W., & Freeman, F. G. (2002). Effects of a biocybernetic system on vigilance performance. *Human Factors, 44,* 654–664.

Morrison, J. G., Kelly, R. T., & Hutchins, S. G. (1996). Impact of naturalistic decision support on tactical situation awareness. In *Proceedings of the Human Factors and Ergonomics Society 40th Annual Meeting* (pp. 199–203). Santa Monica, CA: Human Factors and Ergonomics Society.

Morrison, J. G., Kobus, D. A., & Brown, C. M. (2006). *DARPA improving warfighter information intake under stress—Augmented cognition phase II: The concept validation experiment* (SSC Tech. Report No. 1940). San Diego, CA: SPAWAR Systems Center.

Newell, A., & Rosenbloom, P. S. (1993). *Mechanisms of skill acquisition and the law of practice.* Cambridge, MA: MIT Press.

Parasuraman, R., Mouloua, M., & Molloy, R. (1996). Effects of adaptive task allocation on monitoring of automated systems. *Human Factors, 38,* 665–679.

Prinzel, L. J., Freeman, F. G., Scerbo, M. W., Mikulka, P. J., & Pope, A. T. (2003). Effects of a psychophysiological system for adaptive automation on performance, workload, and the event-related potential P300 component. *Human Factors, 45,* 601–613.

Russo, M. B., Schmorrow, D., & Thomas, M. L. (2007). Operational applications of cognitive performance enhancement technologies. *Aviation, Space, and Environmental Medicine, 78,* Section II.

Schmorrow, D., & McBride, D. (2004). Special section: Augmented cognition. *International Journal of Human-Computer Interaction, 17,* 127–286.

Schmorrow, D. D., Nicholson, D. M., Drexler, J. M., & Reeves, L. M. (2007). *Foundations of augmented cognition* (4th ed.). Arlington, VA: Strategic Analysis.

Schmorrow, D., Stanney, K. M., & Reeves, L. M. (2006). *Foundations of augmented cognition* (2nd ed.): *Augmented cognition: Past, present & future.* Arlington, VA: Strategic Analysis.

Smith, M. (1993). *Neural networks for statistical modeling.* New York: Van Nostrand Reinhold.

Smith, M. E., Gevins, A., Brown, H., Karnik, A., & Du, R. (2001). Monitoring task load with multivariate EEG measures during complex forms of human-computer interaction. *Human Factors, 43,* 366–380.

St. John, M., Kobus, D. A., & Morrison, J. G. (2003). *DARPA augmented cognition technical integration experiment* (SSC Tech. Report No 1905). San Diego, CA: SPAWAR System Center.

St. John, M., Kobus, D. A., Morrison, J. G., & Schmorrow, D. (2004). Overview of the DARPA Augmented Cognition technical integration experiment. *International Journal of Human-Computer Interaction, 17*, 131–149.

St. John, M., Risser, M. R., & Kobus, D. A. (2006). Toward a usable closed-loop attention management system: Predicting vigilance from minimal contact head, eye, and EEG measures. In D. Schmorrow, K. Stanney, & L. Reeves (Eds.), *Foundations of augmented cognition* (2nd ed., pp. 12–18). Arlington, VA: Strategic Analysis.

St. John, M., Smallman, H. S., Manes, D. I., Feher, B. A., & Morrison, J. G. (2005). Heuristic automation for decluttering tactical displays. *Human Factors, 47*, 509–525.

Staal, M. (2004). *Stress, cognition, and human performance: A literature review and conceptual framework* (NASA/TM-2004-212824). Moffett Field, CA: NASA Ames Research Center.

Sy, D. (2007). Adapting usability investigations for agile user-centered design. *Journal of Usability Studies, 2*, 112–132.

Trimmel, M., Wright, N., & Backs, R. W. (Eds.). (2003). Special section: Psychophysiology in ergonomics. *Human Factors, 45*, 523–684.

Van Orden, K. F., Limbert, W., Makeig, S., & Jung, T.-P. (2001). Eye activity correlates of workload during a visuospatial memory task. *Human Factors, 43*, 111–121.

Wickens, C. D., & Hollands, J. G. (2000). *Engineering psychology and human performance* (3rd ed.). Upper Saddle River, NJ: Prentice Hall.

Wikipedia. (2007). Future force warrior. Available at http://en.wikipedia.org/wiki/Future_Force_Warrior

Wilson, G. F. (1993). Air-to-ground training missions: A psychophysiological workload analysis. *Ergonomics, 36*, 1071–1087.

Wilson, G. F., Lambert, J. D., & Russell, C.A. (2000). Performance enhancement with real-time physiologically controlled adaptive aiding. In *Proceedings of the International Ergonomics Association 14th Triennial Congress and Human Factors and Ergonomics Society 44th Annual Meeting* (pp. 3–61 to 3–64). Santa Monica, CA: Human Factors and Ergonomics Society.

Wilson, G. F., & Russell, C. A. (2003). Real-time assessment of mental workload using psychophysiological measures and artificial neural networks. *Human Factors, 45*, 635–643.

About the Authors

Patricia A. Aloise-Young received a Ph.D. in developmental psychology from the University of Florida. She was a postdoctoral fellow at the Institute for Health Promotion and Disease Prevention Research at the University of Southern California. Aloise-Young is an associate professor in the Psychology Department at Colorado State University. Her primary research interest is adolescent cigarette smoking. She has been the recipient of several research grants, including a K01 award from the National Institute on Drug Abuse.

Mark Austin is an associate professor at the University of Maryland, College Park, with appointments in the Department of Civil and Environmental Engineering and the Institute for Systems Research (ISR). Austin is past director of the Program for Master of Science in Systems Engineering at ISR. He is also associate director of the Systems Engineering and Integration Laboratory at ISR and past cochair of ICOSE's Commercial Practices Interest Working Group. His Ph.D. and M.S. degrees are from the University of California, Berkeley. He also has a B.E. (First Class Honors) in civil engineering from the University of Canterbury, New Zealand.

Chris Berka, chief executive officer and cofounder of Advanced Brain Monitoring, has more than 25 years' experience managing clinical research and developing and commercializing new technologies. She is coinventor of six patented and four patent-pending technologies and is the principal investigator or coinvestigator for grants awarded by the National Institutes of Health, DARPA, and ONR, which provided more than $13 million in research funds to Advanced Brain Monitoring. Berka played a key role in the growth of an AMEX public company that patented and commercialized forensic methods for investigating drug abuse, drug addiction, and other health-related concerns. She has 10 years' experience as a research scientist with more than 50 publications on the analysis of the EEG correlates of cognition in healthy subjects and patients with sleep and neurological disorders. She received her B.A. with distinction in psychology/biology at Ohio State University and completed graduate studies in neuroscience at the University of California, San Diego.

Michael C. Dorneich is a senior research scientist in the Human Centered Systems Group at Honeywell Aerospace Advanced Technology. His thesis work (Ph.D., University of Illinois, 1999) centered on the design of distributed collaborative decision aiding, virtual learning communities, and workflow management applications. At Honeywell, Dorneich has worked on multiple NASA- and DARPA-sponsored projects relating to the design of decision support, situation awareness, and mixed-initiative adaptive system applications. NASA programs include the design and evaluation of an integrated cockpit alerting system, integrated weather avoidance optimization for dispatchers, and, currently, the displays and controls design for NASA's Orion. He also developed a collaborative flight diversion decision support tool for airline dispatchers. Dorneich was the cognitive engineering lead for a DARPA MICA project focusing on mixed-initiative control of teams of unmanned air vehicles. He was co–principal investigator on the DARPA Augmented Cognition program, developing adaptive systems triggered by real-time assessments of cognitive state. He has authored more than 55 professional, peer-reviewed papers; holds 1 patent and has filed 16 additional patents; and is an associate editor for the journal *IEEE Transactions of Systems, Man, and Cybernetics*.

Traci H. Downs, Ph.D., is the chief operating officer and principal owner of Archinoetics, LLC, in Honolulu, Hawaii. She has more than 14 years of experience in functional neuroimaging and neuroscience. She is currently involved in the development and testing of several real-time, wearable functional near-infrared imaging (fNIR) systems. Her background includes extensive work with functional magnetic resonance imaging (fMRI) and positron emission tomography (PET). Her doctoral work concentrated on detecting and discerning differences in brain signals for actual movement, imagined movement, and phantom limb movement. Building off her augmented cognition efforts, Downs is exploring innovative ways of coupling user interfaces and emerging physiological sensing. She has authored or coauthored more than 15 publications related to functional brain imaging and has extensive program management experience with government and commercial contracts. Under her leadership, Archinoetics received the prestigious National Tibbetts Award in 2007 for outstanding SBIR achievement. In 2006, she was honored with the Pacific Business News "Forty Under 40" award, which recognizes the best business leaders in the state of Hawaii under the age of 40. Downs received her Ph.D. in neuroscience in 2002 from the University of Virginia and her B.A. in psychology in 1992 from Baylor University.

J. Hunter Downs III, Ph.D., is the chief science officer and co-owner of Archinoetics, LLC, in Honolulu, Hawaii. He has more than 18 years of experience in the analysis and acquisition of medical images. Over the last 7 years, he has been the principal investigator on more than 20 contracts, with projects ranging from telerobotic surgical mentoring to functional imaging post–brain injury. For his work with online training content development, he has served as an advisor to the World Health Organization, has been an invited speaker at a number of international conferences, and has been the focus of articles in *Slate* and *Sun Developer's Journal* online. He has authored or coauthored more than 60 scientific publications, including 3 book chapters, on subjects that include functional medical imaging, telerobotics, human-computer interaction, and advanced medical image visualization. He has also developed a commercially available system for nuclear medicine radioisotope imaging and processing. Under his leadership, Archinoetics received the prestigious National Tibbetts Award in 2007 for outstanding SBIR achievement. Downs received his Ph.D. in medical physics in 1994 from the University of Texas Health Science Center at San Antonio and his B.S. in computer science in 1987 from the University of Texas at San Antonio.

Monica Fabiani is a professor of psychology at the University of Illinois and a cochair of the Biological Intelligence Research Group at the Beckman Institute. She codirects the Cognitive Neuroimaging Lab with Gabriele Gratton and is the author or coauthor of more than 90 publications. She received her Ph.D. in biological psychology from the University of Illinois in 1990. She subsequently was a researcher at the FIDIA pharmaceutical labs in Abano Terme, Italy (1991), a research scientist at the New York State Psychiatric Institute (1992–1996), and an assistant to associate professor at the University of Missouri—Columbia (1996–2001). She is the president of the Society for Psychophysiological Research (2007–2008) and a Fellow of the Association for Psychological Science.

ABOUT THE AUTHORS

Her research involves the integration of multiple brain-imaging methods (including event-related potentials, optical imaging, and structural and functional MRI) and their application to the study of aging.

Sven Fuchs is a senior research associate at Design Interactive, Inc., where he has been involved in several augmented cognition research projects sponsored by DARPA, IARPA, and ONR. He is the principal investigator of an Air Force-funded SBIR effort that investigates approaches of combining physiological measures with real-time performance metrics to improve the context-sensitivity of adaptive systems. Fuchs has authored and coauthored more than 10 papers and 3 book chapters on augmented cognition. In 2006, he was named an "Augmented Cognition Ambassador" and received the Augmented Cognition International Society's "Foundations of Augmented Cognition" award. Other interest areas include usability evaluation, multimodal interface design, and innovative human-system interface technologies. He holds an undergraduate degree in computer science from the Flensburg University of Applied Science in Germany and was a Fulbright scholar at DePaul University, Chicago, where he earned an M.S. in human-computer interaction. Fuchs is a Full Member of HFES.

Gabriele Gratton is a professor of psychology at the University of Illinois. He codirects the Cognitive Neuroimaging Lab with Monica Fabiani and is the author of more than 100 publications. He received an M.D. from the University of Rome, Italy, in 1980 and a Ph.D. in biological psychology from the University of Illinois in 1991. He was an assistant professor at Columbia University (1992–1996), and an associate professor at the University of Missouri—Columbia (1996–2001) before moving to the University of Illinois in 2001. He received the Provost Outstanding Junior Faculty Research and Creative Activity Award from the University of Missouri and the Distinguished Scientific Award for Early Career Contributions to Psychophysiology from the Society for Psychophysiological Research. He is a Fellow of the Association for Psychological Science. His research includes the development of fast (neuronal) optical imaging of brain function, and the use of multiple imaging methods in the investigation of attention and executive function.

Frank L. Greitzer, Ph.D., is a chief scientist in cognitive informatics at the Pacific Northwest National Laboratory (PNNL). He holds a Ph.D. in mathematical psychology with a specialization in memory and cognition and a B.S. in mathematics. His research interests include modeling human behavior, human-information interaction, and augmented cognition mitigation concepts to enhance information processing, decision making, and learning. He has led research on evaluation methods for assessing the effectiveness of tools and methods designed to improve human-system performance. In the area of cyber security, Greitzer serves on the Leadership Team for PNNL's Information and Infrastructure Integrity Initiative in the area of predictive defense. Greitzer also has conducted research to improve training effectiveness by applying cognitive principles in innovative, interactive, scenario-based training and serious gaming approaches. Some of his more than 70 publications and descriptions of representative projects may be found at the Cognitive Informatics Web site, http://www.pnl.gov/cogInformatics. In addition to his work at PNNL, Greitzer serves as an adjunct faculty member at Washington State University, Tri-Cities campus, where he teaches courses in interaction design, human factors, and cognition. He also serves on the editorial board of the *Journal of Cognitive Informatics & Natural Intelligence.*

Kelly S. Hale is the human-systems integration director at Design Interactive, Inc., and has more than 8 years' experience in human-computer interaction, innovative human-system design paradigms, usability evaluation methods of advanced technologies, and cognitive processes within multimodal systems and virtual environments. Hale has been the principal investigator of augmented cognition research and development efforts sponsored by ONR, DARPA, and IARPA; directed Multimodal Information Design Support (MIDS) and Tool for Information Processing Capability Assessment (TIPCA) SBIR product development efforts; and is leading efforts to enhance training using innovative affect induction techniques to optimize emotional experience in real time. She has authored more than 10 papers related to augmented cognition, including "Physiological Sensors Cannot Effectively Drive System Mitigation Alone," which was awarded the AugCog International Society Best Topic Paper Award in 2006. Hale holds a B.S. in kinesiology/ergonomics from the University of Waterloo, Canada, and an M.S. and Ph.D. in industrial engineering from the University of Central Florida.

Joseph Juhnke, CEO and director of delivery for Tanagram Partners, a Chicago-based user experience design firm, has been working in digital interaction since 1996. For the last 12 years, Juhnke has been dedicated to human performance enhancement by enabling humans to interact with data in natural and intuitive manners. He leads a team of researchers, information architects, visual designers, and software developers to produce cutting-edge, commercial applications for companies such as FTD, Nike, Orbitz, Tribune Interactive, and The Nielsen Company. Having worked primarily for commercial-sector Fortune 1000 companies, Juhnke is now leading a new push into defense research and development. Currently, he is engaged in several research efforts for the Defense Advanced Research Projects Agency (DARPA) developing lifesaving mitigation strategies and advanced user interfaces. He received his B.F.A. in art (emphasis in graphic design) from Western Michigan University and also holds concurrent B.A.s in communications and art. Juhnke is a licensed private pilot in the United States.

David A. Kobus is a senior scientist and director of the Medical Systems & Marine Corps programs division at Pacific Science & Engineering Group. He has been involved in human performance research and project management for more than 25 years. His expertise in identifying, evaluating, and correcting environmental problems that cause systems error has been recognized by NASA, the Department of Defense, and the medical community. His research interests include human-machine interactions, sensory integration, human error, augmented cognition, and sustained human performance. He has been involved in several studies that have led to system specifications and requirements optimizing the human-machine interface. Kobus has published more than 70 papers and technical reports. Most of these reports involve evaluating human performance in the completion of real-world tasks. He has also presented more than 80 papers at national or international professional conferences. He has 20 years of teaching experience at both the graduate and undergraduate levels in the areas of experimental design, advanced statistics, cognition, biological psychology, and sensory systems.

ABOUT THE AUTHORS

Blaze M. Keller is a research engineer at the Operator Performance Laboratory (OPL) at the Center for Computer Aided Design. He received his undergraduate degree in electrical and computer engineering and entrepreneurship from the University of Iowa College of Engineering in 2004. He has a master's degree in industrial engineering, specializing in human factors. Keller is working toward his Ph.D. industrial engineering with a focus on human factors in aviation. He has designed and integrated advanced flight deck concepts for flight tests for fixed- and rotary-wing aircraft. He has also integrated advanced systems into automobiles and earth-moving machinery. Keller is the software lead on the Cognitive Avionics Tool Set (CATS) and has extensive experience in developing powerful visualization tools that facilitate the design and evaluation process. These tools have been employed across a number of domains, including advanced flight deck displays, eye-tracking systems, and neural avionics.

Arthur F. Kramer is Swanlund Chair and professor of psychology. He received his Ph.D. in cognitive/experimental psychology from the University of Illinois in 1984. He holds appointments in the Department of Psychology, the Neuroscience Program, the Institute of Aviation, and the Beckman Institute. Kramer's research projects include topics in cognitive psychology, cognitive neuroscience, and human factors. A major focus of his lab's recent research is the understanding and enhancement of cognitive and neural plasticity across the lifespan. He is the director of the Biomedical Imaging Center and codirector of the NIH Center for Healthy Minds. Kramer served as an associate editor of *Perception and Psychophysics* and is currently a member of seven editorial boards. He is a Fellow of the American Psychological Association and American Psychological Society, a member of the executive board of the International Society of Attention and Performance, and a recent recipient of a NIH Ten-Year MERIT Award and a ONR multidisciplinary university research initiative center grant. Kramer's research has been featured in a long list of print, radio, and electronic media, including the *New York Times, Wall Street Journal, Washington Post, Chicago Tribune, CBS Evening News, Today Show, NPR,* and *Saturday Night Live.*

Todd J. Macuda is vice president of product solutions and chief scientist for Gladstone Aerospace Consulting (GAC). He has worked for the National Research Council of Canada and is currently on an industrial secondment program to GAC as director of business development. He is the founder of the aerospace medicine program at the Flight Research Laboratory. Macuda is a graduate of the University of Western Ontario and holds an undergraduate and master's degree in psychology and a Ph.D. in neuroscience. He is an adjunct professor of several universities and a qualified instructor of human factors and related aerospace medicine courses. Macuda's research focuses on UAV and aircraft interoperability, control station design, pilot performance assessment, flight testing, night vision goggles and their effects on human visual performance, visual enhancement systems using sensor fusion, synthetic vision systems, augmented cognition, and workload measurement using physiological methods. Macuda is a member of several technical societies and committees, including Augmented Cognition International (board member), the Canadian Aeronautics and Space Institute-Flight Test Methods working group, and the Society for the Advancement of Modeling and Simulation.

Santosh Mathan is a senior research scientist in the Human Centered Systems Group at Honeywell Aerospace Advanced Technology. His projects focus on sensor-based approaches to cognitive state estimation in real-world application contexts. He is currently principal investigator on the DARPA-funded Neurotechnology for Intelligence Analysts program and leads pattern recognition and signal-processing efforts under DARPA's Augmented Cognition program. He is also exploring the potential for using sensor-based workload estimation techniques in human-computer interaction assessment contexts.

Erin M. Nishimura is an electrical engineer at Archinoetics, LLC, in Honolulu, Hawaii, with a focus on signal processing. Her research interests include biomedical data analysis and algorithm development, and she was involved in the development of commercially available equipment for wireless mesh networks. She has authored or coauthored seven articles on hardware and signal processing of functional near-infrared imaging for assessing cognitive activity for usability analysis and for use in brain-computer interfaces. Nishimura received her B.A. and M.A. degrees in electrical engineering from Stanford University in 2003. She is a member of IEEE and Tau Beta Pi and has held several leadership positions in the Society of Women Engineers.

Colby Raley is an advisory bioengineer with Strategic Analysis, Inc., where she has worked in data fusion, human factors, cognitive psychology, and neuroscience-inspired programs for the past six years. She has supported various projects at DARPA, including augmented cognition, and the Department of Homeland Security Science and Technology Directorate. Raley holds an M.S. in systems engineering and a B.S. in bioengineering from the University of Maryland.

Evan D. Rapoport is an application specialist at Archinoetics, LLC, in Honolulu, Hawaii. His research interests focus on human-centered computer interfaces for deployment in diverse environments and with unique sets of users. One such project involves brain-computer interfaces that enable severely disabled people to communicate and control their environments with only their thoughts. He has authored or coauthored several articles on this topic, specifically on methods for maximizing the effectiveness of limited-bandwidth communications inherent in brain-controlled applications. Rapoport has moe than seven years of experience on research teams developing closed-loop augmented cognition systems for improving human memory and performance. Additionally, he is working on several projects that provide K–12 students with hands-on experiences with cutting-edge technologies for the purpose of broadening science and engineering curricula in public schools. He received his B.A. in cognitive science from the University of Virginia in 2002 with concentrations in computer science and cognitive psychology. He is also an avid photographer.

Thomas Schnell is an associate professor in the Department of Industrial Engineering with a secondary appointment in the Department of Neurology at the University of Iowa Hospitals and Clinics. He is also director of the Operator Performance Laboratory (OPL), which is home to about 17 researchers from the ranks of undergraduate students, graduate students, postdoctoral researchers, and full-time staff. Schnell has an undergraduate degree in electrical engineering and a master's and Ph.D. degree in industrial engineering with a specialization in human factors. He is a jet-rated commercial pilot, flight instructor, and U.S. Air Force auxiliary (CAP) mission-qualified pilot. He is also a research pilot at the OPL. Schnell has conducted human factors studies since 1992. He has specialized in flight testing, eye tracking, vision research, aviation human factors, workload assessment, and modeling. He is the author of 37 papers in refereed journal articles and 60 papers in conference proceedings.

Mark St. John is a senior scientist and director of the Cognitive Systems Division at Pacific Science & Engineering Group in San Diego. He received his Ph.D. in cognitive psychology from Carnegie Mellon University in 1990. He investigates the human factors and psychology of supervision and decision making in complex, real-world environments. St. John received the Jerome H. Ely *Human Factors* Article Award in 2002 for research on how 2-D and 3-D displays afford different types of cognitive and perceptual tasks, and again in 2006 for research on the design of heuristic automation tools to support situation awareness and decision making in the context of naval air warfare. His current interests include the principled design, prototyping, and empirical evaluation of collaboration tools for small, distributed tactical teams; supervision and multitasking tools for command and control; and augmented cognition systems to support sustained attention.

Patricia May Ververs is a staff research scientist in the Human Centered Systems Group at Honeywell Aerospace Advanced Technology in Columbia, MD. She received the 1999 Stanley N. Roscoe Award for the best doctoral dissertation in the area of aerospace human factors. In 2007, she received her Program Management Professional (PMP) certification. Since joining Honeywell in 1998, she has led programs in the areas of high-speed research, flight-critical systems research, the Aviation Safety Program's System Wide Accident Prevention program, advanced primary flight displays, synthetic vision for helicopters, and neurotechnologies for dismounted soldiers and image analysts. Ververs was project manager for the DARPA/U.S. Army Improving Information Intake Under Stress/Augmented Cognition program, focused on developing real-time cognitive state assessment techniques for the dismounted soldier and the project manager for the DARPA DSO Neurotechnology for Intelligence Analysts program. The aim of this program is to improve image throughput of intelligence analysts through the creation of a neurobiologically based image triage system. Ververs is the author of five journal papers, two book chapters, and dozens of conference papers and technical reports. She holds a patent on alerting and notification and has filed 15 other patents in the areas of display design and the human-machine interface.

Stephen D. Whitlow is a senior research scientist in the Human Centered Systems Group at Honeywell Aerospace Advanced Technology. He completed a B.S. degree in psychology (1993) and an M.A in cognitive psychology (1995) from the University of Illinois at Urbana-Champaign as a member of the Beckman Institute Cognitive Neuroscience Group. Whitlow's technical expertise includes experimental design, cognitive neuroscience, usability engineering, user interface design, human-automation interaction, and data visualization. He has worked in several domains, including virtual presence, military operations, unmanned vehicle control, neuroadaptive systems, satellite control, aviation operations support, medical systems, cyber security, building controls, and petrochemical processing. He is currently the principal investigator of Honeywell Labs' Army-funded Augmented Cognition program, which is integrating and testing cognitive state assessment systems for the dismounted solider. Prior to this, he was the principal investigator of Honeywell Labs' DARPA-funded Augmented Cognition program, which developed adaptive systems that managed information flow to dismounted soldiers based on real-time sensed cognitive state. Previously, Whitlow contributed to the design of a monitoring application to provide a single watchstander with shipwide situation awareness of all environmental, machinery, personnel, and structural entities for ONR's RSVP program. Under the DARPA MICA program, Whitlow contributed to the mixed-initiative prototype designed to support human operators at an appropriate level of awareness and control of an advanced unmanned combat air vehicle control system.

Peter M. Young received a Ph.D. in electrical engineering from the California Institute of Technology in 1993 and worked for two years as a postdoctoral associate at Massachusetts Institute of Technology before joining the faculty of Colorado State University in 1995. He has worked extensively on the development of advanced analysis and design techniques for large-scale uncertain MIMO systems, subject to both multiple uncertain parameters and multiple dynamic uncertainties. This work provided a breakthrough in this area, whereas previous tools could only handle small problems because of the computational burden. Young developed computational software packages that were released commercially as part of the MATLAB Robust Controls toolbox.

Index

abstraction, 204
  hierarchy, 185
active learning approach, 163
adaptive automation, 75–78, 187–188, 228
adaptive paradigm, 227–230, 240
adaptive scheduling of communications, 78
additive factors method, 161, 165, 168
Advanced Brain Monitoring, 95–97
  EEG sensor headset, 95
Advanced Tactical Aircraft Simulator (ATAS), 49
Advanced Warfighting Experiment, 92, 96–98,
aircraft-in-the-loop simulation, 50–51, 69
ambient lighting, 135
applications for operational tasks, 225
applications space, 214, 219
architecture space, 214, 219
artifact removal, xi, 3, 43, 55,
  real-time, 237
artificial neural network (ANN), 55, 155, 159, 239
auditory noise, 237
augmented cognition–enabled dashboard systems, 216ff
augmented cognition mitigations, 144–145, 151
Augmented Cognition Program, 134, 151, 167, 237
augmented Personal Viewing System, 114–115
aural-spatial representation, 135

bad situation awareness (SA), 130
behavioral/performance metrics, 154
best practices, 36–37, 72–73, 103–107, 136–137, 167–170, 191–192, 240–241
bias
  of classifier, 91
  variance, and temporal smoothing, 88
Biocybernetics and Pilot's Associate programs, 200
blood-oxygenation-level dependent, 13
Bluetooth, 95

classification metrics, 91
closed-loop system, 1, 6, 163, 178ff, 218, 240–241
  adaptive system, 77
  augmented cognition system, 1, 20, 28, 36–37, 39, 163, 185, 187–189
Cognitive Avionics Tool Set (CATS), 43, 45–46, 48, 52–58
cognitive load assessment, 136, 147–152, 158, 162, 166–168
Cognitive Pilot Helmet, 60
cognitive signatures, 116–117, 119, 122–126, 133–134, 137, 140

cognitive state classification, 78–79, 82–83, 88–89, 94–95, 106–107
cognitive task analysis, 232
cognitive workload assessor, 187
cohesion, 208
combining multiple psychophysiological measures, 238
composite classifier fusion, 87–88
Computerized Airborne Research Platform (CARP), 44–51, 61–62
confusion matrix, 157, 159
considerate mitigation, 113, 125
context-sensitive help, 138
continuous-wave method, 31, 38–39
control system theory, feedback,
control systems, 176–178, 181, 190, 194
  theory, 175
  uncertainty, 181
coupling, 208

Decade of the Brain, ix
decomposition, 204
Defense Advanced Research Projects Agency (DARPA) Augmented Cognition Program, 237
dense-array EEG, 48, 50, 51, 62
deoxy-hemoglobin, 15, 20, 30–33
discriminative classifier approach, 86
dismounted soldier domain, 79
Distributed Mission Operations, 49
DOD architectural model, 213
driver-in-the-loop assessment procedures, 219
dry electrodes, 236
dynamic modeling, 181–184, 190
dynamic, real-time adaptive system mitigation, xi

ecological validity, 235
electrical noise, 237
electrocardiography (EKG), 29, 48, 149
electroencephalography (EEG), 3, 29, 113, 122, 229ff
electroenchephalography event-related potentials (EEG/ERP), 113
electromyogram, 48
electrophysiological measures, 5
engineering control theory, 190–191
error-related negativity, 8
evaluation pilot, 49, 70, 73
event manager, 125
event-based augmented cognition system, 116
event-based cognitive assessment, 112
event-based situation awareness metric, 129
event-related potentials, 4ff, 113, 122–124, 134, 136–137, 235
external validity, 144, 152–153, 155, 168–169, 240

eye-mind hypothesis, 17
eye-tracking module, 53–54, 62, 236–238

Fast optical signal, 11
FAST Track program, 196
Fast-Fourier Transform (FFT), 93
feedback control, 177
    algorithm, 187
fixation durations, 17, 54
flight test engineer, 43, 49, 73
formal models, 204
FRL Fly-by-Wire System, 44
function distribution, 208
function sharing, 208
functional brain-imaging technology, 27
functional magnetic resolution imaging (fMRI), 14–16, 19, 81, 122, 161
Functional near-infrared sensors (fNIR), 27–28, 30–34, 36, 38, 77, 81, 149–150, 166–167, 235
fusion and composite techniques, 87
Future of Augmented Cognition, The, 150

galvanic skin response, 29, 149
galvanic-vestibular stimulation, 135
Gaussian mixture models, 83–88
gaze
    -contingent displays, 18
    -contingent paradigm, 19
    duration, 18
    tracking, 31
good situation awareness (SA), 130
gradient fields, 15
ground truth, 89–90, 93, 96–98, 106–107, 145, 153

head-up display, 55
heart rate sensors, 236
heart-rate variability, 77, 82, 94
hemodynamic methods, 13, 20–21
Hidalgo Vital Signs Detection System, 95–96
high workload, 91
high-key/low-key lighting, 135
human factors engineering, 2, 146
human-computer interaction, 199
human-machine symbiosis, 146
human-systems interaction, 29, 165–166, 169

Improved Performance Research Integration Tool (IMPRINT), 121
Improving Warfighter Information Intake Under Stress Program, 200
inconsiderate augmentation, 113
integral paradigm, 229–230, 240
integrated airborne neural imaging, 42

integrated design, 208
Interface Design Evaluation with Sensors, 29
Internal Paradigm, 229
internal validity, 144, 152
invasivity, 2, 7, 16

K-nearest neighbor, 84

laboratory proxy task, 226, 231
Laplace transform, 183
lateralized readiness potential, 8
lessons learned, 20–21, 34–36, 69–70, 98–103, 133–136, 165–167, 187–191,
Level 1 SA, 117
linear discriminant function analysis, 239
logistic regression, 239
long-term generalization, 89
low workload, 91

magnetic fields, 10–11, 15
magnetoencephalography, 5, 10
Matlab/Simulink simulation, 187
meet-me-in-the-middle process, 199, 210
memory processes, 18
mental chronometry methods, 161
metacognition, 145
MIDS workload equation, 136
missed perception, 134
Mission Awareness Rating Scale, 116, 130
mitigation management, 151, 155, 160–161, 166, 169–170
    framework, 124
mitigation
    decision making, 147
    information processing of, 147
    pacing of, 134
    selector, 126
    scheduling of, 134
    stages, 127
    strategies, 147
        "learning curve" of, 162
        evaluation methodologies of, 144, 154, 165
    training, 147
mobile environments, 75, 81, 88, 92, 94, 98, 114
model brittleness, 76
model predictive control, 186
modular design, 208
module complexity, 208
MOUT (Mobile Operations in Urban Terrain, 92
moving-window or gaze-contingent displays, 17
mu rhythm, 6
Multi-Attribute Task Battery, 228

multi-input–multi-output (MIMO) control task, 177
Multimodal Information Design Support, 121–136, 148–150, 158–160, 167–168
multiple linear regression, 239
multiple resource theory, 128

NASA Task Load Index (NASA TLX), 29, 56, 98, 148–149
National Geospatial-Intelligence Agency (NGA), 154
Near-infrared spectroscopy, 29, 30
neural signatures, overload of, 150
neuronal activity, 5
    measures of, 5
n-fold cross validation, 91
noise artifacts, 81
nonlinear signal-processing approaches, 176

oculomotor behavior
    measures of, 17
open-loop system, 180, 180–181, 186, 188–189
Operational Risk Management, 70
operator functional state, 155–156
ordinary differential equation (ODE), 179
OTIS, 27, 39
oxy-hemoglobin, 15, 20, 30–33

P300, 7–8, 230, 240
Parzen windows, 83–88
perturbed closed-loop system, 181
physiological measures of cognitive state, 77
pi detectors, 12
platform stack, 211
platform-based design, 196–200, 204, 210, 212–215, 219–221
    techniques, 200
positron emission tomography (PET), 14, 21, 77, 81
power spectral density, 100, 101, 105, 123
proportional-integral-derivative control, 186, 189–190
pucker factor, 234
pulse oximetry, 31
pupilometry, 31
pursuit movements, 16

rapid serial visual presentation, 154–155, 165, 168–169
Receiver Operating Characteristic, 91
    Curve, 102, 157–158, 160
recognition-primed decision making, 233
recording channels, or optodes, 12, 16
rule-based control, 186

saccade targets, 19
saccades, 16–18
saccadic movements, 16
safety, 70–72, 166, 203, 236–237
    pilot, 44, 49, 61–62, 64, 71, 73
sensor density, 102, 104, 105
sensor fusion, 87, 100, 102
sensor integration, 41, 46, 97,
Sherlock Experiment, 164
signal detection theory, 156–159, 161, 165, 168-169
signal processing, 93
Signal Quality Tool, 33, 35
signal-to-noise ratio, 6–7, 10, 12-13, 20–21, 33, 35, 82, 95, 104, 149
single-input–single-output, 177–178, 186
Situation Awareness Global Assessment Technique (SAGAT), 29
Situation Awareness Rating Technique (SART), 116, 130
situation awareness, definition, 116–117
spiral model, 200
Stage 1 mitigation, 128
stressors, 93
system identification, 179

Tactical Tomahawk Weapons Control System (TTWCS), 29, 233
tactile seat cueing system, 48
Task Load Index, 98, 148–149
technical compatibility, 3
test card, 43, 61, 69, 71–72
theta rhythm, 6
transfer function, 184
TTWCS, 29, 233
    Tool for Interface Design Evaluation with Sensors (T-TIDES), 29–30, 235

Unified Modeling Language, 202, 210
unmanned air vehicle, 155–156, 231, 233
user-centered design, 29, 146

variance classifier, 91
visual noise, 237
V-model, 200

Warship Commander Task, 149–150
waterfall model, 200
wet electrodes, 236
wireless data network, 89
Workload Index (W/INDEX), 121, 136, 148